Construction in the Landscape

Construction in the Landscape

A Handbook for Civil Engineering to Conserve Global Land Resources

T. G. Carpenter

earthscan
publishing for a sustainable future
London • Washington, DC

First published in 2011 by Earthscan

Earthscan Ltd, Dunstan House, 14a St Cross Street, London EC1N 8XA, UK
Earthscan LLC, 1616 P Street, NW, Washington, DC 20036, USA
Earthscan publishes in association with the International Institute for Environment and Development

For more information on Earthscan publications, see www.earthscan.co.uk or write to earthinfo@earthscan.co.uk

ISBN: 978-1-84407-923-0

Typeset by Composition and Design Services
Cover design by Yvonne Booth

A catalogue record for this book is available from the British Library

Library of Congress Cataloging-in-Publication Data

Carpenter, T. G.
Construction in the landscape : a handbook for civil engineering to conserve global land resources / T.G. Carpenter.
p. cm.
Includes bibliographical references and index.
ISBN 978-1-84407-923-0 (hardback)
1. Building sites. 2. Landscape protection. 3. Engineering geology. 4. Building—Environmental aspects. 5. Landscape construction. 6. Land use—Planning. I. Title.
TA705.C36 2010
624.028'6—dc22 2010023404

At Earthscan we strive to minimize our environmental impacts and carbon footprint through reducing waste, recycling and offsetting our CO_2 emissions, including those created through publication of this book. For more details of our environmental policy, see www.earthscan.co.uk.

FSC
www.fsc.org
MIX
Paper from responsible sources
FSC® C013056

Printed and bound in the UK by TJ International Ltd.
The paper used is FSC certified and the inks are vegetable based.

CONTENTS

List of Photographs, Figures and Tables

Photographs

Figures

Tables

Colour plates

Preface

For many centuries, people have designed functional forms that came to characterize beauty. They were Roman engineers, Renaissance polymaths or modern creators of daring structures using new flexible materials and enhanced mathematical calculation. Less romantically, mankind has also shaped land surfaces to grow crops, to build houses on steep slopes, to divert river flow and to protect property against floods or the ravages of the sea – all activities aimed at making the best use of land resources for increasing human populations.

Aspects of beauty – shapes and appearance that suit landscapes – have long been recognized. Some relate to engineering structures that were explained, for example, by Dame Sylvia Crowe as she advised on British road and power development in the mid-20th century or – more recently – by the Highways Agency in *The Appearance of Bridges and Other Highway Structures*.

By massive earthworks, countries such as the Netherlands have been transformed until they are almost wholly artificial. Elsewhere, as in Bangladesh, man has succeeded only in mildly alleviating the impact of huge natural hydraulic forces. Meanwhile, some ancient fortifications or abandoned railway routes have matured into heritage features. Worked out gravel pits have been converted into wetland reserves or waterside amenities. Today such modern works in well-inhabited regions are controlled by land-use planning authorities. But elsewhere, just as the forests are cut down, strip mining for minerals or increasingly less accessible carbon or hydrocarbon fuels is proceeding with less appreciation of the long-term consequences.

Some of us who practised civil engineering or were involved in planning development projects from the 1950s were fortunate enough to find opportunities in distant countries which had not yet formed their own teams for these activities. We saw great natural and man-made scenery and wide varieties of construction techniques; in more mundane surroundings we could contrast drab monotony with vibrant colourful communities. We were aware of much of the elegance or intrusion which construction could create in the days before environmental impact assessment had been invented. But we did not have current – to us amazing – tools for recording what we saw. Such tools include:

- word processing, for any author the most vital use of personal computers;
- digital photography; and
- the internet, as well as increasingly better-informed information available in the conventional daily and technical press.

So early travel opportunities and recent advances in information technology combine to make this book possible. But, in undertaking it, I am also especially grateful to Edward Hepper for early comments on everything, Peter Hunter for advice on coastal engineering, Bob Clements for final review of a range of issues, my brother Robin Carpenter for transforming photographs and my nephew Paul Carpenter without whom I could never have mastered all the computer applications.

Tom Carpenter
Henley-on-Thames
May 2010

List of Acronyms and Abbreviations

BAA	British Airports Authority
CHP	combined heat and power
CPRE	Campaign to Protect Rural England
DIY	do-it-yourself
EU	European Union
FGD	flue gas desulphurization
FRP	fibre-reinforced polymers
GHG	greenhouse gas
ICE	Institution of Civil Engineers
NGO	non-governmental organization
PFA	pulverized fuel ash
PUD	Planned Unit Development
SCR	selective catalytic reduction
SSSI	Site of Special Scientific Interest
SUDS	sustainable urban drainage system
TGV	Train à Grande Vitesse
UN	United Nations
UNHCR	United Nations High Commission for Refugees

Part I

Land Resources

1

Land Features

1.1 Landscape – an inclusive concept

To most people, landscape is scenery, particularly scenery that they admire and wish to conserve. To some, it also reflects the presence and activities of life forms – from wild nature to cultured humanity – or the opportunities offered by land for production of food, as space to live in or as an amenity for recreation.

The first meaning in dictionaries is indeed scenery – or a painting or other depiction of a view. However, a geographical alternative (in the *Shorter Oxford Dictionary*, 2007) is a 'tract or region of land with its characteristic topographical features', which is perhaps closer to the opportunistic concepts of landscape. Forman and Godron (1986)[1] mention an American (Webster) definition as 'landforms of a region in aggregate'.

Specific definitions of landscape have been proposed by people with interest in particular land characteristics or uses, for example:

- by painters, guidebooks and those who deplore its depredation: as acknowledged scenery;
- by geographers and landscape designers: as the physiographic and environmental aspects of land (for example, Laurie, 1986);[2]
- by ecologists: as clusters of ecosystems (Forman and Godron, 1986);[3]
- by historians: as associations with forests, mountains, rivers or settlements in our individual or collective recollection of past events (Schama, 1995);[4]
- by archaeologists: as surface landforms that recall man's ancient structural achievements (for example, Muir, 2004).[5]

According to Schama (1995)[6] the Dutch word *landscap* once signified a limit of occupation or jurisdiction as much as a depiction of pleasing prospects. In the Netherlands, much of the landscape had already been transformed for productive use; and modern larger scale works have continued its reconstruction to this day.

The Germanic root of the Dutch word is *landschaft* where the suffix *–schaft* was equivalent to *–ship* in English, implying the quality or condition, in this case, of the land. However, by the time the word came to England, first as *landskip*, in the late 16th century, it meant a 'picture of inland scenery' (Jackson, 1986).[7] Man-made landforms or structures have always been elements of that scenery, much of which was once recorded as background in portrait paintings. Then, as the landscape has become more crowded and land resources in greater demand, people have recognized that there are elements in landscape that are crucial to its conservation but cannot be fully expressed in visual terms. Even in viewed scenery, if there is something that jars our aesthetic sense, then perhaps it is because we sense there is something wrong with treatment of the land itself. Thus a popular concept of landscape today is scenery *plus* non-visual attributes such as wildlife habitat and cultural association. In this book we regard it as land resources *including* scenery. Some of the forms of construction addressed in Part II, such as schemes for bulk supply of water for irrigation or the alignment of highways, primarily affect actual land space. Others, such as tower structures or daring bridge crossings, are important elements of scenery.

This chapter introduces basic geologic landforms and the influence of climate and water on them (Section 1.2) and the plant life that clothes the Earth and sustains its animals (1.3). Section 1.4 recognizes the productive and necessary uses of land. The consequent implications for wildlife habitat (1.5) and natural or man-made scenery (1.6) are then discussed as is people's perception of different aspects of landscape and land use and hence their wish and opportunity to influence plans affecting land resources (1.7).

1.2 Landforms, geology and climate

The forms of the Earth's surface are consequences of geological processes – orogenic folding, faulting and thrusting, glaciation and wind or water erosion. The most prominent forms are seen where hard rocks have not yet been eroded down to form flatter softer land.

Climate concerns temperature, winds, humidity and precipitation, all with seasonal variations. Temperatures can be too cold to grow crops in the open or to sustain significant populations of people without support from elsewhere; or conditions may be warm and moist enough to promote luxuriant vegetation and abundant animal life – but also virulent pests and diseases.

Winds bring humid air and water vapour in wet zones, or scorching erosion to arid lands. Precipitation, as seasonal rainfall, falls directly on forest, grassland or cropland, providing opportunity to manage vegetation as crops for people or fodder for animals. Or the rain runs off or the snow melts, feeding streams and rivers or seeping into the ground.

Low temperatures and ice or water cause glacial movement in the high mountains, permafrost in flatter northern lands, and rough terrain with poor soils and irregular drainage in both conditions. High temperatures or diurnal extremes of freezing and thawing can shatter outcropping rock. Where rainstorms are occasional but intense, surface run-off may carry off topsoil. Streams in flood erode their banks or undermine sloping ground. The resulting eroded material – sediment – is transported downstream and deposited as alluvium in less steep territory.

Four representative types of physical geography represent typical landforms in a sequence from hard and chaotic rock geology down to soft plains and deltas.

Mountains, *gorges* and *escarpments*, together with their *lakes* and *cataracts*, are grand scenery. With the exception of certain settlements or storage reservoirs in the valleys, they are best left as such, being less suited to productive land use.

Gently undulating hills are more evenly eroded and rounded, for example in the ranges of the Appalachian Mountains or the less steep regions of Mediterranean Europe. Natural rock outcrops are scarce and, in humid regions, smooth-sided valleys divide often wood-clad uplands of roughly equal altitude. With adequate soil and in suitable climates, these hills may offer slopes, terraces and plateaux suitable for agriculture, grazing land and forestry.

Alluvial plains comprise eroded sediment deposited during past and contemporary floods so that the land is now generally flat. Often the soil is richly fertile, sometimes so uneven as to be suitable only for grazing or so saline that few species can grow. Seasonally flooded land may be better suited to pasture and flood recession agriculture than to permanent settlement; but engineering structures and embankments can do much to influence the flow of flood water over the land or to protect habitation.

Deltas are formed by the deposition of sediment where rivers debouch into the sea, lakes or reservoirs. Their formation is a continuous process as long as the flow pattern of the river is unaltered by constructing dams and control works. The build-up of the delta of the River Nile into the Mediterranean Sea was reduced or reversed in the 1930s and 1940s; one of the causes was construction of barrages in the delta and of dams on the main river further upstream. The delta may have started to advance again from 1954 but its retreat recommenced rapidly after completion in 1970 of the Aswan High Dam prevented any more sediment passing downstream. The Akosombo Dam in Ghana has had a similar effect at the Atlantic outfall of the Volta River.

Other deltas, like that of the Mississippi and the Po rivers continue to advance and the consequences in protecting cities such as New Orleans or Venice are fraught with problems requiring expensive engineering solutions.

1.3 Life forms – animals and vegetation

Living organisms are:

- animals that have sense organs and nervous systems but can survive only by eating organic food, such as vegetation or other animals;
- vegetation, inanimate plants which absorb water and inorganic substances through their roots and synthesize nutrients in their leaves;
- microorganisms, which may fall into either category, are too small to be normally noticeable but are hugely abundant in all animals and plants and critical in the way in which all these life forms interact.

Ecology is the scientific study of relationships between organisms and their physical surroundings (habitat). Flora and fauna interact with other species in complex communities (ecosystems).

Animals fall into four categories as far as land use is concerned:

1 Human beings, who dominate all but the most remote places.
2 Domesticated farm livestock, such as cattle, sheep, goats, pigs or fowl, which require pasture and, sometimes seasonally, extra fodder or weatherproof accommodation.
3 Small domestic, feral or wild mammals that live in human environments such as villages and suburbs.
4 Wild creatures who live entirely naturally and whose survival depends on continuance of their habitat.

Vegetation can be similarly represented by three very broad classes in a world dominated by people:

1 Crops in cultivated fields, orchards and plantations; or commercial forests.
2 Incidental trees, offering shade and green attraction in human settlements, or gardens or lawns managed for amenity and recreation.
3 The plants and trees of wild land – from isolated wilderness on rough slopes or small wetlands to extensive tropical forest – that provide a habitat for wild animals.

Agriculture and forestry are mentioned here because they cover so much space and to observe that certain crops or trees suit certain land resources. Trees as a human amenity are features of settlement, wild animals of nature conservation.

Agriculture is the science or practice of growing food and industrial crops. It uses seeds, plants or livestock that have been selected, bred or modified to yield optimum types and amounts of produce. Types of agriculture can be related to the climate, soil fertility, topography (surface contours) of the land and the way in which the crops are watered. Cold climates favour only those hardy plants that can withstand frost or can grow during a comparatively short warmer season. Dry climates, even where the soils are fertile, cannot be agriculturally productive unless extra water is provided. Temperate climates with timely rainfall are ideal for many agricultural crops. Wet tropical climates engender profuse growth both welcome and destructive.

Soil fertility relates to its grading, chemical composition and nutrients as well as drainage conditions and the thickness of organic topsoil. Fertility can be exhausted by intensive cropping; or it can be restored – artificially by adding chemical fertilizers or organic material, naturally by rotating the primary crops with years of fallow or special plants such as nitrogen-fixing legumes.

In luxuriant growth such as tropical rainforests, the trees and undergrowth are sustained almost entirely by sunlight, rainfall and nutrients recycled through the vegetation itself and not through the soil. The latter is therefore bereft of the sort of nutrients and perhaps water retention or drainage needed for cultivation of ground crops. As a consequence, agriculture on land cleared from tropical jungle tends to fail because:

- the soil is unsuitable due to these deficiencies;
- growth of weeds and soon secondary forest requires arduous effort to combat; and
- pests and diseases are difficult to overcome.

Fortunately rice, the staple food for half of humanity, can overcome much of the tropical soil problems. Paddy rice grows in shallow water, thrives in a warm humid climate and forms an impermeable pan that eludes the loss of water by downward seepage (Young, 1998).[8]

Steep land is difficult to cultivate or harvest mechanically; and it may have to be terraced to retain water. But very flat land may be difficult to drain and can become waterlogged. Crop watering can be entirely by rainfall in wet climates; or water may have to be collected, stored and applied by an irrigation system in drier conditions.

In the past, seeds for agriculture have been selected, hybridized or genetically modified, primarily to raise the yield of crops. These yields are difficult to maintain without considerable inputs of fertilizers and other chemicals. Now, as fertile land is put under greater pressure, more emphasis needs to be placed on development of seed varieties to suit marginal conditions – such as with resistance to drought, salinity and certain pests and diseases.

Woodland includes commercial tree species that grow locally or are imported from countries with similar conditions. Sustainable forestry involves selective harvesting of trees and replanting of seeds or saplings to achieve various products including:

- high grade (hardwood) wood and veneer for furniture;
- structural timber;
- low quality fast-growing timber for pulp (or wood fuel where necessary); and
- non-timber products such as gums, resins and latex.

Mechanized clear felling is not always sustainable. Nor is devoting large tracts of land to beef livestock the best way of growing food. Even arable crops may not be the optimum variety. Most human uses of regetation compete with each other; and wild animals still need wild land.

1.4 Productive and functional land

Directly productive land use includes agriculture, forestry and mineral extraction. Functional land is additional space for infrastructure of human settlement and activity.

Agricultural land is allotted to food production (mainly) as well as to industrial crops such as cotton or feedstock for petroleum substitutes. However, there are shortages of food in some overpopulated countries and others are becoming increasingly dependent on imports, particularly of cereals such as wheat, rice or maize. In many countries – especially in Africa – there is scope for considerable improvement in agriculture including in seed varieties, husbandry of land and crops, and marketing of the produce. In others, food production is already efficient; and on some land it is doubtful whether the high output already achieved can be maintained without impoverishing the soil.

The land area of farms comprises space in several categories:

- Arable fields for planting and harvesting cereals, root and leaf crops on annual or more frequent cycles.
- Similar fields lying fallow or under grass, allowing the soil to replenish its nutrients.
- Water-collecting areas and animal pasture,[9] particularly on steeper or less productive land.
- Shrub and tree crop plantations and orchards: these can also be on sloping ground.
- Water channels, ponds and drainage ditches serving irrigated land.
- Roads, boundary strips between fields, farm buildings, woodland and incidental space.

Thus agricultural activity in well-populated areas requires specific land surface space to be fitted into actual patterns of topography.

Forestry, by contrast, takes place generally in more extensive, sometimes remote, areas where there are a great many trees and few people. Forests occur naturally in a density and diversity that is greatest in the wet equatorial regions.

In the cold northern tundra there are no trees. South of the tundra, 'taiga' forest is widespread but trees are stunted and sparse. This is land that could be kept in reserve for agriculture, forestry or settlement in the distant future, depending on the very uncertain effects of climate change and prospects for new sources of energy. Further south, the forest, generally of conifers or birch, improves in quality, size and speed of growth. Most of the great mainly deciduous forests of Europe, China and the eastern US have long been cleared for agriculture. The remaining reserves of conifers and deciduous trees, still significant in some countries, should be regarded as a valuable resource for many centuries ahead and should be exploited only by sustainable forestry methods.

Tropical forests have been extensively cleared in Southeast Asia and the land is tilled, most successfully for rice. Of such forest as remains – mainly in South America and Africa – there is an urgent need to define ultimate land use in terms of specific zones allocated to:

- conservation of forest biodiversity (in large ecologically viable units);
- sustainable forestry for managed production of timber and by-products;
- agriculture and settlements;
- buffer zones where necessary between the other zones, and forest land to be kept in reserve for needs that may become evident after another century.

Mineral extraction yields a range of useful resources and creates holes in or below the ground surface. The minerals extracted include:

- carbon 'fossil' fuels (coal, oil, gas and more firmly embedded hydrocarbons) and uranium: these are burned and thus non-renewable;
- metal ores: these are also finite in quantity but a proportion of manufactured items can be recycled;
- construction materials including soils and rock: these may be locally scarce but are generally abundant and can also be recycled.

All mining – underground via adits and shafts, quarrying at open rock faces or opencast excavation of softer materials – produces waste and disturbs the ground surface. Restoration of land, after extraction is complete, gives opportunities to form new land shapes or dispose of wastes, and to create or restore surface drainage, rock face and water features as well as agricultural and building space. Such exploitation and restoration are difficult to achieve where there are many thousands of small mining operations, such as in central Africa. In well-governed countries, mining and quarrying are controlled by licence. Permission is given only in those areas where extraction is scheduled in long-term plans to meet regional economic needs.

Human settlement has always concentrated on the fertile areas where the benefits of that land can best be gained. Therefore there is direct competition between agriculture and built development. Building land is very valuable in monetary terms, farm land much less so, however vital its function. Typically, land with permission to build on costs a hundred times as much as farming land.

Wide tracts of the most fertile land have been built over. However, alluvial floodplains are better suited to agriculture than to settlements (because of the floods), while sloping ground is better suited to buildings than to agriculture; and hard rock suits foundations, not vegetation. Most towns are where they are for historic reasons, although planners can now define zones where expansion or new settlements is permitted or forbidden. Factories, power plants and transport infrastructure can be sited taking into account their particular locational needs; and particular areas can be avoided, for example when suited to specific forms of nature conservation.

1.5 Wildlife habitat

This section describes typical types of habitat and the ways in which human activity can affect them. Particular impacts of construction on wildlife are introduced in the next chapter (pp18–19) and examined further in appropriate chapters of Part II.

Wildlife should be conserved for many good reasons including the possible unseen benefits it may yet offer and the impoverishment of our resources that will result from the demise of ecosystems that we hardly yet understand. Animal life is at risk from:

- extermination by human weaponry – the dodo, the American bison – or by capture for consumption as has depleted edible fish; and
- damage and destruction of habitat, hence the diminishing numbers of tigers and giant pandas and the elimination of many smaller species, some even before they have been identified.

Since this book is concerned with land and the effects of construction, it is the space of habitat and its physical attributes that need to be considered when changes to land cover are contemplated. Our wish that habitat should be conserved springs partly from sympathy with and scientific understanding of animals and partly from our association of wildland with scenic playgrounds. The concept of national parks relates particularly to human recreation, that of nature reserves to biological purpose.

Habitat is where vegetation grows or where animals live – that is, where they rest, breed, hunt, feed or move about. Particular species or complete ecosystems depend on certain combinations of physical characteristics and climatic conditions. Habitat can be disturbed or destroyed by anthropogenic action. Fragmentation can reduce both species mobility and the variety of flora and fauna.

Forman and Godron (1986)[10] and others have explained 'landscape ecology' in terms of spatial structure, landscape dynamics, heterogeneity of species and their response to intervention. The structure comprises patches, corridors, matrices and networks, or combinations of these. For a simpler summary, seven types of habitat illustrate typical features and sensitivity to human interference:

1 *Extensive regions* of similar character may cover thousands of square kilometres. Examples include the grassland and water-holes supporting wild African herbivores and the predators who hunt them, or dense rainforests containing rich pristine biodiversity and large enough to perpetuate their own climate. The ecological role of such land is put at risk as soon as roads are constructed through them, allowing agriculture, timber felling and settlement to take place along the route. However, whenever population pressure makes it inevitable and necessary that some of this wild land should be tamed, then occupation should be planned in areas and along routes after zones of suitable extent and character for permanent nature conservation have been defined.
2 *Medium-sized nature reserves* – in woods, scrub or heath – may permit controlled human access but are primarily for protection of particular species or natural communities, probably with a necessity to conserve or create connections to other similar habitat. The most ecologically highly rated sites can be given priority for avoidance by new transport routes or can be sensitively contained within wider areal development.
3 *Wildlife corridors* are vegetated strips of land or water along which animals can move or gain temporary refuge. They are key elements of ecological land patterns because they promote connectivity for species, communities and ecological processes (Bennett, 1999).[11] Their viability depends on their continuity, the needs of different species for safe stopover points or their ability to overcome narrow sections or obstacles. Wildlife corridors include remnant, regenerated or recently planted strips of woodland, grassland or hedgerow; or they may be rural pathways or lanes. There are 'disturbance habitat corridors' that take advantage of verges, earthwork slopes or buffer land alongside railway lines or highways, or are cleared for pipelines or high voltage power transmission lines.
4 *Watercourse beds* or *banks* are habitats for vegetation, fish, amphibians, insects and various land-based predators. There are birds and small mammals in riverside trees, shrubs, earth burrows and grassy space. The vegetation may filter out some of the chemical pollutants in farm run-off. Artificial river bank stabilization can destroy riparian growth although it may result in new habitat formation on bank sections left less disturbed.
5 *Other wetlands* are *marshes*, adjacent to the watercourses that flood or drain them; and *bogs* – poorly drained terrain that receives water only as rainfall. There is also seasonally-flooded land which may be cultivated by farmers, or grazed or gleaned by animals or birds during the dry season. Marshes fed from rivers may suffer from construction of dams, river control and channelization works; and all wetland habitat is at risk from drainage or water-table drawdown unless these are carefully planned to conserve essential seasonal water levels.
6 *Avian stopover places*, especially for migrating wildfowl: feeding and resting zones, for example, at low and high tide respectively

in estuarial or low-lying coastal situations. Attractive feeding grounds may also be seen on post-harvest arable farming land adjoining inland lakes. Significant sites can be protected, adapted or extended in conjunction with river estuary development (for shipping) or coastal protection or management schemes.

7 *Opportunity habitat* occurs where animals or plants take advantage of, or become dependent on, the infrastructure of human settlement. In buildings, gardens, paths and derelict ground, they find secure niches, sources of food and means of access. There are also welcome strips of land from which people are normally excluded.

Land-use plans conserve natural surroundings for animals or ecosystems of recognized value or sensitivity. In semi-natural conditions, precautions can be taken to preserve certain characteristic elements, opportunities taken to enhance others. In towns, suburbs and agricultural or industrial areas, people have a choice as to the domestic animals and plants that they keep or grow and the feral or semi-wild species that they tolerate.

1.6 Natural and man-made scenery

Scenery is the appearance of features in landscape. Recognized elements are:

- soft and hard landforms – the shape and characteristics of soil and rock masses;
- vegetation;
- water – in snow and ice, rivers and lakes, and the sea along coasts;
- evidence of human activity in agriculture, forestry, mining, industry or settlement;
- man-made structures; and
- combinations of these.

Landforms are surviving manifestations of geological processes. A great proportion of the Earth's surface has been flattened by erosion or filled with sediment. In well-populated regions, scenery is mostly man-made. Most fine natural scenery is found in the harder rocks of mountains or along gorges and coastlines.

Mountain ridges are created by huge, often violent, movements in the Earth's crust, outpouring of volcanic lava and more gradual subsequent denudation of all but the hardest strata. Glaciation, where it occurs, has carved sharp summits and pinnacles, cirques of steep high cliffs, knife-edged arêtes and U-shaped troughs. Eroded landforms are shaped by the hardest rock masses within them. In steep, wet or temperate zones, valley slopes and beds continue to be eroded by rainfall run-off and the flow in watercourses. In more arid zones, constant wind or chemical erosion has cut shapes that are sometimes geometric, sometimes weird. Pristine snowfields, sheer rock faces and precipitous waterfalls are glorious seen from afar, awe-inspiring and exciting at close quarters. In storms they seem even more forbidding, under heavy cloud dismal or dangerous.

Away from the recognized mountain areas are isolated monadnocks or inselbergs, abrupt hard rock remnants surviving from erosion across what are otherwise now extensive peneplains. Ayers Rock is a large example in central Australia, Sigiriya a steep one in Sri Lanka while numerous jebels or kopjes are strewn over Africa. More common all over the Earth are evenly eroded rolling hills. Apart from occasional rock outcrops or escarpments, beauty on these gentler slopes is more related to what grows on them, demonstrating how man has fashioned the surface.

Vegetation is the clothing of landscape. It decorates landforms with colour and texture and it harbours animal life. Trees perched on rock outcrops or in steep 'hanging woods' can present astonishing beauty. More often, on less dramatic forested hillsides, deciduous trees tend to be seen from beyond as uniform greenery, attractively mottled in autumn sunshine, as dull as the weather in mist or rain. For wooded hillsides to be admired from afar, there must be clear viewpoints through what may also be forested surroundings. From within, clearings and glades provide a pleasant experience (again in the sunshine).

Terrain is more open where the climate or soils can support only ground cover and bushes, or where growth is prevented from attaining its natural climax state by cutting, burning or grazing. Heath, bog, coarse grassland and chalk downland are examples of ground cover that may dominate scenery – as rare pockets of semi-wild terrain amid commercial agriculture, within a tapestry of colour on a hillside or in huge stretches of more monotonous moorland. In flat farming scenery, there may – in season – be billowing crops or colourful wild flowers in the foreground. Otherwise the eye goes quickly to the nearest green line of hedgerows or copses. Beyond this horizon are only man-made structures rising above what, in reality, is usually much more open ground than distant tree lines seem to indicate.

Water, in lakes, reservoirs and wide rivers, also allows an open view of any rising ground beyond. But water makes a more active contribution to scenery through the energy of its flow descending high waterfalls or white water cataracts and its great value to wildlife habitat on river banks or in pools or wetland.

Human activity in farming can be so intensive as to appear starkly utilitarian. In contrast, mixed farming has created rural 'countryside' in which field boundaries dominate the structure of the scenery. Hedgerows, stone walls, copses, wild pockets, farm buildings, lanes and paths provide a random variety which Crowe and Mitchell (1988)[12] described as 'enriching the mosaic of the Earth's surface by changes in colour and texture'.

Forestry, as a commercial operation, is also utilitarian in the form of strips of young and mature trees or of recently cut hillside. In the Roman era and up to the time of Dante, natural forest was regarded as 'the beckoning antechamber of hell' (Schama, 1995).[13] However, from the time of John Muir in California, forests have been regarded as places for wildlife conservation and human recreation as well as timber resources. Investment in commercial forestry should equally be able to diversify into enhancement of mixed streamside habitat.

Quarries and industrial and residential settlements create their own structured landscapes of occupied, managed and wild land.

Structures, as they affect land resources, are the subject of this book. Their place in scenery depends on how they fit into the surroundings in which they are seen:

- In magnificent natural scenery there is seldom place for any sort of construction that can detract from the natural beauty or for too easy access that could spoil sublime experience.
- Where it is necessary to build dams or bridges across narrow valleys, then these need to reflect honestly the strength and ingenuity of their grand design against that of the surroundings. These structures may exhibit awesome structural simplicity or fit gently into a more modest tapestry of contours and colour. Mountain resorts should be built or extended only at an altitude and in a style proven to be acceptable at existing settlements.
- In comparatively featureless open landscape, there are plentiful precedents for adding man-made towers or spires. Perhaps more contentious are overhead power lines or wind farms where provision should be that routes or layout should fit into other patterns of boundaries, roads or water channels, and that the pylons should be removable if they become obsolete.
- In cities, structures *are* the main landscape.

Combination of scenic elements is the essence of the sort of semi-natural country scenery that most people admire. An archetypal heroic landscape painting (for example, US 19th century)[14] depicts tumbling water, rocky bluffs adorned by precariously perched trees and high rock peaks or denser forest behind. Shafts of sunlight light up some features but the exciting uncertainty of the wilderness is reinforced by glowering storm clouds. A typical quieter pastoral scene (for example, Dutch 17th century) gathers all the interesting features – rough soil and muddy tracks, trees and undergrowth, water features and their bridges or banks, houses or huts, people and animals – in contrast with the dull uniformity of much actual scenery in the Netherlands. More completely anthropogenic scenery includes organized vegetation in lines of trees, lawns and flowerbeds, welcome water features, spacious streets and elegant buildings.

Planning the total composition of scenery must take account of what can be seen from where. Elements of scenic opportunity are:

- viewpoints from which people can look up, down or around;

- vistas giving narrow but intriguing glimpses along watercourses, through trees and between buildings; and
- features, width and compatibility of observed foreground, middle distance and background.

All these must be taken into account when we create structures that will alter the scenery or the opportunities to appreciate it.

1.7 People's perception of landscape and its alteration

People's appreciation of scenery depends on their emotional response, cultural upbringing and actual involvement. Their opinion about how land should be used depends on their perception of comparative values, for example as to priorities for economic production, social harmony, human recreation or nature conservation. The need to alter landscapes arises from popular or corporate pressure to build structures and change land usage. Opposition comes from those who prefer what already exists. Solutions should be chosen from a number of options posed by those technically familiar with the issues and concepts concerned.

Perception of scenery

Appleton (1996)[15] describes two related theories that seek to connect landscape with the emotions that it excites:

1. What he calls *'habitat theory'* relates to pleasurable sensations in the experience of landscape as well as environmental conditions favourable to biological survival.
2. His *'prospect and refuge theory'* relates that same pleasure and biological survival to 'the ability to see without being seen'. A 'prospect' is an unimpeded opportunity to see; it may be a view. A 'refuge' is an opportunity to hide; it may also be a viewpoint. A 'hazard' is the danger, which may lurk in the prospect or from which one can hide in the refuge. All three terms may be real in primitive human or animal conditions or may be symbolic in our own wider interpretation of scenic features.

Hazards that are more potential than likely can give 'sublime sensations of delightful horror' recognized by 18th century writers such as Edmund Burke, Jean-Jacques Rousseau and, on his Scottish tour, Samuel Johnson. Typical is the cautious thrill of precipitous places (Colour Plate 3). Wilson (2002)[16] observes that people's free choice of environs 'would combine a balance of refuge for safety and a wide visual prospect for exploring and foraging'. He also recognizes general preferences concerning depths of view, vegetation and water features.

Cultural upbringing determines a person's preferences as to what is important, what is remarkable or mundane in one's surroundings, and what exhilaration or depression this generates. English 19th century milords exploring the Alps were in no doubt as to the beauty of raw nature, whereas the people who actually lived in the Swiss valleys and acted as guides or porters to the foreigners were much more concerned about arduous survival. Guides lucky enough to be invited to visit cities such as London marvelled at the fine buildings and industrial activity they saw. So too can today's observers, well-travelled or familiar with television images of the world, still express strong feeling about the splendour or brutality of prominent structures or architecture. Technical interest is as much a stimulant to enjoyment of artificial landscape as is aesthetic pleasure in the wilderness.

Involvement in landscape is experienced by living in it, exploring it or travelling through it. *Local inhabitants* have a strong familiarity with their environment in which scenery is a significant but not the only element.Their normal reaction to any suggestion of change may be negative although, after the event, this attitude often softens. Concern of local inhabitants is probably most significant in residential situations – where familiar buildings and adjacent open spaces either constitute the local landscape or are foreground to more distant scenery. In working areas, such as industrial estates, people are likely to be less conservative and to welcome change if it implies improvements in employment conditions or opportunities. *Tourists or visitors* come to explore and enjoy scenery and its related cultural or ecological heritage, or to relax in more intimate comfort. Energetic enjoyment can encounter a variety of surprises, glimpses or wide views whether walking or riding, climbing or skiing, or participating in open water sports or golf.

Weather is an important determinant of the potential attraction of landscape. Rainstorms or blizzards and low cloud conditions are common occurrences in mountain or steep woodland scenery. Bad weather deters the less dedicated visitor but charges the watercourses and can change rapidly to sunshine, clear air and brilliant views that exhilarate everyone. As Spirn (1998)[17] points out: 'Weather is the ephemeral expression of wind, river, heat and cold from minute to minute, day to day, season to season whilst climate is responsible for long-term variations; and deserts, prairies and forests express enduring patterns.'

Travelling through landscape affords views that are enjoyable in themselves and often whet an appetite to return for closer acquaintance. Travel can be:

- on foot or bicycle, experiencing a landscape by passing slowly through it;
- by train, including on routes renowned for fine scenery or more fleeting glimpses of abrupt interest;
- on canals, since they became leisurely and picturesque; on rivers or lakes, or by ship into fiords or harbours;
- by bus, particularly for gregarious people who want easy access to popular scenery in an environmentally efficient manner;
- by car, by far the most common means of leisure transport at the beginning of the 21st century: offering great ease and flexibility, it is a mode that involves great use of fossil fuels and may have to be more constrained in the future;
- by aeroplane, even more consumptive of fuel: on many airline routes, window seats give astounding views of huge tracts of outstanding scenery – mountain ranges, lakes, jungles, terraced fields, coastlines – much more comprehensive, if more momentary, than any view available at ground level.

People's experience of scenery and their opinions of its value engender more professional assessment – as to what are the most significant features, how viewers' enjoyment may be enhanced, and what will be the effect of changes by construction or in land use. The same applies, perhaps with more objective analysis and explanation, to the significance of the non-visual attributes of landscape.

Priorities for land use

Development entrepreneurs do not doubt social and economic needs to build houses, construct new roads, or generate electricity; nor do people who are looking for a home, driving on congested roads or expecting that power should always be available at the throw of a switch. Other people are more concerned that the countryside should not be built over, that traffic should not increasingly disturb them, that valleys should not be flooded behind hydroelectric dams or that British power stations should not deposit acid on Scandinavian forests. Some commentators, after the abrupt stall in rich country economic growth in 2008, cast doubt on the necessity or desirability of ever-greater production or facility to meet seemingly inevitable, often hedonistic, demand. Yet the ever-increasing numbers of people on Earth require more land and more irrigation to grow enough food, expansion of cities to accommodate the burgeoning population in reasonable harmony, and transport to distribute that food and enable all these people to gain livelihoods; and we must still find enough land space for essential nature conservation and what can then be fitted in for human recreation.

People may be persuaded to modify their views and, in due course, compromises may result. Houses can be built or employment generated on disused land space or, if there must be greenfield development, on land less strategically needed for production and in a manner that will create a pleasant environment without spoiling the resources that favour the region already. Roads bypassing towns or trunk inter-city highways are welcome in countries where economic development has only recently taken off, several decades behind those where there is too much traffic already and construction of new highways tends towards futility. Damming of rivers is vital for water storage in many countries. Generating electricity is then a secondary benefit. Reservoirs necessarily flood the wider more fertile and settled parts of highland valleys, but – provided the storage and control remains operational – the benefits to much more fertile and settled lands downstream are so considerable that they should adequately afford necessary upstream compensation. There has to be political persuasion to make sustainable solutions publicly acceptable.

Sustainable solutions

People or their governments strive for equitable solutions to meet their logistic and social needs by:

- setting up institutions whereby land and other resources are assessed, mapped and allocated to optimal use or to conservation;
- public sector investment programmes, which follow official policies and rules in built development judged to be best afforded; and
- entrepreneurs or companies that take prime responsibility for specific projects on the land, hoping to make a profit within the rules.

Public authority or private companies identify the need and formulate a preliminary concept for a land development project. Professionals in the sciences and techniques concerned are then appointed to draw up a number of options fulfilling the functional needs of the project whilst meeting at least the stipulated levels of environmental protection. However, sustainable use of land and related resources requires much broader strategies than relate to particular projects and, in the long term, the best-explained technical solutions are likely to gain well-informed democratic support.

Notes and references

1 Forman, R. T. T. and Godron, M. (1986) *Landscape Ecology*, Wiley, New York.

2 Laurie, M. (1986) *An Introduction to Landscape Architecture*, Elsevier, New York.

3 Forman and Godron (1986) as Note 1, p594.

4 Schama, S. (1995) *Landscape and Memory*, HarperCollins, London.

5 Muir, R. (2004) *Landscape Encyclopaedia*, Windgather Press, Macclesfield.

6 Schama (1995) as Note 4, p10.

7 Jackson, J. B. (1986) 'Vernacular landscape', in E. C. Penning-Rowsell and O. G. Lowenthal (eds) *Landscape Meaning and Values*, Allen & Unwin, London, p79.

8 Young, A. (1998) *Land Resources: Now and for the Future*, Cambridge University Press, Cambridge pp12 and 132.

9 Livestock, for meat and dairy products, graze a lot of land, for example, in Argentina, Texas or (precariously) the Sahel region of Africa. On fertile land, meat is a less efficient form of food than crops.

10 Forman and Godron (1986) as Note 1, p83.

11 Bennett, A. E. (1999) *Linkages in the Landscape: The Role of Corridors and Connectivity in Wildlife Conservation*, International Union for Conservation of Nature, Cambridge.

12 Crowe, S. and Mitchell, M. (1988) *The Pattern of Landscape*, Packard, Chichester.

13 Schama, S. (1995) as Note 4, p227.

14 As illustrated in Wilton, A. and Barringer, T. (2002) *The American Sublime: Landscape Painting in the United States 1820–1880*, Tate Publishing, London.

15 Appleton, J. (1996) *The Experience of Landscape*, Wiley, Chichester, pp63 and 262.

16 Wilson, E. A. (2002) *The Future of Life*, Little Brown, London, p134.

17 Spirn, A. W. (1998) *The Language of Landscape*, Yale University Press, Newhaven and London.

2

The Impact of Construction

This chapter introduces construction processes (Section 2.1) and the consequences for land-related resources (2.2). It then describes the human skills and disciplines that have been developed, historically and recently (2.3), and the ways in which they are applied in land-sensitive location (2.4), layout (2.5) and design (2.6) of construction works.

2.1 Structures and construction processes

A *structure* can be defined broadly as any whole constructed unit. A narrower structural engineering definition is the load-bearing framework of such a unit. *Buildings* are structures with roofs to provide protection from the weather for people, animals or equipment.

Construction is making things or putting them together. In this book it is closely related to, and often synonymous with, *civil engineering*, which is the design, construction and maintenance of a wide range of large structures. It is generally accepted that the range of civil engineering structures includes those that perform certain roles of which these are the principal ones:

- Railways, roads, bridges and tunnels; junctions, stations and depots; airports; parking and servicing facilities for vehicles, rolling stock and aircraft.
- Harbours and port structures; coastal and estuarial engineering; sea defence or coastline management schemes.
- Dams for water storage reservoirs; river engineering, canals and hydraulic control structures; flood containment or re-routing; and flood protection.
- Foundations, structures and access required for hydroelectric schemes, wind and solar power systems and all types of thermal electricity generating stations.
- Foundations and frameworks for large or special buildings.
- Fixed services for towns, suburbs and industrial premises: surface drainage, water supply and treatment, sewers and sewage treatment facilities.

That is a functional list of civil engineering works. Such works can also be classified according to the materials and processes used in their construction. The materials include soils and rock – in excavated form, as graded for particular use or as incorporated in concrete or bricks; timber; iron, steel and other metals; and polymers (plastics) derived mainly from petroleum. The processes include demolition and site clearance, earthworks, placing concrete, assembling structural frameworks, constructing paving, pipelines and drains, and organizing construction sites.

Demolition of existing structures may be necessary to make space for new ones. Brick buildings have traditionally been knocked down by a heavy ball swung on a chain from a crane jib. More spectacularly, large concrete-framed buildings can be demolished by carefully placed explosives. Precautions to protect neighbouring buildings and to prevent excessive dust are necessary. But it is better if structures are designed and erected in the first place so that each element can eventually be dismantled and reused. Recycling of construction materials is an essential element of civil engineering.

Site clearance is the removal of extraneous matter necessary to make space available for construction. Some of this matter will be reusable, such as topsoil, fencing and certain building materials. Some will be an amalgam of organic and inorganic matter that is difficult to separate. The area to be cleared will have to include any extra space that is needed for construction activity that cannot be accommodated within the boundary of the new structures. This space may include that needed for storage and processing of materials (for example, to make concrete), parking and servicing of vehicles and machinery and accommodation of construction personnel. On an extensive site, extra land may not be necessary if construction can take place on each part of the site successively. But, where a river is to be dammed, it is often necessary to find space for materials, workshops and accommodation on land beyond the dam area; the great expanse of space within the reservoir may not be available because it usually has to be flooded before the project is more than about half complete. But the reservoir area may have to be partially cleared, for example of vegetation that might cause eutrophication of the water after impounding.

Earthwork is removal, transport and placement of soil or rock – to provide a new surface shape suitable for foundations or as level ground, to strengthen formations to withstand new loads or to construct a new landform, such as an embankment. In labour-intensive earthwork, men and women dig up soil, carry it on their heads and place it where required; or they get animals to carry it further. This process can be selective as to what dug earth is suitable for what purpose as fill. Mechanical earth-moving equipment – excavators, scrapers, dump trucks, conveyors – cannot so easily discriminate; but their use is essential for large scale work that has to be accomplished quickly, such as closing a river channel and diverting the water along a new route. As for sorting earth into specific particle sizes, this can be achieved on large projects by mechanical sieving equipment.

Tunnelling is underground excavation, often in conditions of geological complexity. In hard formations, constructors are at risk of accidental rock falls or sudden water inflow. In softer conditions, precautions may have to be taken to prevent collapse and subsidence.

Pouring concrete is casting artificial stone. Concrete is a remarkable material in that:

- it can be poured in fluid form at most ambient temperatures into any shape for which formwork has been prepared;
- it can be made into beams capable of taking tension as well as compression, either with steel reinforcing bars to withstand the tensile loads or by prestressing cables that hold the entire beam in permanent compression; and
- with the right constituents, it may be lightweight or self-compacting, retain particular thermal properties or even consume pollutants that might otherwise disfigure its surface.

Possible drawbacks to the use of concrete may arise:

- if there are shortages of suitable stone aggregate (gravel or crushed rock);
- because mass concrete in foundations and heavy walls is difficult to remove when a structure becomes obsolete and has to be replaced;
- if the concrete surface becomes discoloured, stained or cracked; or
- if there is not enough space at the construction site for the storage and mixing of materials, in which case it has to be mixed elsewhere and delivered quickly.

Precast concrete is used in elements, handleable by lifting equipment, which can be mass-produced at casting yards for repetitive inclusion in structures. Commonly these elements are beams but they may be piles, pipes, wall sections or architectural features. While it is difficult to move very heavy precast concrete elements on land, there have been many large-scale waterborne lifting applications such as for bridge foundation caissons or deck sections, immersed tube tunnels under estuaries and oil production platforms. In much simpler technology, lightweight concrete blocks are a common substitute for bricks, especially when brick clay is not readily available or for walls whose surfaces are going to be covered.

Structural frameworks take advantage of the strongest tensile materials available. Greek, Roman and Renaissance constructors used timber beams, stone masonry blocks and arch and buttress forms to construct buildings, bridges, aqueducts and pavements. Later, cast and wrought iron, then steel and reinforced concrete enabled stronger more slender elements of generally rectangular frameworks. More modern high tensile steels, other alloys and non-metals such as fibre-reinforced polymers (FRP) can be so light or flexible as to enable their incorporation in spectacular and seemingly daring structures.

A major advantage of lightweight high strength materials has been that they can be incorporated in very tall structures with a high capacity/base area ratio; so they are economic in use of city land space. Their construction has been possible because of the invention of very high self-erecting tower cranes, in parallel with precise calculations and measurement, control of mechanisms and training of skilled operatives.

Infill of floors, walls and windows can be completed in a range of materials – light aggregate concrete, bricks and facing or insulation panels designed for heat or cold conservation, fire resistance and visual appearance. There remains considerable skill in fitting these pieces accurately, maintaining them and when necessary renewing them – all to suit the structures and their ancillary access stairs, lifts, electrical and heating equipment and ventilation ducts and cables.

Paving, buried pipes and *open drains* have to fit into the final layout for space between buildings. In layouts, for example for new residential or industrial areas, civil engineers usually plan roads and drainage together. Other pipes and ducts have then to fit alongside roads and across ground leading to the buildings they serve. Also to be determined is the sort of vegetation that will grow or have to be managed on any fertile land still exposed; this and all paved or other ground surfacing must be considered in determining how storm run-off will be retained or flow. Note that subsequent building extensions or changes to surface space may have to be planned by agencies other than those who undertook the original design and construction.

Planning of *construction sites* has to arrange that construction can take place within the precincts of the final layout. The location of various permanent features will affect how the works can be tackled and particularly where the temporary facilities and activities can fit in, for example:

- space, that is to be paved or grassed over later, can be used to locate cranes and other construction equipment;
- wider spaces, where available, can be used for storage and processing of materials, preparing reinforcing steel or manufacturing precast concrete beams;
- final roads or drainage routes may or may not be suitable for construction traffic or site run-off; certain equipment or heavy machinery may require special access, such as by barge or helicopter; storm run-off through concrete aggregates or fuel residues may need special treatment before release to local drainage systems;
- boundary fences around the site, primarily for security, can incorporate public viewing points enabling interested people to view the processes and progress.

On large remote construction projects, such as dam building, complete townships may be built for the construction and supervisory personnel and their families. The main features of these are often designed for permanent settlements after the construction teams have left.

2.2 Consequences of construction

The effects of construction on land-related resources are both intentional and incidental. Dams and irrigation schemes can greatly increase production from arid lands; but structures in well-populated areas can displace farming territory. New transport routes can bring economic benefits and reduce congestion; but access to some communities may be severed and some heritage buildings may be threatened. Adverse impacts can occur to both wildlife and scenery; but some projects can be adapted to enhance both aspects. The nature and implications of the impacts of construction can be assessed as part of the planning process – for industrial or social sectors, for geographical regions, or for particular projects.

Impact on productive land

The most extensive productive land is that devoted to agriculture. Farm products, particularly food, are needed in proportion to populations; yet farming land is always threatened by development of human infrastructure. While it is inevitable that some building takes place on fertile land, there is often opportunity to build on less fertile or steeper ground and to avoid construction on flood plains where buildings are unnecessarily at risk. A positive and generally substantial role in increasing agricultural production can be control of rivers and construction of canals bringing water to otherwise marginal or arid lands. However, poorly managed irrigation or inadequate drainage of fields can damage soil quality, as can inappropriate cropping practice.

Particularly useful are pieces of land strategically suited to certain types of structures such as bridges or dams. Often on rock or infertile soil, such sites are much less extensive than the space needed for agriculture or building; but, because of the finite number of sites that are suitable, the best sites should be reserved for the most vital needs. This might mean, for example, that a bridge should not be built until the maximum likely and acceptable traffic demand for the whole district can be predicted, and a long life – or simple replacement – of the structure can be assured. The site may then be used to full advantage and there should be no likelihood of another location being sought subsequently. Similarly, no dam should be built, regardless of the need for river control, unless it can be built and operated such that it is not likely to fill with sediment, eventually rendering the site useless.

Impact on existing structures

Construction in already well-populated regions occurs as:

- new development or extension to existing built-up areas, where rural land has to be taken;
- new building where there are structures that are redundant and – if they cannot be adapted – may be demolished; and
- urban redevelopment in which some of the existing buildings, roads and services may be retained but others will be removed to make way for new buildings or street layouts.

In any of these cases there may be certain buildings, groups or situations that are regarded as cultural heritage. The fruits of construction that we have inherited include:

- buried ancient structures that have been covered in alluvium or interred in lava; protruding remains may be evident on open ground or in cleared woodland; buried walls, foundations and paving can be revealed by archaeological excavation;
- monumental stonework – placed as megaliths, cut into mother rock, or fashioned from quarried rock;
- castles and fortifications dominating the surrounding landscape and often affording good viewpoints;
- temples and shrines – pagodas, churches, mosques or mausoleums – less robust than castles but often architecturally or structurally splendid;
- country houses and their adjacent rural landscape – parks, gardens and agricultural or forest estates; villas or palaces, built as aristocratic opulence but since enabled by public access to become popular landscape heritage;
- architecturally historic buildings, of which at least the basic shell has survived and the remainder may be conserved or reconstructed;
- city heritage comprising buildings (their facades, domes and towers) and layouts (of streets and parks) with consequent foreground and vistas;
- industrial landscapes – associated with mining, factories, warehouses and transportation – in so far as their adaptation or demise has provided opportunities for landscape improvement or industrial archaeology; and
- major civil engineering structures if their continued existence remains essential (as in the Netherlands) or if they can become admirable features of the landscape (as are some ancient bridges).

The problem of heritage structures blocking new development or transport routes can be tackled by a number of options:

- It can be accepted that the structures and their setting are sacrosanct and that any new work

should be relocated, realigned or foregone altogether.
- A structure can be dismantled and rebuilt piece by piece in an appropriate setting elsewhere.
- The entire superstructure can be lifted onto skids or wheels and moved to a new location.
- It can be decided that the structures are not worth the cost of conservation because they are neither regionally scarce nor particularly fine examples.

The last option – demolition – is not necessarily disastrous if the best elements of workmanship, decoration or historic association are recovered. Many types of structure can be faithfully reproduced in a state in which subsequent maintenance will be less costly than for the crumbling original and with greater sympathy for contemporary surroundings.

Impact on wildlife

Construction and related development can:

- destroy or fragment biotic communities such as by clearing woodland, draining marshes or upsetting patterns of seasonal flooding;
- block land routes or watercourses which serve as wildlife corridors;
- disturb wildlife habitat, for example by traffic, dust or artificial lighting; and
- pollute water or land.

Nature conservation avoids this damage by resisting or mitigating harmful forms of development and preserving or improving the state of natural resources that still exist.

Conservation – for some already defined biological purpose – can be incorporated in construction planning that:

- avoids the most sensitive areas altogether;
- protects or adapts any habitat that can be retained at a construction site;
- creates alternative or new wildlife habitat, corridors or crossings; and
- after construction, continues to manage the vegetation, perhaps the flow of water and possibly the animal communities that result.

Successful conservation is only likely to result from careful application of vital information gleaned from practical research (ARC, 1992)[1] and in the light of guidelines prepared for particular species in particular situations. There also needs to be scientifically rigorous monitoring of the success or failure of habitat creation schemes.

Avoidance of sensitive areas has to be practised in the light of priorities set by some sort of objective analysis. Thompson (2001)[2] mentions classifications of habitats in Europe 'in terms of their sensitivity to perturbation' and using 'that classification in formulating mitigation measures'. Risk analysis[3] may be appropriate for determining, for example, the ability of salt marsh plants to recover from oil spillage near proposed port development, or how land drainage can be most effective for agriculture or least destructive of wetland. However, Forman and Godron (1986)[4] point out the difficulties in practising risk assessment in varying conditions and suggest simpler recognition of contrasts, for example between dry and moist years.

Protection of habitat is afforded by some sort of seasonal or permanent barrier. These confine some species within the protected zone as well as keeping others out. *Adaptation of habitat* could involve managing a watercourse ecosystem, for example, by an altered pattern of flow. *Adaptation of constructed works* could be modification of the banks of a new watercourse to promote growth of particular species.

Creating habitat may be attempted in mitigation of what is being destroyed. However, habitat cannot be reconstructed precisely. It is impossible to replace the complexity or maturity of an ecosystem or to replicate the microclimate, topography, soil conditions and boundary features which existed before disturbance. It may even be impossible to replace identical seeds. Nevertheless, a wealth of semi-natural habitats has been created after interference. So it is possible to create conditions in which natural ecological processes may – in time and among new structures, landforms, waterways and drainage – create habitats perhaps similar to those previously existing and certainly suited to the new situation. Completely new semi-wild areas can also be created in space made available by construction. Wild flowers planted in motorway cuttings are a welcome addition, provided that the species are suitable for the conditions and contribute positively to regional ecological capital.

Management of vegetation may be necessary to achieve a semi-natural equilibrium that provides security and rough edges for wildlife habitat and functional space for human activity. Advantage can also be taken of any regulating mechanisms introduced with engineering works – such as control of water levels.

Impact on scenery

New construction can affect existing visual opportunities:

- by obstruction, usually at close proximity;
- by intrusion, mainly into longer distance views; and
- by its visual impact on the character of the place at which construction takes place, in a range from sympathy to discord.

The first arises or is avoided in the layout of structures, the second in their location, the last in their design.

Visual obstruction occurs when a significant part of a particular view is obstructed by the construction of a structure or earthwork. Obstruction may also comprise exclusion of light or introduction of gloom where people live or work. Where obstruction of a fine broad view is the inevitable price of essential construction, then there could be an obligation to replace the viewpoint. This might be adjacent to, on top of or within a new structure; or foreground could be cleared elsewhere – whatever solution maximizes people's opportunity to take advantage of it. Similarly, where a narrow vista – such as along a passage – is to be blocked, it may be possible to provide space for a new one on a different alignment. Simmonds and Starke (2006)[5] suggest that the best view is not always a full view and that many views are enhanced by a compatible close framework. However, the view in Photo 2.1 is not one that welcomes partial obstruction by a new office building beyond the established building line at Porthmadog Harbour railway station (Photo 2.2) or the front of the wharf (Photo 2.3).

Photo 2.1 Classic view of the Moelwyn Mountains and the Glaslyn estuary since the latter was reclaimed by construction of an embankment (the 'Cob') in 1810. This view was obstructed, more than 150 years later …

Photo 2.2 ... by construction of an office building, partially blocking the view from the Porthmadog Harbour steam railway station at the end of the Cob ...

Photo 2.3 ... and from the front of the wharf, on which slate shipment sheds have been replaced by holiday homes.

Photo 2.4 Although the neutrally coloured textured concrete is perhaps not, at close quarters, a blatant feature of this new upper storey converting a former chapel into a civic centre ...

Visual intrusion occurs when new structures or activities adversely affect the perceived quality of a view without materially obstructing it. Intrusion can take place anywhere within a visual envelope encompassing all places from which the structure can be seen. Photos 2.4 and 2.5 show close and distant views of a structure that intrudes into an otherwise historic setting.

In wild country, almost any form of construction is likely to be considered intrusive. In semi-natural rural landscape, the form of structures or the alignments of roads intrude when these fail to blend with existing buildings or linear features. In views over cities from high ground or buildings, intrusion may be incursions into islands of greenery, tall buildings into skylines where there were none or, where there were many, profiles radically different from the rest. Photos 2.6 and 2.7 illustrate distant and closer aspects of structures that some viewers consider sharply intrusive.

Measuring the severity of visual intrusion is as difficult and contentious as assessing the value of scenery itself. Comparison with similar effects of construction that has already taken place may be the most objective approach. The nature of intrusion can be demonstrated by depicting particular views before construction (by photographs) and after (by photomontage or virtual reality representation).

Visual impact on site character depends on whether the new development:

- dominates its site: for example, a power station is so large as to constitute a major feature on its own; in the flat part of the Trent Valley in the English East Midlands, the landscape *is* the power stations and their cooling towers; the fields are merely their foreground;
- is intended to appear subordinate to the existing scenery: a visitor centre, an electricity substation or a water utility pump house may be designed to blend with the historic and semi-natural features that already typify the scene; or the new buildings may be screened by trees or natural topography to hide them from the main viewpoints.

In either case, the visual impact depends on what sort of balance or imbalance is struck between the essential purpose of new construction and the characteristics of the existing site landscape. That landscape can be the surroundings of a dominant structure or the setting of a subordinate one.

The colour plates at the end of this chapter illustrate various effects of construction on scenery. They range from splendid natural features that dwarf any cautious human approach (Colour Plates 1–4), through various dominant structures (Plates 5–12) to a range of colourful features in semi-natural or built-up landscapes (Plates 13–16). A feature of many colour images of structures in landscapes is the varying effect of green vegetation.

Impact assessment

Assessment of the incidental consequences of construction and recommendations as to action

Photo 2.5 ... its shape does intrude strongly on the view, from the town walls, of the three bridges, castle and high street of Conwy, North Wales.

Photo 2.6 Montreux, Switzerland. The early 20th century lakeside and hilltop buildings and the rural hinterland were intruded upon more recently by the slender viaduct of the bypass motorway and by an apartment building of exceptional height ...

Photo 2.7 ... Closer to the dominant and elegant viaduct, it can be seen that the bottom of the valley was already occupied by a less than elegant factory.

that should be taken to remove or mitigate adverse changes can be undertaken:

- as a formal environmental impact assessment;
- as an inherent part of planning and design, with the advice of specialists;
- as a long-term iterative process thereafter.

The first is partly a check that design is being undertaken within accepted norms. The second is the way in which large construction projects were planned in the past, often successfully, occasionally with brutal indifference, nowadays perhaps with apprehension lest some specialist or social interest is overlooked. In any case, it is easiest if the planners of a project are aware from the conceptual stage of any issues described as environmental. Certain consequences always attract comment – such as whether a new structure will enhance or blot the landscape. Sometimes local interests have been recognized at an early stage, such as in the provision of fish ladders around Scottish dams. Today more precise analysis tends to be expected of the effects on all species – from great crested newts to floral indicators of ancient woodland. Assessment of impact on land resources normally encompasses:

- recognition of any protective land classification that has been applied, such as a nature reserve or area of outstanding natural beauty;
- description of the existing situation in terms of land use, buildings, natural resources or scenery; and calculation and presentation of how these will be affected by construction (including with any proposed mitigating features); and
- some sort of evaluation as to the suitability and compatibility of the project, preferably by comparison with changes that have resulted from similar land development elsewhere.

Impacts can be assessed at three levels:

1 Sector, such as for power stations development in a well-populated industrial region, for dams

throughout a highland area or for mineral extraction within a county.
2 Regional, covering the effects of all forms of built development on all land resources, balancing needs against opportunities and offering optimal strategies for the region as a whole.
3 Project impact assessment, covering more precise planning and design for a particular development.

We must plan, design, construct and manage projects with the aim of fair assessment. It is worth considering first how the professions to do this have evolved.

2.3 Skills, artistry and ingenuity

Throughout history, building and construction have combined:

- skill, exemplified for many centuries by stone masons and still evident in fine brickwork or manipulation of composite materials;
- artistry, including painting, sculpture and architecture, but also the design of elegant bridges, towers and their adjacent landscapes; and
- ingenuity, for example, in the way in which structures are erected.

Ingenuity and engineering can be synonymous. The first is cleverness at inventing or constructing. The Latin word *ingenium* meant a talent or device, as did the Old French derivative *engin*; thence came the English word 'engine', generally a powered mechanism, while 'engineering' became the application of science to the design, building and use of machines and structures. An 'engineer' is a person qualified in a branch of engineering.

Construction and structural design

Giotto di Bondone (c.1266–1337) has been credited as an originator of new qualities in Italian painting that heralded the start of the Renaissance. However, a close relationship between art, architecture and engineering was fundamental at that time. Artists were regarded as artisans, albeit very skilled ones, at a time when master masons, also artisans, were building cathedrals. Giotto himself spent much time in the production of frescoes – huge works in plaster involving skills in placement as well as brilliant aesthetic sense – but such was the impact of his work that he and later painters became esteemed members of society. In 1334, Giotto was appointed chief master of all the city works of Florence, a role which covered everything from construction of the cathedral campanile to the city fortifications and probably public utilities – indeed what came eventually to be classified both as architecture and engineering.

Subsequent Renaissance polymaths included Filippo Brunelleschi (1377–1446), an architect, designer and constructor of the remarkable dome of Florence cathedral as well as a skilled goldsmith and sculptor; Leonardo da Vinci (1452–1519) – artist, scientist and mathematician – also prepared designs for military engineering, vehicles, bridges and hydraulic works; the great sculptor and artist Michelangelo (1475–1564) was also a leading architect and engineer.

The term 'civil engineer' was introduced centuries later by John Smeaton (1724–1792) to distinguish his work from that of military engineers.The Institution of Civil Engineers (ICE) was founded in London in 1818 and given a royal charter in 1828. Later the mechanical engineers founded their own institution as, in due course, did electrical, chemical and other engineers. Civil engineers continued in the fields defined at the beginning of this chapter.

In practice, further specialities developed including:

- geotechnical engineering specializing in foundations, earthworks and adaptation of landforms; and
- structural engineering using a variety of traditional and new materials for structural frameworks and with considerable mathematics/ mechanics-based advances in design, described by Chen (2009).[6]

In the 19th century, ICE defined its profession as directing the great sources of nature for the use and convenience of man, more recently[7] as the practice of improving and maintaining the built and natural environments to enhance the quality of life of present and future generations.

Civil engineers and *construction companies* design and make structures, particularly of a non-building type and often on a large scale. Certain engineering concepts and techniques are applied in special purpose or large buildings and in services for medium and large communities. Construction also makes a substantial contribution to agriculture where the transfer of large quantities of water is needed for irrigation. Industries supporting construction include mining and quarrying, processing of raw materials, design and manufacture of mechanical equipment, production of steel, plastics and cement, and prefabrication of standard structural units.

Architects and *building firms* lead the design and construction of most buildings – from standard housing units to city skyscrapers. Architects also often handle aesthetic appearance of engineered structures and cater for the people in them. Teams of architects and engineers have achieved major structural triumphs, whether by individual polymath leadership (Renaissance builders or modern architect-engineers such as Santiago Calatrava) or through joint effort.

Landscape architecture

Forman and Godron (1986)[8] defined landscape architecture as the arranging and modifying of natural scenery over a tract of land for aesthetic effect. Going further than this traditional 'scenic' concept, Hopper (2007)[9] says it mediates between (human) art and (untamed) nature or, in less abstract terms, 'is concerned with the design of the external environment from the drip-line of architecture to the project property line'. But many landscape architects and engineers believe that aspects such as the scenic setting and the land drainage system extend well beyond that line. Both disciplines are involved, often jointly, in the ways that Hopper suggests landscape architecture is undertaken, 'by manipulating the materials of the Earth and the products of human industry to modify and create outdoor space'.

The first prominent practitioners in adapting extensive outdoor space for aesthetic effect were probably the English 'landscape gardeners' Humphry Repton and Lancelot 'Capability' Brown. Brown later practised architecture while architects such as William Kent or Sir John Vanbrugh also designed the parks around their stately buildings. According to Hopper (2007),[10] the French title for a landscape designer (from 1804) was *architecte-paysagiste*. *Paysage* is usually translated as landscape and *paysagiste* as a landscape painter. 'Landscape architecture' was used in connection with painting as early as 1828 (Turner, 2001).[11] But the American F. L. Olmsted may have coined the term 'landscape architect' after he visited France in the 1850s (Laurie, 1986;[12] Hopper, 2007).[13] However, Clark (1955)[14] maintained that Olmsted did not like the term, which was probably introduced by his partner Calvert Vaux in the title of a document concerning their appointment as landscape architects to design Central Park, New York. In any case, Olmsted was prominent in bringing landscape design into urban settings. He 'argued that the growth of cities was inevitable and fundamentally beneficial to society, and that incorporation of parks and natural landscapes into the urban fabric could counter many of the negative effects of this growth' (Sendich, 2006).[15]

Laurie (1986)[16] defined four types of landscape architectural practice:

1. Landscape evaluation and planning, in terms of ecological and natural science as much as visual quality.
2. Site (landscape) planning, giving that quality to the constructional aspects of site planning.
3. Detailed landscape design – a 'process through which specific quality is given to the diagrammatic spaces and areas of the site plan… selection of components, materials and plants and their combination in three dimensions as solutions to well-defined problems such as entrance, terraces, amphitheatres, parking areas and so on'.
4. Urban design.

City planning is perhaps a separate discipline, albeit one with support from architects in determining what are viable built communities, civil engineers in planning the provision of roads, drainage and other infrastructure, and landscape planners in spatial arrangement. City skylines can be 'total works of art' (Bacon, 1978)[17] and are therefore important elements of both architecture and landscape design.

Planning

A plan can be a scheme or a proposed method of proceeding, for example, for designing and building a structure. A plan can also, in an engineering sense, be a drawing made by projection on a horizontal plane, which, in large-scale geographical terms, is a map. In the context of this book, a plan for construction is both a scheme for defining where and in what form construction shall take place, and a proposed method for detailed design. Maps, showing different topographical features and other characteristics of land, are important tools in formulating a scheme.

Planning is the process of making plans and has specific connotations in the control of land development by local government authorities. Planning permission is formal licence for building. Various people participate in land and construction planning, such as:

- Professional planners, many of whom serve or advise government authorities in control of land use. By training they may be city or rural planners, architects, municipal engineers or development economists. In cities there may be community and transport planners. In developing countries there may need to be expert advice from soil scientists and agronomists on agricultural land layouts and from anthropologists for designing settlements.
- Civil engineers – for assessment of foundation and tunnelling conditions, concepts of superstructures, earthmoving requirements and the layout of drainage, roads and services – all with specialist advice where necessary from hydrologists, hydrogeologists and geotechnical engineers.
- Landscape designers, who analyse the natural elements and visual opportunities of the site, participate in the layout and earthwork plans, and formulate solutions for surface finishes, planting and other treatment of outside space – to be adopted later at appropriate stages of construction.
- Other specialists in outdoor planning, particularly ecologists.

The main components of interdisciplinary construction planning are location, layout and preliminary design. The first determines how land resources shall be used. The second includes consideration of the created landscape, the third what structures will look like.

2.4 Suitable locations

A location is the chosen site at which a project may be built.

Location to suit regional plans

A region may be a geographical area (such as a river catchment) or an administrative division (such as a county, state or province) or the whole of a small country. Within a well-inhabited region, it is likely that government may have already set out plans or rules as to where new development may take place, for example:

- regional master plans defining specific areas or zones within which certain types of development are permitted or encouraged;
- designated areas where certain or most development is *not* permitted; these include heritage or nature conservation areas, green belts and recognized islands of tranquillity;
- basic intentions that new road, rail or power connections may be needed between certain towns;
- policies, guidelines or regulations regarding the location of particular activities or land uses, which developers have to satisfy.

Where official plans do not exist, prospective developers may need to devise their own, at least in outline, so as to satisfy whatever authorities are in power or may be in the future. Maps, descriptions and measured data can be drawn up – of the whole region or the environs of possible development sites, showing:

- any prescribed restrictions or zones;
- physical characteristics of the land, including geographical features, liability to flooding, nature

of settlements, and possibly land ownership data; and

- intrinsic suitability of areas of land for specific types of development.

These enable selection of locations to be undertaken in phases of exclusion, technical assessment and identification of aggregate land suitability (McHarg, 1992).[18] Modern urban planners tend to give emphasis also to the location and capacity of existing infrastructure and its necessary extension.

Functional requirements and topographical opportunities

At broadly indicated locations, specific areas of land have to be delineated that can satisfy the functional requirements, for example, for space permitting particular configurations, for ground conditions suitable for certain types of structure, or for access to particular resources or already existing infrastructure. At the same time, advantage can be taken of the prevailing topography. Hillsides offer fine views for residential areas. Buffer zones, necessary for operations or safety around airports or petrochemical plants, can offer extra opportunity as conservation, recreation, storage or special activity space. Particular obstacles must also be recognized – such as key strategic or heritage structures that must be left in place, wildlife reserves that must not be disturbed, or satisfactory compromises whereby these built or natural assets can be modified without destroying their viability.

Thermal power stations need adequate supplies of cooling water; so do many heavy industries. Either may require surface or underground storage space for hazardous and other wastes. Before road transport became ubiquitous, factories had to be sited adjacent to wharves or railways, and houses close to work places. There still has to be longitudinal space for railway sidings, parking areas or container yards; and passenger transport must serve the workforce's living areas. Topographical opportunities include narrow valley constrictions suitable as dam sites with wider valley space for reservoirs upstream. Hydroelectric power schemes can be located where there are steep drops in river level. Viewpoints can enhance open air amenity areas.

Once a project site is selected, ensuring the compatibility of structures with their surroundings can be completed:

- in the layout of individual elements so as to provide suitable lines of sight, foreground and skylines; and
- in the design of each structure to make its appearance acceptable.

2.5 Appropriate layout

Planning a layout for a functional project involves appreciation of site characteristics, assigning positions for certain main elements and then allocation of remaining ground space and ancillary structures.

Site characteristics

Features of a stretch of ground are determined by topographic survey and geotechnical investigations as well as recognition of landscape features such as vegetation or items of built heritage. Contour mapping records the shape of the ground surface and its drainage pattern. In relatively flat country, even gentle slopes are critical in planning layouts for irrigation schemes or drainage. Sloping ground is relevant to views upward and downward and to the character of foreground.

Geotechnical investigations – by boring and sampling or electrical or seismic probing – determine typical or specific subsurface conditions. Where is there sound rock or what are the properties of softer material? This sort of information is necessary in locating where there are suitable conditions for particular structures and what sort of foundations or ground strengthening are appropriate.

Vegetation and man-made features on the ground can be depicted on maps or illustrated directly on photographs. Further, high resolution photographs (such as satellite imagery) can be combined with topographical and geotechnical information in 'remote sensing'. For example, growth of certain types of vegetation can be linked to possible occurrence of particular soil conditions and even the likelihood of subsurface minerals.

Positioning of main elements

Positions can be determined at key points, as fixed lines, or by less place-specific requirements concerning proximity or isolation, connection or orientation.

Key points could be the offtake location for a gravity-fed aqueduct from a river; or it could be that of any existing structure which is to be incorporated in the new works. A *fixed line* could be that of a dam, a waterfront, a railway track or a vista. *Proximity* is required for transfer of goods or for transactions between people; or comparative *isolation* may be needed for hazardous storage facilities. *Connections* could be road junctions or electricity substations, each with certain space requirements – on preordained routes or adjacent to specific structures, such as bridges or underpasses, power stations or factories. *Orientation* relates to light and microclimate (catching or eluding the sun or weather), to the direction of runways for aircraft to take off or land into the prevailing wind, or to views available from buildings.

Having determined the position of each main structure, its actual extent on the ground must be delineated. For example, electricity generating sets, petrochemical towers and tanks, or blast furnaces maybe disposed in blocks or rows; or the land space required in a building may be reduced with multi-storey construction or basement levels. Thus the permanent positions will be fixed for the main functional structures – such as dams, power stations, quays, railway tracks, buildings or monuments. For minor roads, small ancillary buildings, overhead cables or buried pipelines, there may be more options as to routes and for separate or combined use of the remaining space.

Allocation of remaining ground space

Open space can be functional – for storage or parking areas. It can be semi-functional – such as a buffer zone – or purely for amenity or decoration. It can be paved over, gravelled, grassed or planted with thicker vegetation and managed or adapted over time to suit changing circumstances.

Open ground can be allocated for specific activities, for example outdoor processes; but there may be a choice as to where and in what pattern the required space is delineated. Open storage space may also be required for stockpiles of raw materials, for mechanical equipment or for vehicle parks. In the configuration of most of these, it may be possible to ensure that any 'spare' ground left over is not fragmented; such space can then be shaped with a view to possible nature conservation, human amenity or subsequent extension of economic activities.

Preliminary land space allocation can take account of the various functional needs, environmental opportunities and dictated priorities. Iterative adjustments can then be made to total layouts and any adjustments made to plans for earthworks and drainage features. Visual issues include simplification of over-complexity, relief of stark structural appearance, the arrangement of vistas or fences and the role of vegetation and its management – from mowing lawns to thinning or pollarding of trees.

2.6 Functional and aesthetic design of structures

Beauty and function

Elegance and utility were united in Renaissance engineering. Beautiful bridges tended to reflect functional simplicity, based on structurally effective design. Sufficiently massive stone piers supported more daring masonry spans in a way that expressed the magic of arch action and created a form that became structural elegance in subsequent bridges and in new materials.

Wrought iron and steel, developed in the 19th century, produced some less than elegant truss and girder solutions as spans and loads increased but before the materials' properties were fully understood. Isambard Kingdom Brunel's railway crossing of the River Tamar combines a suspension bridge with a high bow girder and stiffeners, a belt-and-braces approach compared with the simple lines of the road suspension bridge built beside it a century later. However, Brunel's bow girders have a massive elegance, as do his triumphal suspension towers. Modern high tensile steel and a capability to produce non-uniform beams and sections enables daring and thus aesthetically exciting engineering, providing that the material is manufactured under the strict quality control needed to guarantee its performance.

In the 1920s, the German Bauhaus school of architecture saw the 'idea of physical function… elevated to the dominant principle of design, almost to the exclusion of any other. The idea of beauty was not derided, but was thought to reside in fitness for physical use, in function itself' (Preece, 1991)[19]. Frank Lloyd Wright took a completely different approach, making a house just part of the scenery in which it is located. Perhaps Bauhaus concepts are most appropriate to the town, Wright's to the country?

By 1996, Britain's Highways Agency[20] found it 'generally accepted that expression of function is the basis of good design, and that any adjustment or addition required to improve the appearance should exploit this functional basis and not run counter to it'.

Summarizing structural approaches to conforming with nature, Wines (2000)[21] finds that 'a great part of the solution is technological, but filtered through a study of the way nature solves its own engineering problems and how resourcefully energy and materials are converted to function'. So part of this affiliation between beauty and function relates to the parallels between natural processes and construction techniques, for example, the flexibility of trees in the wind compared with that of tall buildings resisting earthquakes, or the similarities in the properties of timber with those of artificial fibre-reinforced polymers. Also relevant are natural solutions to engineering problems, for example, in directional river training, passive coastal management or the use of vegetation in slope stabilization.

Form and engineering

To meet any performance requirement, there can be different structural options. For example, dam walls or bridges may be of various forms that suit the width of a gorge and the strength of its rock. Civil engineering design options in determining conceptual forms concern:

- the type of foundations, determining what structural loads can be borne;
- structural framework, determining the bulk or grace of the basic shape;
- methods of construction, with different demands for materials, machinery and labour, working space or access, and any seasonal climatic or hydrological limitations.

A number of choices may arise in the adoption of particular forms – for the shape and proportions, colour or texture of surfaces, or for embellishment.

Appearance of structures

Shape reflects the massiveness of a structure or the form of its more slender elements. For example, in stonework, there is mass in a pyramidal monument or a defensive bulwark but slenderness in tall pillars or flying buttresses. The clean curving face of a concrete arch dam is its outstanding feature, not least because of the evident ingenuity in its construction and the audacity with which it holds back a reservoir. The shape of a tower can be equally inspiring. The fascinating curvature in power station free-draught cooling towers or the cables of suspension bridges became familiar sights to an older generation who saw them as more graceful than the straighter profiles of forced draught cooling towers or cable-stayed bridges.

Proportions relate the dimensions of part of a structure to those of the whole. In bulky buildings, different proportions are devised for different options for providing the required volume of internal space. In more complex frameworks, such as those of bridges, proportions – of heights to spans, of struts to decks and of different sections of long viaducts – can be analytically pleasing or arbitrarily clumsy. To those who see them, shapes and proportions can seem threatening or comforting, whether of man-made structures, the shape of planted woodland or natural rock skylines. Cloudy weather can be dismal while sunshine draws out any brightness.

Colour can be that of the basic material, such as in rock revetment, brickwork or poured concrete; or can be added by exterior cladding, stucco or paint. Rock weathers, often sympathetically, occasionally detrimentally; cladding may need renewal to conserve its more extrovert character. White is a colour well-suited to massive isolated structures, particularly when the sun is shining. White walls reflect light and keep small buildings cooler in warm climates. But the brightness and cheerfulness of surfaces can deteriorate if they are not kept pristine. Where brightness is not appropriate, rocky grey seems natural in many materials – indeed, so far as to confuse the distant onlooker as to whether construction is in steel, concrete or masonry.

Yellows, blues and bright reds are not well-suited to most natural backgrounds but offer welcome contrast among closely-built houses or other urban structures. Green is for vegetation; it can turn to gorgeous lighter hues in deciduous autumn or be darkly sinister in evergreen forestry. Buildings, if they are to be subordinated in green surroundings, are perhaps better partially concealed by the trees. Browner camouflage colours may be appropriate in more open arid sparsely vegetated scenery.

Texture can be as expressive on hard walls or paved surfaces as on cut hedges or mown lawns. Stark plainness can be striking where modesty is not needed and if inaccessible to graffiti artists. Texture can be that of plain or patterned bricks, dressed stone or concrete impressed by patterned formwork. Carved ashlar or moulded Babylonian ceramic tiles show up in brilliant relief, tracery has been evident in ceilings or walls since medieval times, and decorative ironwork for more than two centuries. Glass is a special surface in that it reflects as well as transmits light; it is not strong unless laminated with tough transparent polymers.

Decorative embellishment of castles, palaces and municipal buildings – such as battlements, spires and Moghul or Gothic windows – became such symbols of splendour that they were later incorporated in 17th century Persian bridges, baroque architecture, Victorian castellated water intake towers and imperial railway stations.

Such decoration is seldom fashionable today unless it is specifically functional or in shapes, such as spires or domes, that remain traditional in regional culture. Petrochemical works, large cranes and industrial chimneys can provide bizarre skylines, often of a temporary nature; but it is not unknown for their configurations to be designed to present a profile that is at least intriguing.

Architectural licence accounts for classical (Greek or Roman style) columns, even friezes, in 19th century public buildings and railway stations. More modern functional features that can nevertheless be decorative are exemplified by impact blocks in hydraulic stilling basins and plates on the exposed end of anchorage or tie-bars strengthening retaining walls or bridge spandrels. Works of art can be incorporated in a range from paintings commissioned along pedestrian passageways, through the patterns of motorway overbridge abutments to concrete grids stabilizing hillsides. If function is normally the dominant theme, decoration can be a welcome enhancement. The main way in which functional design can affect structural appearance is in the choice of form and construction materials.

Design for sustainable construction

The technical and economic justification of a project should be based on an anticipated lifetime. If the structural durability or functional need is to be short-lived – say for less than a human generation – then project design should take into account either its eventual replacement or restoration of the land resources involved. However, for many large structures, it is evident that their productive or aesthetic function continues for centuries. So means have to be devised for maintenance or renovation far into the future. A dam has to continue to provide a reservoir unfilled by sediment. Harbour walls have to provide protection against tides and storms even if channels are deepened. Buildings, which accommodate people or their machines or products, are less critical as to their location and form but have to be readily capable of adaptation or dismantling when lifestyles or industrial processes change.

Particularly for modest-sized locally-constructed systems in difficult ground or drainage conditions, structures may better be adaptable than robust. Referring to mountain valley irrigation channels, Yoder (1994)[22] observed:

> Designing *for sustainability does not necessarily mean designing for permanence. Whether the structure lasts for one season or for a thousand years is independent of the concept of sustainability. Sustainability in the concept of hill irrigation refers to the ability to mobilize resources to meet expected needs on a continuing basis to keep the systems operating within* tolerable limits.

The design life of structures and the extent to which they should be repaired or adapted during their lifetime can be determined by economic, technical and risk assessments of the life-cycle performance of each design option.

Inherent in both the endurance and adaptability of construction materials is the way in which they are used. Some principles are suggested:

- Massive constructed landforms should be designed to last forever. But sufficient adjacent space and adaptability should be allowed so that the landform can be extended or modified if this is necessary to meet continuing, perhaps unforeseen, future needs.
- More modest earth, rock or concrete forms should be capable of being broken up and recycled in other construction; and specifications for new work should permit such reuse.
- Preformed elements – such as steel sections, precast concrete beams, even facing and roofing units – should where practicable be recoverable and of standard dimensions.

Sustainable civil engineering makes optimal use of resources. In densely populated countries, the actual area of land surface is a scarce resource that should be conserved by planning policies. The way in which the land's space, stability or fertility is altered can affect any subsequent use for which it may be needed. In less crowded regions, much of the value of land resources continues to lie in conservation of the more natural and scenic aspects.

Notes and references

1 ARC (Amalgamated Roadstone Corporation) and the Game Conservancy (1992) *Wildlife After Gravel*, Game Conservancy, Fordingbridge, Hampshire.

2 Thompson, S. (2001) 'Natural Habitat' in Carpenter, T. G. (ed) *Environment, Construction and Sustainable Development*, vol 1, Wiley, Chichester, p74.

3 Risk analysis – as a comparison of the consequences of alternative options – is referred to again with regard to floods (pp41–42), slope stability (p259), optimum road maintenance (p60) and water control works (p11). Risk has been defined as the hazard times the potential worth of the loss, where hazard is the probability of a particular damaging phenomenon which causes a certain degree of loss or damage (Yoder, R. (ed) (1994) *Designing Irrigation Systems for Mountain Environments,* International Irrigation Institute, Colombo).

4 Forman, R. T. T. and Godron, M. (1986) *Landscape Ecology*, Wiley, New York, p524.

5 Simmonds, J. O. and Starke, B. W. (2006) *Landscape Architecture: A Manual of Environmental Planning and Design*, 4th edition, McGraw Hill, New York, p188.

6 Chen, W-F. (2009) 'Seeing the big picture in structural engineering', *Civil Engineering*, vol 162, no 2, pp87–95.

7 ICE (Institution of Civil Engineers) (1992) *Members' Guide*, ICE, London.

8 Forman and Godron (1986) as Note 4, p594.

9 Hopper, L. J. (2007) *Landscape Architecture Graphic Standards*, Wiley, Hoboken NJ, p5.

10 Hopper (2007) as Note 9, p5.

11 Turner, T. (2001) 'HyperLandscapes', *Landscape Design*, no 304, pp28–33.

12 Laurie, M. (1986) *An Introduction to Landscape Architecture*, Elsevier, New York, p1.

13 Hopper (2007) as Note 9, p5.

14 Clark, H. F. (1955) 'Landscape architecture' in *Chambers's Encyclopaedia*, vol 8, Newnes, London.

15 Sendich, E. and the American Planning Association (2006) *Planning and Urban Design Standards*, Wiley, Hoboken NJ.

16 Laurie (1986) as Note 12, p11.

17 Bacon, E. N. (1978), *Design of Cities*, Thames & Hudson, London.

18 McHarg, I. L. (1992) *Design with Nature*, Wiley, New York.

19 Preece, R. A. (1991) *Designs on the Landscape*, Belhaven, London, p10.

20 Highways Agency (1996) *The Appearance of Bridges and Other Highway Structures*, Her Majesty's Stationery Office (HMSO), London, para 1.5.

21 Wines, J. (2000) *Green Architecture*, Taschen, Cologne, pp233–234.

22 Yoder, R. (1994) *Designing Irrigation Systems for Mountain Environments*, International Irrigation Institute, Colombo, p11.

Plate 1 Yosemite, California. The cliff of El Capitan can be viewed from the comfort of the valley floor. The tourists' cars are hidden among the trees.

Plate 2 The Rhine Falls near Schaffhausen, Switzerland. The falls can be viewed from terraces on both banks and from the railway and pedestrian bridge just upstream (photo by Robin Carpenter).

Plate 3 Les Galeries, Champéry, Switzerland. A modest timber paling secures the thrill of walking across the face of the precipice. The chalets seen in the valley below …

Plate 4 … also enjoy views of high peaks and pristine snowfields.

Plate 5 Pont du Gard, France. Great structural elegance constructed in local stone.

Plate 6 Ely Cathedral, England, constructed of imported stone, retains a rural setting on its southeastern side.

Plate 7 Edinburgh, highly scenic capital of Scotland. The castle stands aloft on its own crag, the railway takes the lowest route through what was once a marshy lake, the picture galleries sit on the man-made Mound, and most of the rest of this view is the public Prince Street Gardens.

Plate 8 Near to the city centre of Fribourg, Switzerland, are wild cliffs, thick natural vegetation, partially trained river channels and historic and modern structures.

Plate 9 A transport and development corridor near Montreux, Switzerland – in sequence from the bottom: lakeside promenade, railway, streets and bypass motorway. Higher up, the forest and crags are still dominant natural grandeur, except where there has been quarrying (towards the right) ...

Plate 10 ... This section of the same motorway viaduct cannot be concealed. But its slender deck and piers offer a clean contrast with woodland splendour.

Plate 11 The M3 motorway cutting through Twyford Down near Winchester, England. The stark whiteness of the freshly cut chalk mellows with time ...

Plate 12 ... as has happened where the older M40 route cuts through the Chiltern Hills.

Plate 13 Rolling farming country in southern Spain. Visual attraction is in contrasting colour of grassland, crops, fallow, wildflowers, orchards and the village.

Plate 14 Each apartment in this terraced block at Montreux appears to have its own garden in the wood.

Plate 15 The light grey bridge over the River Douro blends well with the yellow buildings of Porto, Portugal (photo by Robin Carpenter).

Plate 16 Bright yellow trams running on the freshly mown track bed were perhaps one of the happy features inherited from the East German regime in this 1996 shot in Dresden.

Plate 17 A shared railway/canal corridor at Heyford, Oxfordshire. The 'narrow boats' bring colour and interest to the scene ...

Plate 18 ... as do passing trains (both photos by Robin Carpenter).

Part II

Man-made Forms and Structures

3

Landforms and Their Modification

3.1 The creation and substance of landforms

A *landform* is defined by the Landscape Institute (1995)[1] as a 'combination of slope and elevation producing the shape and form of the land surface'. Natural landforms are the result of geological processes. The processes occur over millions of years in the formation of sedimentary rocks, as a series of cataclysmic orogenic events, or as a continuing response to climatic and hydraulic forces. However, physical processes can be modulated or accelerated by man's interference, such as through agriculture or construction. For thousands of years, man has moved rocks and earth to create fortifications, flood protection banks, canals, dams, roads and harbours, or to win ores, fossil fuels and building materials.

The tectonic or erosive forces that create natural landforms are far more powerful than the efforts of men or machinery. But human ingenuity may take advantage of stable forms, for example as foundations; it may influence gradual processes such as local erosion and sedimentation, attempt to control powerful forces such as those of river flow, or take protective measures to reduce risks of damage from earthquakes or tidal surges.

Landform engineering is the application of earth science in adapting or creating landforms to accommodate various forms of activity or structure. The science involves soil and rock mechanics – the performance of these materials, in situ or transposed, and the consequences of structural and hydraulic loads on them. The application is in earthworks – removing and placing material to form new land shapes

The *constituents of landforms* are rock and soil. These occur in sometimes complex mixtures and need to be classified or sorted to ensure that they are suitable for construction. *Rock* properties concern its strength, the form in which it can be removed as a structural material, and its stability in freestanding cliffs. Problems arise where there are discontinuities, where rocks are soluble or where they have decomposed to soft material. *Soils* are rock particles that have been eroded and removed by hydraulic action (alluvial), have been weathered in place by wind forces and temperature changes (residual) or have fallen by gravity (colluvial). Soils are classified according to properties of which the most fundamental is grain size distribution. The diameter of grains varies from pebbles and gravels, through sand and silt to fine clay particles. The latter are too small to separate by sieve and have to be differently identified, for example by measuring their rate of settlement in water.

Sand and coarser particles are permeable and can provide a free-draining mass or layers for drainage between less permeable materials. Dense sand, contained laterally, is capable of withstanding considerable vertical pressure. Performance of clays is much more difficult to predict and depends on the situation in which they lie. Because of their small grain size, clays are permeable only very slowly even though there may be considerable porosity (pore space between grains). The natural condition of clay depends on:

- the amount of consolidation which has taken place, perhaps over thousands of years; and
- whether the clay is in a saturated, unsaturated or constantly changing condition; clay expands as its moisture content increases.

The condition and properties – such as shear strength – of consolidated clay can be completely altered by disturbance, exposure or removal of overlying material. Change in hydraulic conditions or more sudden seismic vibration can affect the properties of fine-grained soils, including their liability to liquefaction. Soft rocks also vary in the way that they react to being exposed to the atmosphere.

Geological complexity is a feature at many sites of large-scale construction. Complications arise:

- because of the random way in which all sizes of alluvial soils are deposited in flood events in constantly changing river channels;
- when mountain-building activity has produced folds, dykes and faults in rock strata; and
- where there are seams of weak material behind cliffs or beneath slopes, such as layers of clay that can form slip surfaces, particularly when lubricated or brought under pressure by the load of groundwater.

The nature of complexities can be partly revealed by comprehensive site investigations and testing but seldom can these reveal all subsurface anomalies. Therefore, recourse has also to be made to empirical judgement in the light of experience of similar soil conditions elsewhere. Additional investigation, measurement of soil parameters and appropriate design adjustment can be made as full excavation takes place. Engineering should, above all, take account of the implications of changes to drainage in complex landforms.

The rest of this chapter concerns:

- natural changes that occur to landforms and the ways in which human action should accept these changes or can aggravate or alleviate them;
- reconstructing landforms to provide useful (especially flat) space; and
- visual and landscape impacts of altered landforms including their vegetative cover.

Chapter 4 describes structural earthwork – adaptation of land resources by cutting the earth to shape, using it as a foundation, depositing earth materials in embankments and stabilizing slopes. Chapter 5 deals with the means and effects of extracting rock and earth materials for use elsewhere, including as construction materials.

3.2 The frailty of landforms

Long-term wearing down of the Earth's surface

Denudation is the general lowering of land surfaces over millions of years; it takes place through weathering of rock, transportation of the resulting debris and erosion by that same moving debris. Figures have been estimated for the average net rate of lowering of the world's terrestrial surface; but the great variation in climatic and geological conditions and the influence of different anthropogenic activities make such data less significant than, for example, that concerning the transfer of material eroded from mountainsides and its deposition on the plains within particular river basins.

Theoretically denudation continues until the land surface is totally flat. However, throughout geological history, the process has been interrupted by periods of rapid and powerful movement in the Earth's crust, creating new mountains and starting afresh the denudation process. In the millions of years between such interruptions, lesser manifestations of these crustal movements have occurred; and these continue as occasional yet comparatively violent natural events.

Weathering is the in situ breaking-down of rocks by mechanical or chemical processes usually related to climate. Mechanical weathering results mainly from temperature changes, for example, in freezing water in rock pores or expansion and fracture of rock surfaces under extreme diurnal variation. Resulting features in the landscape are rock debris below cliffs, 'onion-skin' exfoliation and formation of smooth 'boiler plate' slabs of exposed rock. Chemical weathering processes include that caused by growth in rock pores of crystals of salt evaporated from solutions in arid basins. In somewhat wetter limestone landscapes,

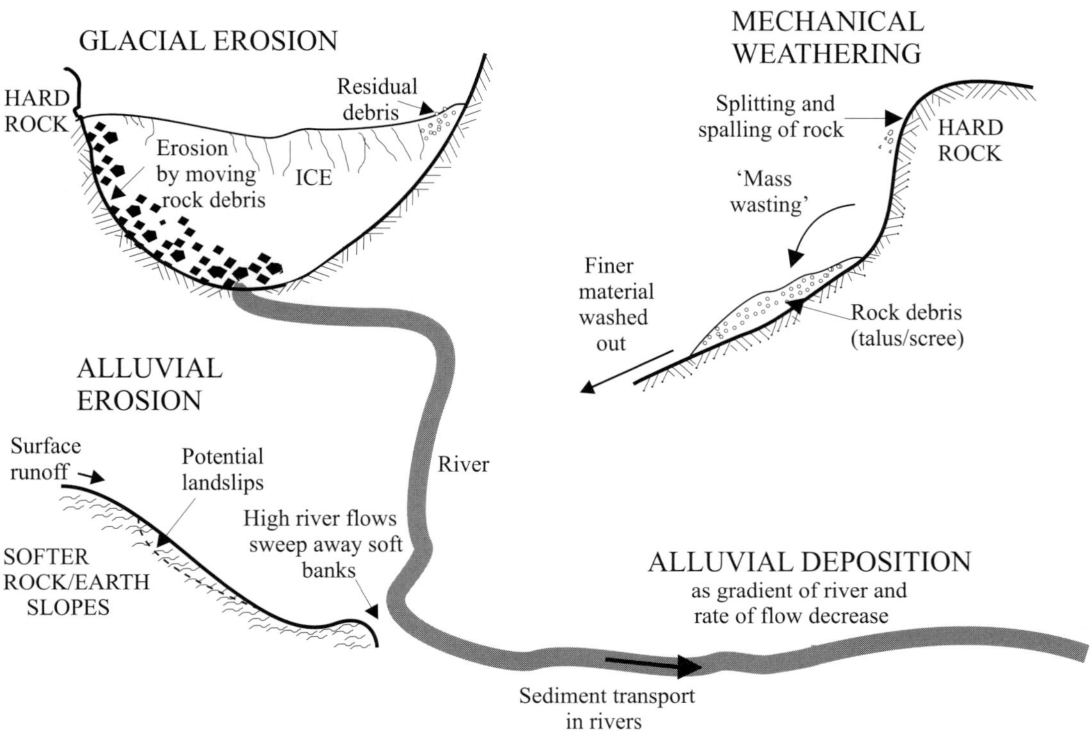

Figure 3.1 Transportation of weathered material

caverns and subterranean channels are created by dissolution of calcium carbonate ($CaCO_3$) in water containing dissolved carbon dioxide (CO_2). In some tropical conditions if land is cleared, intensive leaching of the soil causes the formation of hard laterite surfaces and a barren landscape unsuitable for the agriculture.

Choice of construction materials – stone, brick, concrete – has to consider susceptibility to weathering. Tunnelling into limestone, or founding on it, has to be undertaken with awareness of its permeable or cavernous nature.

Transportation of weathered material is shown in Figure 3.1 and takes place:

- by gravity ('mass wasting') down slopes – slowly as in soil creep, rapidly as pyroclastic flow, rock avalanches, mud flows or mass sliding along a shear plane;
- as sediment transport in watercourses (mainly as fine suspended alluvium), glaciers (coarser moraine) or coastal currents (littoral drift).

Creep and more abrupt mass sliding are evident symptoms of instability, which can be overcome only by retaining walls or other measures to resist or divert gravitational or hydraulic loads (see Section 4.4). Sediment transport in rivers has important implications in the planning of dams and other hydraulic control structures (Chapter 6).

Erosion is a term generally used for denudation caused by the scouring away of rock by air, ice or water, or by the debris which these carry. *Wind erosion* is the result of strong winds that readily pick up fine weathered material, particularly in arid conditions, and use it to abrade and loosen further material. Deposits of aeolian (windblown) sediment can form sand dunes, which are deposited in the direction of the prevailing wind, or loess, often in thick unstratified but vertically jointed layers of silt accumulated over thousands of years. Wind erosion removes topsoil from fields causing uncultivable dust bowls. The effects can be reduced by planting windbreaks and sowing hardy grasses on moving sand dunes – if there is

Photo 3.1 Morainic debris, Scotland.

sufficient rainfall to support the vegetation and unless the winds are too strong and continuous.

Ice erosion by glaciers has shaped much of the world's fine cliffs, cwms and arêtes. Yet the eroded debris creates moraine, revealed after the snow and ice has retreated, as starkly rough desolation. Eventually vegetation may soothe its bleak surface if not relieve its rough topography (Photo 3.1).

Water (alluvial) erosion is not only the principal and most continuous means of mass transport of material from the mountains to the plains, it is also the most common hazard resulting from neglect of land management or ill-planned land clearance, whether for agriculture or construction. Alluvial erosion takes place as sheet, gully or stream erosion. Its prevention should be planned in conjunction with all aspects of development including optimum use of land and its drainage. Alluvial transportation and deposition occurs principally in rivers and coastal waters. Photos 3.2–3.4 show examples of actual or potential erosion by water.

Photo 3.2 Very severe erosion in the Upper Rhine Valley, Switzerland.

Sediment transport and deposition by rivers

Erosion in the higher and younger immature mountain ranges is often severe and rapid. Rivers such as the Indus flow profusely in the summer when the snow in the mountain basins melts. In relatively arid catchments, deposits of erodible material are washed away at high river levels. Thus, during the flood season, there is not only much

Photo 3.3 Colluvial and alluvial eroded material is able to support modest vegetation and human settlement in this steep valley in the Elburz Mountains, Iran.

Photo 3.4 Potentially erodible highland where forest has been cleared and if tea garden land structure is not adequately maintained, Sri Lanka.

more water discharge but each cubic metre of water carries a heavier load of sediment. Waterborne sediment is often referred to as 'silt', that being the dominant fraction of the earth constituents concerned. In the lower flatter stretches of river basins, seasonal high waters spread out over the flood plains bringing fertile (nutrient-carrying) silt to benefit the irrigated fields or flooded meadows. However this deposited sediment also raises the bed levels of rivers and forms natural banks (levees) beside their courses.

Human interference in the natural cycle of river flow and sediment transport can take various forms including:

- clearance of vegetation in the catchment, reducing the ground's capacity to temporarily absorb heavy rainfall and increasing the peak discharge and erosive strength of run-off;
- paved surfaces and direct drainage from built-up areas, with a similar effect;
- construction of dams and entrapment of sediment in reservoirs, progressively reducing storage capacity unless the sediment can be sluiced or flushed out through the dam; and
- flood protection works – artificial levees to contain the river across flood plains, providing many areas with positive protection but sometimes increasing risks during exceptional floods.

It is sediment transport and deposition by rivers, as much as their flow characteristics, which determine valley bottom land character and create problems in river control.

Coastal erosion

Whereas inland rainfall and stream flow have roughly seasonal characteristics, coasts are subject to precisely regular tides and irregular but often violent storms. The consequences of continuing assault by the sea include both hard rock scenery and softer more fragile coastlines. Inland, mother rock is usually buried in soil. Rock is exposed only where the superficial cover has been naturally removed, for example by glaciation, or excavated by man. But by the sea, the onslaught of waves has long ago gouged out any softer material revealing the underlying hard formation, forming stacks, headlands and coves; and the erosive process continues along softer coastlines.

The debris at the bottom of eroding cliffs either becomes an instrument with which the waves continue to batter the cliffs, or it accumulates as a beach protecting the bottom of cliffs. The combination of frontal and angled assault of the waves and their retreating backwash can sort the material into distinct zones of shingle, sand and mud. These zones can form sequences down the slope from high tide level to deep water. The sand and shingle that form the intertidal shore zone are propelled gradually along the coast in the direction of the prevailing current. This 'littoral' or 'longshore' drift is essentially eroded debris being transported and deposited as the energy of storms develops and is dissipated. Much of the deposition occurs in shallow water where the coastal strip widens, forming shingle spits or sand banks. On coastal lands above the tidal zone, the direct force of the wind becomes effective and builds up the sand into dunes; these continue to move unless stabilized by vegetation.

The pattern of coastal erosion has to be taken into account in attempting to strengthen the shoreline or in building outwards from the shore into deep water. Shore protection removes a source of eroded material that would drift along the coast to replenish material eroded from other sections of beach further along the shore. There, the absence of such replenishment probably results in increased erosion. Groynes and breakwaters block the littoral drift with similar consequences.

Sudden natural hazards

Most upstanding natural landforms are stable. Where they are not, the tectonic or erosive forces that change them are far more powerful than man's constructive efforts. These forces cannot be conquered; but perhaps, if they are understood, people can take precautions to reduce the risk or magnitude of damage. Rapid natural events changing landforms and threatening land use comprise sudden, generally unpredictable events, such as volcanic explosions or earthquakes, and partially predictable events, such as exceptional floods.

Earthquakes occur, as do volcanoes, in the currently active zones of the Earth's crust near to the borders of tectonic plates. The majority of recorded seismic events are minor in character. In areas of greater risk, foundations and superstructure can be designed to reduce damage and loss of life from major incidents.

Volcanoes constitute striking scenery and their occasional eruption is an awesome sight. Fuji and Demavend are cones of quieter distinction. Many Atlantic islands encompass rough landscapes of active and dormant craters offering cliffs and lakes amid profuse vegetation, fertile fields, still hard lavas and manifestations of geothermal energy. Construction in these circumstances may comprise:

- villages and small towns in attractive locations where risks of devastation are less significant;
- roads gaining access to these settlements and to scenic viewpoints and paths; and
- steam power stations converting geothermal energy into electricity.

Tsunamis result from severe earthquakes as well as from coastal volcanic explosions and landslides. Millions of people may be remotely at risk from tsunamis but their rarity of occurrence is such as to render structural safeguards impracticable beyond those taken against more likely events – such as wind-generated sea level surges combined with extreme high tides.

Semi-natural sudden landform changes

Landslides are abrupt manifestations of slope failure. A soil mass may collapse under increased weight or hydraulic loads that may overcome inherent internal friction; or rocks may slip along discontinuities or layers of softer materials such as shales or clay. Slope failure can result from earthquakes, exceptional rainfall and other natural causes on fundamentally unstable slopes, or from anthropogenic activity that alters the drainage pattern and subsurface flow or excavates into slopes that would otherwise remain stable.

Landslides can create dams, often unstable, that block river valleys as has happened many times in Central Asian massifs (Photo 3.5). Landslides have destroyed communities in the Andes, China and Turkey. Travellers have been entombed on a Canadian highway. Bad siting for mining waste has caused catastrophic slips; and the construction of the Vajont Dam in the Italian Alps created a reservoir whose freeboard capacity was insufficient to contain the huge amount of water displaced by a landslide resulting from an altered groundwater regime. Fuller prior investigation and consideration of subsurface drainage should lead to construction solutions avoiding such risky situations.

Photo 3.5 Site of an ancient landslip of perhaps 2km^3 into the valley of the Karkheh River, Zagros Mountains, Iran.

Landslides may also create dynamic, that is unstable, 'undercliff' landscape that is scenically intriguing, attractive to particular types of flora and wildlife habitat, and less suited to human construction. At the comparatively few places in the inhabited world where landforms or their slopes are known to be naturally unstable, these are best conserved as scenery or nature reserves or for limited agriculture.

Storm runoff and floods

Periods of excessively high rainfall occur only occasionally in any particular basin, several times a year somewhere or other on Earth as a whole. It has been argued that the occurrence of violent storms is increasing in the current era of climate change. Whether this is correct or not, human action such as clearing land of vegetation or paving it extensively can increase – at least locally – the rate of rainfall run-off, the severity of flash floods and the instability of steep soil slopes. Civil engineering works should be designed both to minimize such disturbance and to make what allowance they can for events beyond their full control. For example, in planning roads and bridges, a balance has to be sought between the risks of their destruction during severe storms and the costs of reconstruction and subsequent maintenance and renewal for various levels of robustness.

River floods are common natural phenomena. Compared with volcanic or seismic activity, the hydrology of rainfall, run-off and stream flow is easily comprehensible and measurable. In large catchments, the occurrence of flood flows can be predicted approximately on an seasonal basis; and their volumes can be forecast more definitely within a few hours of the occurrence of heavy rainfall upstream. In smaller catchments, there is no time to forecasts floods such as inundated Lynmouth or Tintagel. Again, a balance has to be sought between the engineering cost of containing or diverting the water and the risks of damage.

Less predictable even than flood flows are the volumes of sediment that they move. Especially in arid sparsely vegetated basins, high stream levels erode considerable quantities of soft bank material that normally lies above water level. This eroded sediment is carried downstream and deposited as the valley bottom widens, the gradient lessens and the speed of flow decreases. First the coarser material is dropped, leaving numbers of branching and rejoining (braided) channels, then the increasingly fine fertile material as the floods spread out and slow down over wider plains. The pattern of landform and river course change can be capricious because of the variations in flow and uncertainty as to where the river spills over its banks and where it drops its sediment load. Thus the effects of river floods on the geomorphology of their courses are complex and seldom precisely predictable.

Photo 3.6 Fields terraced for rice cultivation in northern Iran.

3.3 Reshaping landforms

Man's most fundamental economic activity was always farming; and agriculture remains the main use of fertile land. The flatter land tends to be the most fertile and the easiest to crop. So where such land is available, steeper country is for sheep, forest or scenery. However, human populations soon spread onto more difficult land where they have constructed terraces of flatter ground on which they can pond up water, till the ground and transplant or harvest their crops. Thus terracing of slopes and subsequent maintenance of the terraces became vital elements in the economy of hillside communities (Photo 3.6).

At first, people settled on or near the land they cultivated; but, especially as the flatter more fertile land became more intensively used, they sought sites for settlement on any higher ground, if it existed, to avoid flood hazards, or on the slightly rising ground at the foot of hillsides where springs might be a perennial source of water. As populations rose further and fewer people were engaged in agriculture, so it became increasingly necessary to build homes and other buildings on sloping ground. Such ground is less fertile and thereby less suitable for agriculture; but it is usually harder and therefore more suitable for foundations of permanent structures. Landforms are reshaped to make surface space suitable for buildings and their functional surroundings, through engineering approaches suited to the circumstances.

Creation of useful space

On gentle slopes, houses can be built on one-level foundations. These have to be excavated below ground surface level anyway, formed horizontally but slightly deeper at the uphill end. Roads can run directly up the slopes.

On steeper slopes, homes may be constructed to a split-level design and road formation has to be cut into the hillside. But for more spacious buildings, wider roads, parking areas and tennis courts, horizontal terraces have to be created on the hillsides as shown on Figure 3.2 or Photo 3.7. The terraced space is then more useful than the original natural hillside; but there is less of it because of the intervening, now steeper ground (that may have to be stabilized or supported as described in Section 4.4).

The purpose of more modest land reshaping – which Laurie (1986)[2] described as 'grading' – is 'remodelling of existing land form to facilitate

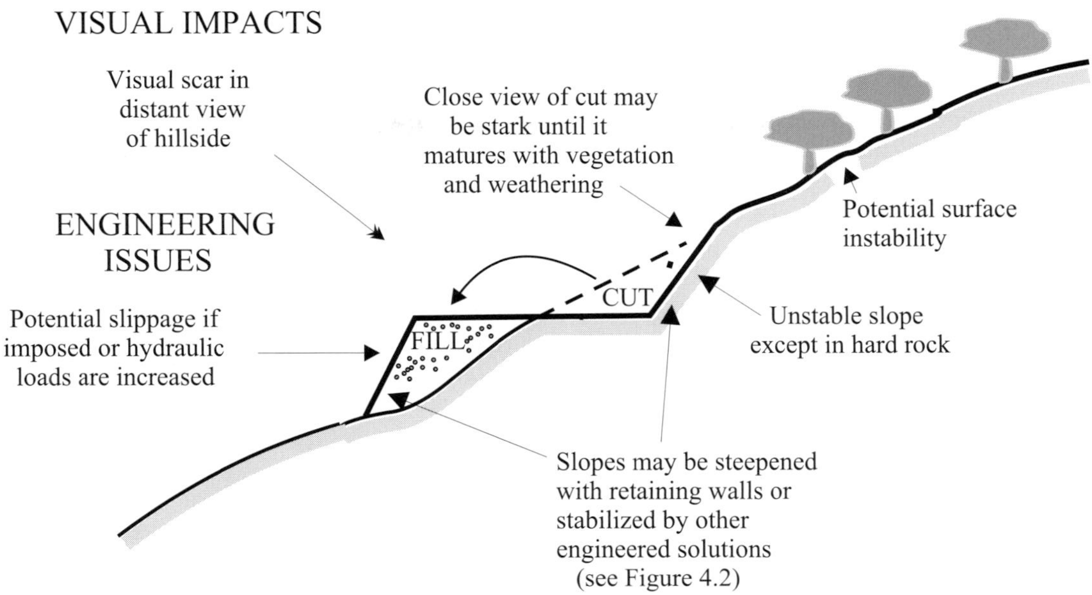

Figure 3.2 Creating level ground

Photo 3.7 An industrial estate occupies a flat area originally cut out of chalk hillside for Ventnor Station, Isle of Wight. The station, high above the town, was the terminal of a railway line that emerged from a tunnel (centre right).

the functions and circulation of the site plan and to ensure adequate drainage'. One of the main objectives of a site plan is to enable building to take place on a suitable foundation. Circulation can be the way people and vehicles move on paths or roads between the buildings as well as what Laurie describes as 'connections between architecture and landscape, between indoors and out'. Adequate drainage has to prevent build-up of water within the disturbed slope; but too rapid drainage may cause problems of erosion or flooding downstream.

Land reshaping creates level space for new construction with steep banks or walls at its borders and new boundaries around any still undisturbed but possibly fragmented ground. On the newly levelled ground, the layout of buildings suits their function but also determines the composition of the outside space. In this composition, space may be conceived as volumes, especially visibly, rather than just ground areas. 'It is a matter of designing the volumetric enclosure and spatial interconnections to suit the use' (Simmonds and Starke, 2006).[3] Any undisturbed space may continue as a feature of the overall landscape, as managed vegetation or for any other role which does not require ground reshaping.

Engineering issues

The type and amount of earthwork, drainage, slope treatment and remedial work undertaken determines how much useful space is available for functional development. The cost of providing each area of useful space has to be balanced against the benefit of subsequent development.

A fundamental factor in balancing engineering costs against spatial benefit is scale. What proportion of the original area should be altered and what should be the extent of each earthmoving or slope stabilization operation? Solutions have to be sought, at the conceptual planning stage, in light of economic and environmental impact of different sorts of development and the engineering properties of the rock and soils concerned.

Two geotechnical questions have to be answered in detail before designing any restructured landform:

1. What are the characteristics of existing landforms that will remain in place, albeit under new loads and surface or drainage conditions? This has to be resolved mainly by site investigations.
2. What performance is expected from the soils to be placed in new formations? This will indicate, based on experience and the principles of soil mechanics, what available soils are suitable, how their characteristics should be specified to meet design parameters, and what safe assumptions are appropriate.

Factors of safety cannot be a catch-all for inadequacies; they can allow only for limited variations from what is assumed or specified. Furthermore, an increased factor of safety against failure in one respect may increase the chances of failure in another, especially where different conditions apply at different times, for example as regards hydraulic loads or moisture content.

Site investigations include survey of surface conditions and drainage; but they need, in particular, to determine the nature of the underlying ground and any variations or faults within it. There is a range of choice between:

- detailed and intensive subsurface investigation, analysis, design and construction techniques with high confidence of future performance; and
- minimal investigation, use of traditional – probably empirical – design concepts and local construction resources, accepting that adaptation may be needed during construction or remedial action subsequently.

A design strategy for landform reshaping then depends on the nature of the earth materials and the methods of excavating and placing them, which in turn depends on the scale of the project and the costs of supplying earthmoving machinery or specialist structural techniques. Laurie (1986)[4] suggests that ground should be reshaped 'at the scale of the machinery. Grading machinery is by definition gross in nature and subtle details are difficult to achieve except by hand labour'. Whatever the scale of the total operation, there needs to be intensive – often skilled – human effort in detailed earthwork, conserving topsoil, installing ground-strengthening and drainage measures and managing vegetation.

3.4 Visual aspects of landform engineering

Design for open space has developed from that of deliberately organizing land shapes, water features and vegetation to create admirable views to that which takes opportunity to create pleasant or interesting sights in the consequences of functional engineering. Lancelot Brown estimated the 'capability' of each park landscape on which he was consulted. Civil engineers and landscape planners can do the same by recognizing landscape opportunities in the functional landforms they create. This they can do at the stages of:

- preparing layouts of structures and adjacent open space at the different levels dictated by the topography and its modification for function and drainage;
- adaptation of functional earthworks;
- construction or modification of water features; and
- planning and management of vegetation.

Layout of structures in open space

Recognition of site characteristics and positioning of structures in relation to each other has been introduced in the last chapter (pp27–28). The same planning, applied in three dimensions where landforms are being altered, can take account of vistas possible between buildings or through woodland and the views that can be seen over these or toward rising ground.

New high points may be created or others opened out. Monuments commemorating battles at Marathon in Greece or Waterloo in Belgium are high mounds that also serve as viewpoints. Perhaps spoil heaps at mines could do the same.

Of course, not all structures – or even some untidy pockets of natural habitat – are welcome sights. What we do not wish to see can be a matter for contention – according to different opinions as to what features are visibly offensive or ugly. Practicable concealment strategies may also seek to protect areas that are secret, a category that could include very sensitive wildlife conservation areas as well as military installations.

Adaptation of functional earthworks

The shape, colour, texture or orientation of man-made features may be made sympathetic as well as functional. While high steep rock faces are scenic features in any surroundings, so can be structures built on top of them (Colour Plate 7, Photo 4.2). Weathered stonework in Hadrian's Wall tones well with outcrops of the Great Whin Sill scarp on which it stands.

Excavated quarries may resemble the entrance to a limestone gorge or ape a natural escarpment. Indeed people admire Salisbury Crags behind Edinburgh or the gritstone 'edges' of the Peak District without necessarily being able to differentiate which features of the cliffs are entirely natural and which have been quarried.

In softer surface geology, there may be more subtle ways of blending new with old forms. A low embankment dam seen from downstream may have likeness to a grassy hillside. A mature reservoir can be mistaken for a mountain lake. Landscaped mine waste heaps may not differ greatly from grassed colluvial or morainic debris. Mining subsidence is equivalent to dolines (sinkholes), hillside lynchets are like natural soil creep or sheeptracks. Agricultural terracing and tiered land development have some affinity with ancient river terraces. On the other hand, neatly trapezoidal 'landscaped' spoil heaps do not necessarily blend with U-shaped valleys or rough slopes at the foot of escarpments. Mapped contours give a guide as to what are natural shapes locally.

Optimizing water features

Water features in contrived scenery have included:

- lakes, as foreground to scenery;
- watercourses that have been adapted or rerouted; and
- waterside strips comprising pathways, hard edges and softer beaches for human recreation, or rougher growth forming habitat for terrestrial, aquatic or amphibious wildlife.

Man-made lakes, constructed as aesthetic features of leisure grounds, differ from those created for water storage although – except in certain worked-out quarries – both require dams. The water level in ornamental lakes can be maintained by moderate inflow from a surrounding catchment and any outflow escapes over a fixed level spillway or through a concealed pipe. To a casual visitor it may not be obvious whether such a lake is dammed or is a natural geological feature. But, for deeper storage reservoirs, timely demand for release of water can only be managed by providing full outlet control facilities to release or impound water. This involves varying the level of the lake and consequent difficulties in maintaining riparian habitat or beaches.

Watercourses are pleasant features of gentle parkland but can be particularly scenic where they flow in rapid rocky sections or through a steep-sided valley. Jesmond Dene, near Newcastle, is a small example of such a valley. It has seemingly natural attributes of beauty and nature conservation although the waterfalls and rocky character of the gorge are largely the result of stream diversion and other more cosmetic works undertaken for the local industrial magnate, Lord Armstrong. He presented the gorge for public enjoyment.

More often, rerouting of less torrential rivers across flatter landscape is necessary to suit extensive land development projects. Environmental mitigation is then likely to be necessary in terms of restoring or relocating the natural fauna and flora that constitute the ecosystems and micro-landscapes inherent in watercourses. Relocation of habitat can be arranged in the following sequence:

1. Construction of the new river section in the dry, preserving, as far as is practicable, the gradient, width, depth and line characteristics; provision of berms at the banks for waterside plants and to aid the subsequent establishment of suitable river's edge conditions.
2. Collection of plants from the condemned section of the river (or elsewhere); temporary placement of these in wet conditions, such as behind gabions in a nearby section of river.
3. Replanting in the new section when water is diverted into it.

Waterside strips are significant for their riparian vegetation, which may have hydraulic as well as ecological functions.[5]

Taking advantage of vegetation

Vegetation is planted and managed in agriculture, horticulture and forestry for the production of food, fibre crops, animal fodder or timber. Vegetation can also be relevant to built development for reasons that are aesthetic. Greenery is a main element of scenery, and can be functional, such as in windbreaks or grass planted to stabilize banks or to delay storm run-off (see Figure 3.3).

Botanical expertise is essential in determining appropriate plant species or seed materials, the necessary conditions for cultivation, and the programmes for planting and subsequent growth management. The associated role of construction planning lies in:

- location and layout of vegetated areas so as to best enhance the landscape within or around the construction project;
- measures to support planting in the design of earthworks and drainage for the project; and
- actual processes in the preparation or implementation of vegetative works that may have to be included in construction contracts and programmes.

To fulfil its role, vegetation has to be planned both for establishing it – as grass and low plants or as trees and woodland – and then managing its subsequent growth, all to a pattern that fits the construction and function of the structures, adjacent land use and development of the landscape as a whole.

Maturing landforms and their associated scenery

There is always a time element to be considered. Geological erosion, accumulation and formation of sedimentary strata take millions of years, but sudden tremors or landslides occur in seconds. After a hectic year or two of actual construction, a few more years required for consolidation of deposited material in an earthwork seems comparatively slow, as does the maturation period of some planted vegetation. Yet hydraulic loads can build up rapidly and landslips can occur suddenly under the new conditions created by landform construction or alteration. Some earthworks have to be undertaken slowly giving time for each layer to settle; but sometimes

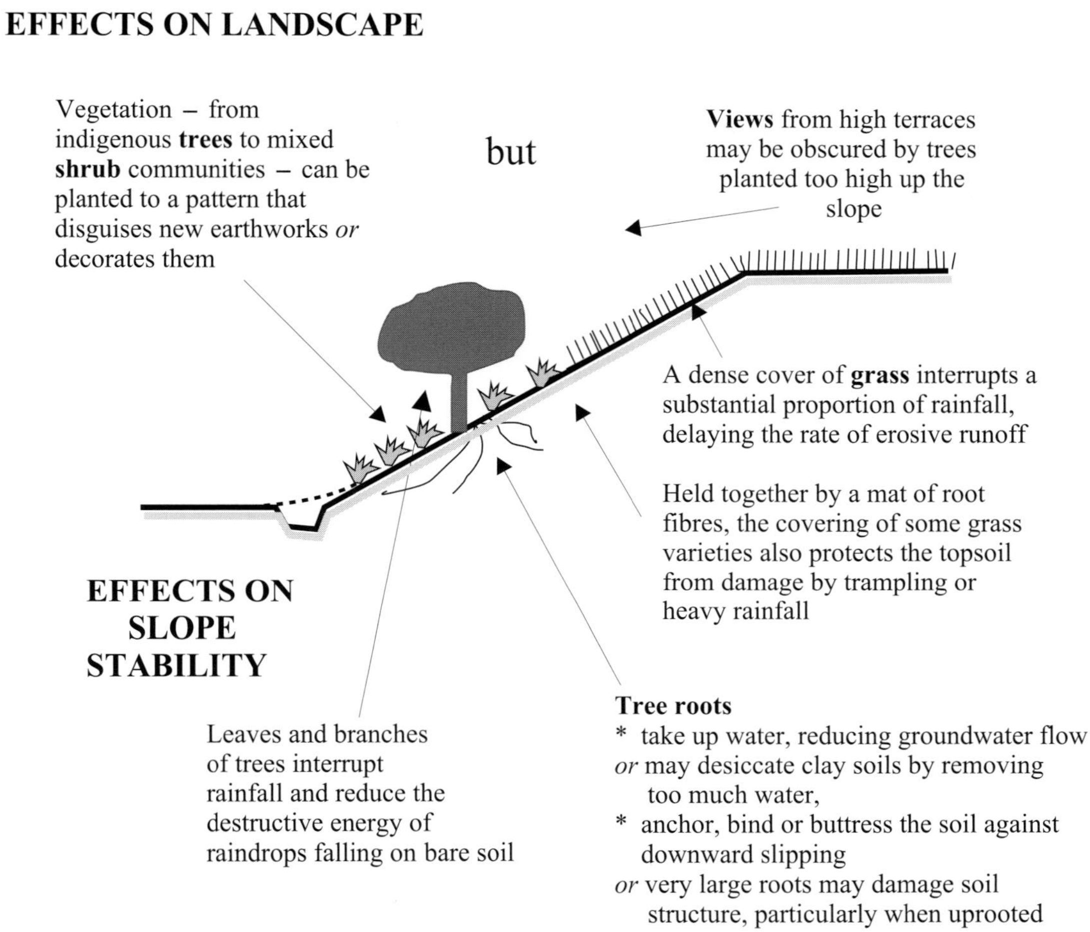

Figure 3.3 Effects of vegetation on landscape and slope stability

changed circumstances compel adjustments and timely additional safeguards.

Landscape itself needs time to mature after it has been remodelled. Just as new growth arises and animals find habitat gradually in the chaos of a landslide, so vegetation reaches a controlled climax on an embankment; or the colour of structural stone or concrete assumes a hue less stark in comparison with its surroundings.

Notes and references

1 Landscape Institute/Institute of Environmental Assessment (1995) *Guidelines for Landscape and Visual Impact Assessment*, Spon, London.

2 Laurie, M. (1986) *An Introduction to Landscape Architecture*, Elsevier, New York, p12.

3 Simmonds, J. O. and Starke, B. W. (2006) *A Manual of Environmental Planning and Design*, 4th edition, McGraw Hill, New York, p157.

4 Laurie (1986) as Note 2, p221.

5 The hydraulic function of wetland generally is to store storm run-off and alleviate river flood flow. The biological features of riparian habitat have been explained by others, for example Sendich, E. and the American Planning Association (2006) *Planning and Urban Design Standards*, Wiley, Hoboken NJ, p118.

4

Earthworks for Structures

Earthwork comprises the removal of soil or rock from one location to another. Excavation ('cut') may provide a level surface or subsurface on which construction can take place; and placement as 'fill' elsewhere may create a raised foundation or a structure in itself. If excavated material is taken to a placement area, that is 'cut and fill'. If the cut material is not needed it is 'wasted' as 'spoil'. If there is not enough suitable cut for necessary fill, then more is excavated from 'borrow pits'.

Chapter 3 explained the variability of soils and the need to investigate them for use in engineering. Meanwhile, in recent decades, advances in earthwork design have resulted from the incorporation of geosynthetics. These are polymeric products manufactured as rolls of sheet (geotextiles or geomembranes), as less flexible geogrids or in combinations as geocomposites. Within earthworks, they separate, contain, strengthen or protect different soils. They may allow seepage or can act as impenetrable barriers.

This chapter examines the constructional and land resource issues in the various aspects of earthwork:

- excavation (4.1) and any other preparation to make ground suitable for foundations (4.2);
- construction of embankments (4.3); and
- stabilization of cutting, embankment or natural slopes (4.4).

4.1 Excavation

Shallow excavation

The construction of buildings and paved areas starts with site clearance, usually including removal of organic topsoil and any loose or contaminated surface material. It may then be necessary to excavate deeper to reach base level for the chosen type of foundation. Shallow excavation is undertaken by bulldozers that push the soil into a heap on the perimeter or, over extended space or to slightly greater depth, by tractor-hauled scrapers that transport the stripped soil farther away. Consequently, the material becomes mixed, aerated and bereft of many of its in situ properties. Sometimes there are ground features that could be lost to indiscriminate excavation and may call for more selective, careful, even manual removal. Examples are:

- fossils, in sedimentary, often relatively soft, rocks and evident only as a result of active erosion or excavation: therefore, since the formations are to be destroyed, careful excavation or preparation of casts has to be undertaken so that valued fossils can be removed;
- buried archaeological structures or evidence: these also have to be revealed delicately and a plan devised for removing, conserving or recording them; if ancient timber or brickwork structures are revealed, the consequences of exposure to the air and changes in the water table have to be considered; and
- particularly fertile topsoil or special botanical turf or shrub habitat.

It is possible that fossils or archaeological remains may emerge unexpectedly. Extra time and funds are then likely to be needed to deal with them.

Topsoil intended to foster growth elsewhere needs to be handled so as to conserve its fertility. Precautions mentioned by Emberton (2001)[1] include:

- use of relatively light machinery for excavation so as to avoid compacting and breaking up the structure of the underlying soil; use, if necessary, of temporary tracking laid across susceptible surfaces;
- grading, placement and shaping of stockpiles to make them secure against incidental rainfall (even by grassing long-term stockpiles) and to allow aerobic activity to continue in the soil; and
- avoiding over-compaction of placed soil.

Special botanical habitat may justify the removal and reconstruction elsewhere of particular vegetation and its substrate. To achieve close replication of conditions it may be necessary to excavate and transport each of several soil horizons separately so that they can be replaced in the proper sequence.

Open cut

Deeper excavation takes place:

- along hillsides for road routes;
- through high ground in two-sided cuttings, commonly for railways and major highways, sometimes for waterways; and
- for foundations or basements of tall buildings or to provide terraces for groups of buildings.

Excavation in soil is usually undertaken by scrapers or bucket excavators and in rock by drilling, blasting and loading to dump trucks. In soft material, excavated shallow slopes may not differ greatly – when mature – from the surrounding landscape. In rock, the steeper slopes and outcropping strata create a sharply contrasting sight. Aesthetic judgment can be combined with practicability in the way that benches (ledges) are incorporated as the slopes are created, as geological strata are revealed, as vegetation is managed or as loose surfaces are treated.

Cut into rising ground can make a wide impact on landscape character. Any skyline notch will affect the view of the hillside from afar. Within a cutting, the subsurface geology can be seen together with any vegetation that the new steep surfaces attract. Colour Plates 11 and 12 show cuttings in chalk before and after vegetation has become established.

The profile of cuts along or into hills depends on:

- the propensity to slip of the exposed material, restrained by friction (up to the natural angle of repose in granular material), by cohesion (of clay), by inherent massive strength (in rock) or by structures such as retaining walls;
- drainage conditions behind slopes or walls, particularly in cohesive materials or where there are potential slippage layers sensitive to moisture;
- provision of horizontal benches, between geological strata, as a further precaution against slippage or for aesthetic effect; and
- vegetation or other means of protecting surface material against slope erosion.

Hillside scars can result if excavated material is tipped over the edge of road or terrace formation, creating a chaotic and unstable mass that does not easily weather or permit growth. However, if loose spoil can be removed, the slopes beneath mountain roads may be planted with trees sufficient to hide the road from afar but sparse enough to afford views from the road itself.

Tunnelling and underground works

Underground excavation creates metro railway routes and storage tanks or caverns that might otherwise take up scarce surface space. The effects on surface landscape are secondary, being those of possible subsidence, spoil disposal and any surface structures related to the underground works. The planning of underground excavation has to ensure:

- stable walls or means of preventing collapse; and
- reasonable watertightness or, in the case of caverns for storage of hazardous materials, absolute isolation.

The first concerns the method of construction and provision of any necessary walling or support. The second is primarily a matter of proven tight geological surroundings. For hazardous storage, such as of radioactive materials, very long-term means of ensuring adequate surrounding drainage and of access and monitoring become vital. Geological or hydrogeological conditions suitable for secure underground storage constitute a valuable land resource. Most storage capacity is created by excavation. However, fluids may also be stored in the pores of natural formations.

Disposal of surplus materials

Excavated material can be used, if suitable, for fill on the same site, or it can be transported for structural earthwork elsewhere. Industrial use can sometimes be made of materials such as clay and chalk. Soils that are surplus to these needs may have to be placed in designed spoil heaps. Spoil excavated for city tunnels or foundations may have to be carried through crowded thoroughfares – unless waterways are available. In rural areas mined spoil left near the entrance to adits or the top of shafts has for long shown up in surface mini-topography.

Large-scale excavation from large caverns or very long tunnels may provide great amounts of surplus material. That from the English half of the Channel Tunnel was used for construction of a wide strip of new coastal land (Photo 4.1).

The objective in disposing of surplus excavated soil, or indeed of any construction waste, should be to make resourceful use of it, not to treat it as refuse. If it is fertile it might be mixed to achieve a suitable consistency for organic topsoil. If it is infertile, then it can form the core of earth embankments, laid in zones according to its grading, permeability and strength. If it is contaminated it may be treated (see p194) or incorporated in a suitably isolated zone of fill.

Banks of surplus soil, built up above ground level:

Photo 4.1 Material from Channel Tunnel excavation was used to create new land adjacent to the existing railway at the foot of Shakespeare Cliff.

- can be planted to provide green background (or foreground to already higher land beyond);
- can enable views, through gaps in the greenery, from on top of the new raised ground;
- can block off views (or noise) of structures or activities which are judged to be unsightly – if good views in the opposite direction are not thereby impaired; and
- should be formed in sympathy with the ground contours, geology and drainage pattern of the area and with variations, random or functional, to give a general impression of informality; the functional variations can suit the movements of earth-moving machinery during construction or perhaps define pathways.

4.2 Ground suitable for foundations

Foundation depends on the load of structures to be supported and their height, shape and land space needed per unit of capacity. Many soils offer low resistance to point loads. Saturated silts may be particularly weak; confined dense sand is much more resistant. Beneath weak soils there may be resistant layers that can support the points of bearing piles; or loads on friction piles can be transferred to cohesive clays or silts around them. Depending on the material penetrated, piles of reinforced concrete or steel can be driven (hammered, jacked or vibrated) into the ground; or a hole can be augered into the soil and the void filled up with concrete; or cement grout or special resin can be injected under pressure through boreholes to displace weak soil layers.

On softer formations or on reclaimed land, soils can be artificially consolidated, for example by dynamic compaction, sand drains, vibroflotation or calcining clay; or concrete rafts or grids can be constructed to enable structures to float. Inadequate foundations have led to collapse of structures or their expensive resurrection. Landscapes can be devastated by distorted or abandoned structures inappropriately founded – for example on weathered rock, in waterlogged or permafrost conditions or because of differential (unequal) settlement.

In cities, where land surface space is scarce and costly, a number of possibilities arise:

- Underground storeys, providing extra space, may form part or all of a watertight structural foundation.
- There may already be a maze of earlier foundation blocks or piles from previous buildings as well as underground tunnels. Sometimes it may be practicable to remove the obstacles or extract piles (Wheeler, 2003).[2] Occasionally blocks or pile caps can be adopted for new structures. More often, new piles or drilled and grouted foundations have to be threaded among the debris. The design of the foundation grid or slab has to be adjusted accordingly (Chapman et al, 2001).[3]
- Ancient heritage relics such as mosaic paving may be found at earlier (lower) ground levels. It may be possible to design structures to bridge over these (Bolton, 1992).[4]
- As part of foundation construction or as a separate operation, buildings can be heated by circulating fluids (ground source energy), for example through piles (Flynn, 2007).[5]

Hard rock outcrops determine the location and configuration of Edward I's castles in Wales (Photo 4.2). The robustness of the rock had been proven by resistance to glaciation and erosion over millennia; yet, even over the 750 years of Harlech Castle's existence, the softer coastal landscape below it has extensively changed.

Where nominally hard strata are concealed by vegetation or topsoil, their integrity is less guaranteed. Geological maps may indicate the dominant presence of hard underlying rock but this does not necessarily imply a readily suitable foundation for heavy structural loads. Around Hong Kong harbour, the parent rock is predominantly coarse granite. However – except in a few exposed features such as along the narrow ridge north of Kowloon – the granite has been deeply weathered. The scenic consequences are hillsides supporting thick vegetation but a general lack of cliffs or surface rock. The engineering consequence is great difficulty in constructing roads or hillside structures in respect both of foundations and slope stability. In southwest England, granite is hard at Land's End and in Dartmoor tors but completely decomposed where it occurs as china clay (kaolin). In intermediate situations, the depth

Photo 4.2 Harlech Castle stands on a prominent rock boss, to the bottom of which there was once sea access.

of material to be excavated is investigated by borings, sampling and testing; but ultimately only expert risk assessment can determine the degree of compromise to be adopted between very sound foundations at great depth or more flexible higher-factor-of-safety design on less perfect shallower material.

4.3 Embankments

Embankments carry transport routes, dam rivers or protect low-lying land from floods. Road and railway embankments are the commonest linear earthworks; dams and flood protection banks are typical of those that have to withstand water pressure across them. In each case, embankment stability is the paramount engineering requirement. The most direct issue for land resources is the space occupied by the base of the embankments including its sloping sides; the wider consequences concern land areas that may be protected or inundated. The main visual impact of embankment construction is obstruction of views along valleys; but some embankments provide welcome viewpoints over flat country.

Stability and strength

Strength characteristics of raised earth fill are very different from those of the parent undisturbed material in a hillside. The compressive and shear strength of layers of selected material may be more uniform than in some random geological formations; but centuries of consolidation of the parent material cannot be promptly reproduced by compaction of swiftly laid earthworks. Designs for earthwork above ground level require that loads to be superimposed must be borne both by the earth fill and by the natural formation (subgrade) below; and any side slopes should remain stable, securely containing the fill.

The *strength of fill* – its capacity to bear loads – depends upon its density, which in turn depends upon the degree to which it is compacted in place. Sands are usually compacted by vibration, cohesive soils by rolling at optimum moisture content.

Particularly in clay, special application of heavy weights (pre-loading) or allowance for subsequent gradual settlement may have to be considered in the design. If the subgrade is not of adequate strength, then either some densification process must strengthen it, the width of the embankment must be increased to spread the loads, or the weight of fill may be reduced by using lightweight material such as polystyrene or pulverized flue ash.

The *stability of slopes* depends on containing embankment material and placing the materials so as to preclude slippage. In temperate climates, free-standing slope surfaces are usually grassed. Today grass binding may be augmented by geotextile mesh. But for steeper slopes the soil must be contained, supported or strengthened as described in Section 4.4.

Road and railway embankments

Settlement of uncompacted fill was seldom a serious issue in non-water-retaining earthworks before the 20th century. Railways in the 19th century were mostly constructed by tipping excavated material directly from the end of the completed embankment onto the next section. Either the soil in the finished bank was allowed time to settle under its own weight or the level of the railway track was adjusted later by varying the depth of broken stone ballast on which it lay. However, with the onset of motor vehicles, road pavements needed to be smoothly based and surface undulations and cracks were regarded as failure. After the invention of heavy earthmoving equipment, the placing of the earth fill in carefully compacted layers became standard practice, anticipating complete settlement before construction of the pavement or track (Photos 4.3 and 4.4).

Special precautions are necessary where existing highway embankments are widened. New earthwork has to abut on to already well-consolidated material. If the overall width is restricted it may also be necessary to strengthen the base or to add retaining walls.

Viaducts are alternatives to embankments for construction across valleys. Although usually more expensive, comparative advantages of viaducts are that:

Photo 4.3 M20 motorway embankment under construction, as seen from the adjacent A20 road in 1989…

Photo 4.4 …and in use, amidst concealing growth, 20 years later.

- they occupy less ground space, are less obstructive to river flood flow and can be sensitively constructed to cross wetland without damaging its long-term characteristics; and
- they intrude upon, rather than obstruct, scenic views along valleys.

Water retaining embankments

How watertight an earth fill or rock fill embankment is depends on the head of water that has to be contained on one side and on the impermeability and strength of impermeable material placed within the embankment or inherent in its foundations and abutments. Figure 4.1 shows various types of embankment for containing water on one side.

Thus *embankment dams*, for permanent storage of large volumes of water, need to incorporate:

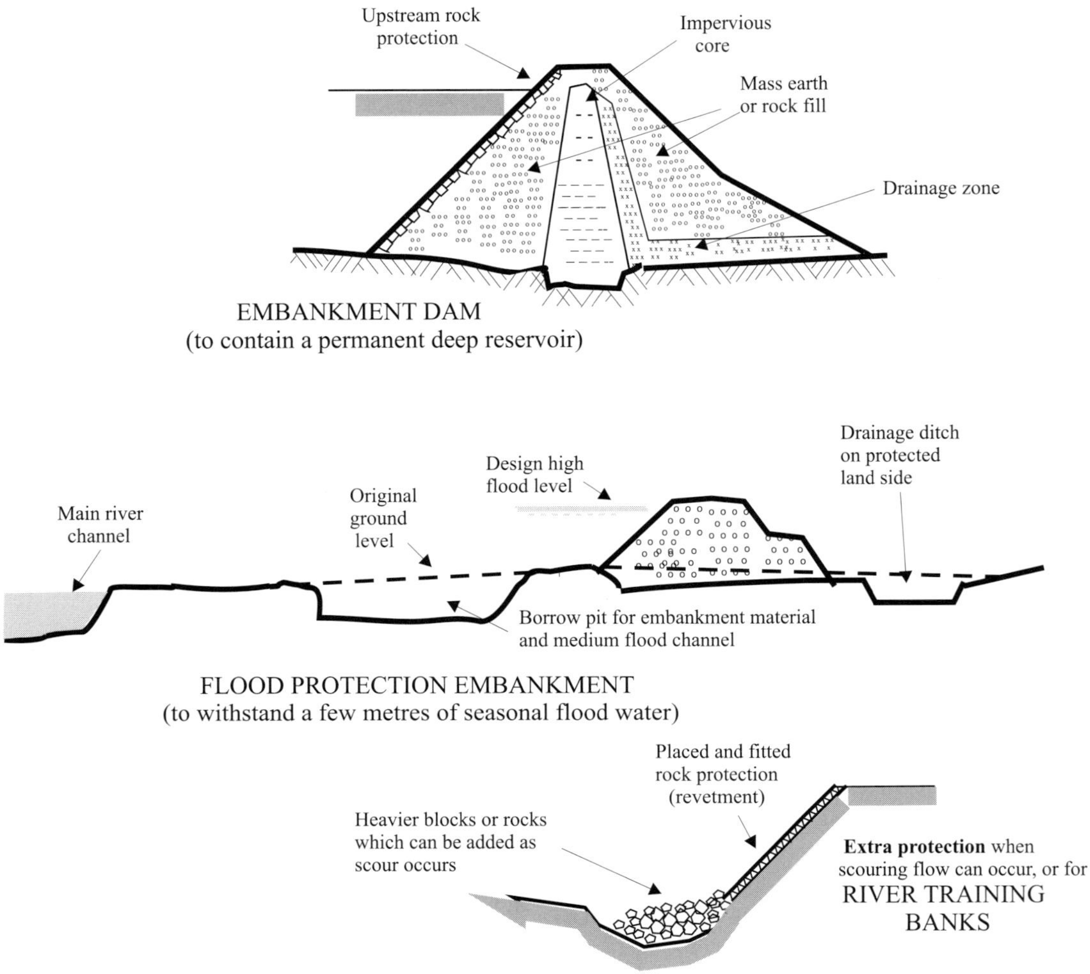

Figure 4.1 Cross-sections through water-retaining embankments

- an impervious zone of impenetrable material (such as clay or a concrete, bituminous or geosynthetic membrane) in the centre or near the upstream face to prevent flow through the dam;
- drainage and filter zones to protect the other zones of the dam by channelling out any flow which does penetrate; and
- rock or block protection (revetment) against wave erosion on the reservoir side and against overtopping, if that is conceivable, on the downstream slope.

The actual cross-sectional design of fill in an embankment dam is very dependent on the characteristics of the material that is available, as well as the scale of the project. For the Tarbela Dam on the River Indus, more than 270 million tonnes of earth and rock fill were placed in embankments. Thirteen screening and crushing plants were set up to provide the required proportions of eight types of fill from a variety of material, excavated both for structures and from borrow pits. Probably such a complex system cannot be justified for much smaller projects although there is no reason why quality control cannot be rigorous at any scale.

Flood protection embankments do not need the stringent precautions that are necessary for the safety of dams. The benefits of alleviating occasional flood damage are not so great as to afford sophisticated construction of banks many miles long. Therefore, flood embankments are made at a uniform cross-section using whatever material is at hand. However, banks must be founded at sufficient depth to key into the natural ground formation, certainly below the bottom level of nearby waterways. According to Nicholson and Silva-Tulla (2008),[6] the levees that failed at New Orleans in 2005 were too shallowly founded on soft clay. Extra provision has also to be made for:

- hard revetment, protecting sections where scour is likely to occur;
- structures to permit timely release of runoff impounded on the non-river side of the embankment; and
- deliberate or automatic flood release into emergency channels – or even breaching of embankments at suitable points – to ensure that, when overtopping is inevitable, flooding takes place where it is least disastrous. Such strategies may be considered unacceptable by some river basin authorities but an essential planned precaution by others.

The alignment of flood protection banks determines precisely what flood plain land shall be seasonally flooded and what can be devoted to more secure economic activity. The banks themselves afford wide views otherwise not available in flat country. Some banks carry roads.

River training banks guide river flood flows in a particular direction, for example under bridges. If the banks are constructed mainly of soft material, their erodible sides must be protected by revetment – hard covering of loose or tightly fitting rocks, precast or cast in situ concrete or various forms of flexible mattress. Photos 4.5 and 4.6 show the construction of revetment for river protection works in Pakistan, while Photo 4.7 shows heavy rock toe protection for a railway/road cross-estuary causeway in North Wales.

Canal banks impound flowing water for prolonged periods. As Goudie (1993)[7] pointed out, many irrigation distribution channels are aligned along natural levees, dunes or terraces from which they can feed the flood plain fields by gravity; and these elevated alluvial landforms are composed of silt and sand prone to seepage loss. Therefore, earth canal banks need to be carefully constructed and preferably lined with some relatively impervious

Photo 4.5 'Rip-rap' rock protection being constructed for training works for a cross-river structure on the Sutlej River in Pakistan. On top of the smaller-sized under-layer, the medium-sized rocks ...

Photo 4.6 ... are cut and fitted by hand. The largest rocks will be placed with machinery for a 'falling apron', resisting scour at the foot of the bank ...

Photo 4.7 ... as do these that protect this railway/road causeway across the Glaslyn Estuary at Porthmadog in Wales.

soil to avoid excess leakage. Even if construction of a more robust impermeable lining can be afforded, some water will accumulate outside the canal banks, running off the adjoining land after occasional rainfall or as excess irrigation water. The implication for the landscape is that moisture is available along the side of canals, accumulated or drawn up by capillary action. This allows growth of vegetation, especially planted lines of shade-giving trees that do so much to enhance the canalside environment and views across flat country.[8]

4.4 Retaining walls and slope stabilization

Stabilization strategies

'The best solutions to retaining the ground are natural ones: rock cliffs and planted embankments' (Highways Agency, 1996).[9] However, steep earth surfaces need to be supported or strengthened when they have been cut back to a steeper than secure angle to provide more level space at the bottom; or the surface or deeper layers of an existing cliff or slope may have become precarious due to weathering or changed load conditions.

Stabilizing measures have to be devised taking into account:

- the type of rock or soil concerned;
- the useful space being created; also the available techniques and space on the site for constructing walls or strengthening soil; and
- the appearance of the wall or treated surface as seen beside a hillside road or behind a residential development.

The characteristics of soils and their investigation have been introduced in the last chapter, as has the creation of level ground. Available space on site may determine the type of stabilizing measures that can be constructed, particularly if excavation and subsequent backfill with well-draining material is envisaged. As to the appearance of structures like retaining walls, Laurie (1986)[10] points out that these are architectural elements linking structures to the landscape. The Highways Agency (1996)[11] suggests that 'retaining walls are best minimized in size and extent. This can be done both physically and visually.'

A number of support or strengthening strategies are illustrated on Figure 4.2. These include retaining walls, anchoring devices and strengthening the soil itself.

Sometimes surface treatment of a loose rock face may obviate the need for more penetrative solutions; and such treatment – together with the appearance of wall faces, panels or bolt heads – determines the total visual impact.

Retaining walls

Retaining walls hold back soil or rock formations so as to prevent them collapsing or sloughing onto level ground or structures below them. They can comprise simple but regularly maintained or rebuilt earth or stone walls for terracing of agricultural land, or permanent structures in poured reinforced concrete, precast or manufactured segments, rock masonry, brickwork or gabions (wire cages filled with rock or boulders).

Agricultural terracing is an important element of rural scenery in some hill country. Photo 4.8 shows lines of retaining walls that stabilize and slightly reduce the steepness of vineyards in Switzerland. More globally significant are level terraces flat enough to retain a depth of water such as for growing rice paddy (Photo 4.9).

Particularly magnificent achievements are terraced fields on steep hillsides in tropical East Asian countries. However, it is becoming increasingly difficult for farmers to earn enough from their crops to justify continuing arduous maintenance of these terraces and the hillside channels that water them (Cruickshank, 2003).[12]

Permanent retaining walls have long been common features of railways – and latterly top category highways – through undulating country. Retaining walls are designed to withstand overturning and to prevent structural failure under direct loading, or sinking or lateral movement of foundations. A 'gravity' wall resists the loads entirely by its own massive weight. More often conventional retaining walls are now cantilevered with a toe usually extending into the soil beneath

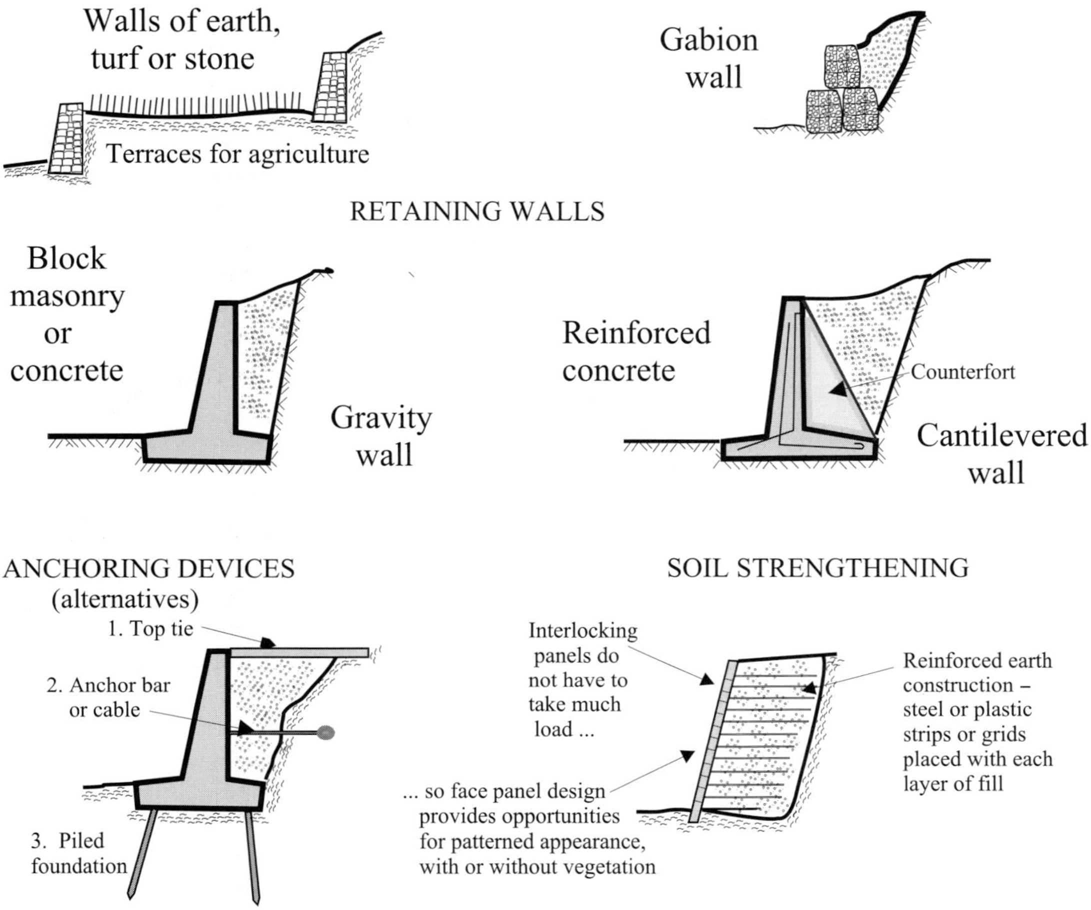

Figure 4.2 Slope support and strengthening measures

the slope. That soil has to be excavated temporarily to construct the wall.

In soft material, the difficulties of excavating behind a wall may be removed by constructing a wall out of sheet piling before any earthwork takes place. But sheet-piled walls are seldom handsome structures and are more often used for temporary containment. An alternative wall form is to insert concrete panels between vertical H-beams driven as piles at intervals into the ground.

The clearly visible functionary element is the wall itself, vertical or steeply battered with a plain or decorated face. Unseen are those structural elements which support the face – the cantilevered toe, perhaps counterfort wings, possibly a tie rod at the top or, in soft subsoil, vertical and raked pile foundations. The wall may be faced in stone, reflecting local geology or vernacular structures; it may be of brickwork or manufactured segments perhaps similar in appearance to the facades of adjacent buildings. Some high walls may have stepped faces that can break up the monotony of an unbroken face (Photo 4.10). Older walls may be supported by front buttresses where it was not practicable to construct adequate back counterforts (Photo 4.11). All are likely to have drainage outlet holes. There may be a narrow earth slope at the bottom on which shrubs may be planted to hide part of the wall.

More recent, less steep, non-load-bearing walls can present a variety of facing panels, the retained earth mass being anchored or strengthened as appropriate (Photo 4.12).

Photo 4.8 Vineyards stabilized by retaining walls above the shore of Lake Geneva, Switzerland.

Photo 4.9 Terraces flooded for rice cultivation, Sri Lanka.

Photo 4.10 A high gravity retaining wall with stepped face separates a former railway line at the bottom from a roadway above, Caernarfon, North Wales ...

Photo 4.11 ... and a lower section of the same wall with front buttresses.

Photo 4.12 A stabilized slope with pervious partly vegetated face panels.

Anchoring devices

Anchors tie walls to robust rock or soil layers behind. This was formerly achieved by excavating the ground so as to construct the inner anchor block and connecting cables, just as was done for the cables at the abutments of suspension bridges. Today the space for the block can be created by inserting a mechanical widening device or injecting concrete under pressure into a drilled cable hole to form a bulb at the inner end of the hole.

In rock bolting or soil nailing, slopes, cuttings or retaining walls are tied back by steel bars drilled into the ground and filled up with grout (cement slurry). Rock bolt capping can be discreet or ornamental according to aesthetic choice and to the width of capping needed to stabilize the rock surface. Japanese prize photography reveals what many regard as expert artistry in the patterns of such capping (Shikati, 1994).[13]

Strengthening the soil

Soil strengthening reduces the earth pressures acting on a retaining wall or other slope facing. Techniques include artificially reinforcing the earth by means of steel or plastic strips, geosynthetic mats placed with each layer of backfill or three-dimensional geocells filled with soil that is given extra strength by its containment. Depending on the nature of the fill, the surcharge or any lateral loads and the bond between the soil and the reinforcement, the width of soil to be excavated behind the proposed face may have to be greater than would be necessary for a free-standing wall.

Surface treatment and appearance

Cliffs or cuttings in rock may be fundamentally stable but liable to weathering or other surface deterioration. Rockfalls or similar hazards may threaten people or structures below. Spalling of face material can be prevented for a time by spraying gunite (cement/sand mortar) or can be contained by a wire or geosynthetic mesh fixed over the surface. Possible detachment of large blocks can be forestalled by drilling horizontal holes through the rock and inserting and cementing long rock bolts into the mother rock.

Aesthetic appearance is influenced by the shape, colour, reflectivity and texture of the treated face and its role in imitating, assimilating or contrasting with the its surroundings (Photo 4.13). Cemented coating casts a dull veneer over any geological splendour but may in time peel off. The most drastic solution is to cover a cliff in structural concrete – at worst a brutalist solution (Photos 4.14 and 4.15), at best an imaginative sculpture in its own right.

Vegetation can be conserved, planted or managed as an integral part of the structural forms, stabilizing the surface of sloping ground and absorbing run-off and resisting erosion. Turf can be laid or seeds planted in conditions where they can thrive without unwelcome competitors. In addition to the stabilizing effect, greenery usually lightens heavier artificial features where sharp patterns are not appropriate. Surface mesh can be largely hidden by vegetation that grows through or around it.

It is possible to install total cover on steep faces. Coverings resist erosion and may resemble high retaining walls; but generally they are non-load-bearing walls, used mainly in connection with anchoring or soil strengthening measures. Precast concrete facing panels are commonly provided with soil strengthening techniques and may be able to tolerate differential movement. However, in the opinion of the Highways Agency (1996),[14] proprietary reinforced earth systems using precast concrete panels are 'almost invariably hideous'. On the other hand, with facings in masonry or shapely geogrids, there may be scope either to create unashamed large scale works of art or to adapt the slope protection structures as giant or extended flower pots for whatever natural or cultivated vegetation can thrive in these lofty locations.

Sustainable geotechnical solutions

Techniques for stabilizing or covering slopes have developed over centuries. Today fresh approaches are still being made:

- through increased use of geosynthetic materials as these are developed in new forms; and
- through risk analysis of each situation to determine how much should be spent on

Photo 4.13 Two railway tunnels, a highway and more recent, less discernible higher protection complement fine scenery beside the River Rhine in Germany ...

Photo 4.14 ... whereas, at this point, clumsier cliff stabilization has taken place.

Photo 4.15 Massive concrete buttresses and a retaining wall at Tremadog, North Wales, seem excessive for the modest properties they protect.

constructing slopes and how much on subsequent maintenance.

Geosynthetic meshes, sheets and grids need to be proven in terms of their performance and longevity so that the most cost effective long-term solutions can be applied.

Lloyd et al (2001)[15] describe how risk assessment was undertaken on slope stability in landslide conditions along an existing 116km-long highway across peninsular Malaysia. The input to the assessment was digital video mapping of the entire route and a ground-based data base survey at each of 467 cuttings up to 80m deep and 578 embankments up to 100m high. Risk criteria were established by the use of integrated slope-hydrology slope-stability software. The benefit of this planning was a move from a remedial to a preventive management philosophy for slopes along the route. On average, preventive action cost only one-fifth that of full post-failure remedial works; in four years, maintenance costs nearly halved and landslides became rarer events.

Similar approaches can be adopted for earthworks generally. Synthetic materials, if they can be produced cheaply, may have a role to play in making more robust earthworks out of soft soils, for example, in establishing more safe havens in flood-prone Bangladesh. Risk assessments could determine how continuous construction, maintenance and reconstruction programmes could be implemented in such regions.

Notes and references

1 Emberton, J. R. (2001) 'Green engineering', in Carpenter, T. G. (ed) *Environment, Construction and Sustainable Development*, vol 1, Wiley, Chichester, p372.

2 Wheeler, P. (2003) 'Second helpings', *New Civil Engineer*, 11 December, p25.

3 Chapman, T., Marsh, B. and Foster, A. (2001) 'Foundations for the future', *Civil Engineering*, vol 144, pp36–41.

4 Bolton, A. (1992) 'Roman handle', *New Civil Engineer*, 29 October, p12.

5 Flynn, S. (2007) 'Heated exchange', *New Civil Engineer*, 13 December, pp28–29.

6 Nicholson, P. and Silva-Tulla, F. (2008) 'Reconnaissance of levee failures after Hurricane Katrina', *Civil Engineering*, vol 161, pp124–131.

7 Goudie, A. (1993) *The Human Impact on the Natural Environment,* Blackwell, Oxford, p141.

8 Vegetation, particularly trees, can have a variety of effects on how stable and watertight are the banks. In the Netherlands, it has been found necessary to provide a thick layer of clay on the outside of certain peat dykes to prevent the banks drying out in hot weather.

9 Highways Agency (1996) *The Appearance of Bridges and Other Highway Structures*, Her Majesty's Stationery Office (HMSO), London, para 29.12.

10 Laurie, M. (1986) *An Introduction to Landscape Architecture*, Elsevier, New York, p227.

11 Highways Agency (1996) as Note 9, para 29.2.

12 Cruickshank, J. (2003) 'Trouble on the terraces', *New Civil Engineer*, 17 July.

13 Shikati, T. (1994) photographs in Y. Nakamura *Terra*, Toshishuppan, Japan

14 Highways Agency (1996) as Note 9, para 29.22.

15 Lloyd, D. M., Anderson, M. G., Hussein, A. N., Jamaludin, A. and Wilkinson, P. L. (2001) 'Preventing landslides on roads and railway: A new risk-based approach', *Civil Engineering*, vol 144, pp129–134.

5

Mines and Quarries

Minerals are removed from the ground by mining them directly from the underground deposits where they are found, or they are quarried in open pits after removal of any overlying material. Traditional subsurface mining is undertaken from shafts or adits giving access to the underground deposits. Fluids such as petroleum or natural gas can be released more directly, under their own pressure or pumped, through wells drilled from the surface.

Open pit quarrying takes place:

- in deposits that lie near the original ground surface, as in steep-sided high or deep rock quarries, or in softer shallower often alluvial layers of gravels, sands and clays; or
- at greater depth into layers, for example of coal or phosphates, that can be reached by opencast mining after removal of the intervening overburden.

This chapter is primarily concerned with the extraction of construction raw materials such as rock, gravel, sand and clay. Indeed, these form the major proportion of all material removed from the ground. However, the mining of ores and solid carbon fuels, which are more regionally concentrated and in the long-term are scarcer and more valuable, has its own implications for local landscapes.

Landscape and land use issues in general subsurface (5.1) and opencast (5.2) mining are reviewed first followed by those in rock or gravel quarries (5.3, 5.4). Strategies are then discussed for restoration or enhancement of land at worked out quarries (5.5). The final section (5.6) recognizes the balances that have to be maintained between economic use of minerals, their rate of depletion and the long-term value of the land from which they are extracted.

5.1 Subsurface mining

Means of underground mining and consequences for overlying land

Coal and ores were first exploited where seams outcropped on the surface; but outcrops were soon exhausted and it became necessary to dig below the surface. Underground mining proceeded through adits (horizontal or gently sloping tunnels) into hillsides to extract coal or ore from seams that had been evident on the surface. Later, geological deductions indicated that seams may be found at depth, including beneath flat or gently undulating ground. Access can then be gained by vertical shafts, generally in larger scale operations. Sooner or later, most subsurface mining encounters groundwater; this has to be drained by gravity through adits or pumped up to ground level through shafts.

Impacts on the landscape of underground mining can result from:

- construction of surface works related to mining or processing of minerals;
- waste disposal; and
- subsidence, interference with drainage and contamination of water.

There are opportunities to create storage or other useful space in excavated caverns.

Surface works

Structures erected on the surface above mines include winding towers (headframes and hoists), ventilation equipment (extractor fan outlets) and water pumping stations. These are components of industrial scenery or, in cases such as Cornish steam pump 'engine houses', industrial archaeology. More extensive land surface areas may be required for railway sidings, cleaning and grading plants, or settling or leaching ponds for processing ores or wastes.

Waste disposal

The waste products of deep mining are materials that have been removed while excavating shafts and adits to give access to the workable seams. There may also be inferior coal or ore, or unwanted remnants from coal sorting at the surface, as well as ash or slag from power stations, metallurgical works or other heavy industrial processes historically located near mines.

Most early coal mines were in hilly country. In some regions, steep valley mining continued well into the 20th century. Waste had to be disposed of in valley bottoms, which might block or pollute watercourses, or in heaps on the slopes. Although relatively low proportions of waste emerge from underground seams following excavation, the quantities become considerable over decades; spoil heaps may become dangerously steep and unstable. Tragedies have occurred.[1] Safeguards include wider heaps, further transfer (including in landscaping measures and backfill of old workings) and drainage measures. In flatter country, typified by many North European coalfields, colliery waste is tipped by conveyor belts onto conical heaps, long ridges or flat-topped mounds, all of which soon become prominent landscape features (Photo 5.1). The actual shape depends on the

Photo 5.1 Colliery waste heap, Upper Silesia, Poland.

orientation of the conveyors; so if the heaps are to be permanent then their ultimate layout and profile should be borne in mind from the start; it is costly to undertake substantial reshaping later.

Even during the active operation of a mine, strategies be devised for planting on waste tip slopes and preparations for any structural foundations or access roads needed for their ultimate use. Hackett (1971)[2] mentions a 'British coal-mining area where the surrounding landscape is a dusty coastal plain, stripped of the vegetation it must have had at one time; here, the cleanliness of the conical mounds at least give character where it does not otherwise exist, and the ugliness is due more to the poor relationship of building groups to the waste heaps than to any fault of the cone form'. Often the nature of the tipped material may be difficult for planting or natural generation of vegetation, for example in slate, shale or ash. A compromise may have to be adopted such as selective placing of more fertile strips of waste.

Subsidence, disturbed drainage and water contamination

Subsidence occurs during mining or after a seam has been worked out. If redundant caverns are not backfilled with waste or supports are inadequate or no longer maintained, then roofs and overlying material may collapse. Structures at the surface may be damaged or displaced. On a larger scale, extensive lakes have been formed in holes that have appeared on the surface after the abandonment and collapse, for example, of Cheshire salt mines. However, some of these lakes are now recognized as wetlands of international importance.

Interference with surface drainage systems results from both subsidence and tipping of waste. Random pools and hillocks and blocked or polluted watercourses are typical of mining and industrial land devastation. However, as an 'optimistic environmentalist', Ridley (2000)[3] sees the southeast Northumbrian coastal region as a place to see wildlife or to experience a landscape 'getting better not worse. A great deal of this is due to the coal industry, past and present… land above has slumped into numerous pitfalls… little flashes of marshy water rich in aquatic life… dead remnants of woods, hedges or fences… full of wildlife… A landscape that was once monotonous fields has… been made greatly more interesting by the coal mines beneath it'.

South Lancashire was once a region of innumerable ponds and pitfalls but these have been progressively filled in since deep mining ended. Colliery waste tips have been levelled and housing estates built. When the coal that is still left in the ground is needed again, possibly in the present century, it may be the harder to find surface space from which to construct new shafts or make stockpiles, or to pump out, treat and dispose of groundwater.

Contamination of water at deep mines occurs by seepage or surface flow of run-off carrying coal dust and process waste, or by discharge of hot, saline or even radioactive water which may be pumped from deep workings. Much of the suspended sediment and all of the heat can be dissipated in settlement ponds if ample space can be allocated for these. There are also ways of reducing or treating polluted effluent to the extent necessary to improve downstream quality to an acceptable level. Whether the treatment processes can be afforded depends on the commercial value of the mined product.

Underground storage in mined caverns

Serious consideration can be given to making use of the more secure mined caverns, for example, for storage of liquid or gaseous fuels or for water reservoirs at the lower level of pumped storage hydroelectric schemes.

For underground storage of hazardous or radioactive materials, it is more likely that caverns will be specially excavated in rock masses that are geologically and hydrogeologically secure. There need to be adequate access routes to the caverns, monitoring of their watertightness, and measures to seal their walls more securely when necessary. Essential safeguards for storage of long-life nuclides include some means of locating the caverns should they be lost or forgotten thousands of years hence, even if the marker is a modest obelisk offering the only surface evidence that mining has ever taken place.

5.2 Opencast and high waste mining

Opencast mines include those where considerable quantities of unwanted overburden material has to be removed in order to reach deeper productive seams. Coal and lignite have been gained largely by opencast mining in Europe and North America since large earth-moving machinery powered by abundant cheap petroleum became widely available and coal seams under comparatively soft material were identified. Large quantities of waste also arise in the initial processing of the commercial minerals. Typically copper and bauxite (aluminium ore) yield copious waste although the amounts concerned may be exceeded by the quantity of overburden that has already been removed.

Opencast mining generally takes place along long deep strips (furrows) and the overburden is placed in equally long heaps (ridges) – sometimes devastating the land, sometimes as new deliberately reshaped landforms (Photo 5.2).

Black coals still tend to be found in regions where they were first exploited and which are now more densely inhabited. European opencast mining of black (bituminous) or brown (lignite) coal is currently undertaken on a medium scale, although recently in Germany this was primarily to give subsidized employment; but some deposits, even those underground, may be extended or reopened as fuel demand/supply patterns change later in this century. In the Rhineland, lignite is mined down to 500m depth while overburden thicknesses can be up to 300m.[4] Such operations cause very significant landform changes. So the process of permitting and defining the scope of any mining operation should be an integral part of any long-term regional land use plan.

Deposits of iron ore or uranium in Australia, bituminous sands or shales in Canada, copper in Africa or nickel in the US are generally in remote locations. In these conditions, the ideal eventual land use may be less apparent and final landform grading may be deferred for years. In any circumstance, adequate drainage has to be maintained throughout operations and after, if land is not to risk permanent spoliation.

If a mined area is one of natural beauty, then the scenic value will be damaged or destroyed. If

Photo 5.2 Overburden from opencast coal mining placed, perhaps over-tidily, at the foot of a natural escarpment, South Wales.

mining is in unexceptional landscape, then the consequence depends upon how the land surfaces are restored – how excavated holes are refilled from adjacent overburden, and what scenery is contrived as mining is completed. Not all the furrows need necessarily be filled and not all the overburden ridges need to be flattened. Strategies adopted can be those that are economic but balance high and low level features in a topography that is pleasing without being out of character. High waste ridges can be fashioned to provide hill features offering viewpoints; and furrows can be connected to suit an appropriate surface drainage pattern.

In Cornwall, china clay is hydraulically separated from large amounts of sand and clay (Photo 5.3). These wastes outweigh the kaolin product by about nine to one (Swann, 1995)[5] and have been conveyed to white conical waste tips beside the roughly mined basins. Some recent workings in the same region have been shaped into more sympathetic landforms and hydroseeded with grass. Meanwhile the older works are of a nature that characterizes industrial archaeology with some chaotic botanic recovery. More modern large-scale operations and earthmoving techniques are better suited to more rational economic land use restoration. The Eden Project is perhaps an excellent compromise providing exotic structures in their own large but secluded hole (Photo 5.4).

There was no need for either import or export of earth materials to or from the disused China clay pit that was dramatically transformed into the Eden botanic garden. Stability of both the surrounding rock cliffs and the soft clay residue was achieved with 'almost no assisted engineering such as mesh or gunite (sprayed concrete)' (Cole, 2003);[6] and topsoil was made up of China clay waste 'sand' (decomposed granite) with green waste to provide the humic content. Elsewhere, in various parts of Cornwall, 750ha of heathland has been restored on land devastated by clay mining.

Photo 5.3 Land devastated by china clay mining, here still in progress ...

Photo 5.4 ... may, after cessation, be adapted to form a new semi-natural landscape (right background) or a completely new structural and landscape development – the Eden Project, Cornwall, in 2001 before much of the outside planting had been completed.

5.3 Rock quarries

Uses of rock in construction

Most quarried rock is blasted out along faces formed by lines of drilled holes in which explosives are then filled and detonated. The loosened rock is cut, cleaved or crushed for construction materials; or it may be used directly in massive more random excavated form.

Cut stone can be dressed as a facing or used as a principal structural element of prestigious buildings or vernacular architecture. Sources of

Photo 5.5 Extracting large blocks of marble at Carrara, Italy. Rock is cut from the face by diamond-dust-coated cables running over pulleys to avoid the high wastage that would result from blasting (photo by Robin Carpenter).

high quality rock in homogeneous, unfaulted, uniformly coloured condition are relatively few. Exceptionally high quality products, like marble (metamorphosed limestone) are cut from a limited number of long-identified quarries such as at Carrara in Italy (Photo 5.5). Other rocks with surfaces that can be polished are more widely distributed; so are granite and some other hard but rough-faced igneous rocks used for block masonry. Some quarries produce particular local characteristics. English Cotswold architectural style has developed from – and been limited by – the strengths, weaknesses and subtle differences between limestone sources (Young, 2000).[7]

Cleaved rocks are slates, widely used for roofing – once worldwide, now mainly in the regions where they are produced. Thin and smooth surfaced when split, they are derived from a fine grained pressure-metamorphosed shale.

Massive angular rocks are delivered for use in the state in which they are blasted from the quarry face. The blasting pattern may be varied to maximize the proportion of a certain size of rock that may be specified for a particular purpose. For example, rock armouring may protect the beds or banks of rivers or coastal structures against water flow and scour (see Photo 4.7), or skilful hand shaping and insertion of smaller pieces of rock can produce handsome, firm, hydraulically smooth surfaces (Photo 4.6).

Crushed rock is much the commonest product of quarrying. It can come from a wide variety of geological types, principally of igneous origin but also of limestone or the harder metamorphic rocks. Indeed new quarries may be opened specially for single large projects at whatever is the most accessible site. Crushed rock is used for concrete aggregate, railway ballast, drainage courses and surfacing. Rock for crushing is usually blasted by the means most economic at the face, producing fragments at the most manageable size. These can then be handled by mechanical shovels, lorries and crushers.

Among softer but homogeneous rocks, chalk is a soft fine-grained limestone that can stand almost vertically at a face and is therefore easily quarried. Of no structural value in its loose extracted condition, it can be a main ingredient of cement. Other crushed limestones are extensively used in chemical industries.

Quarry situations, processes and appearance

Many rock quarries are cut sideways into open hillsides, some originating where rock was exposed in natural outcrops. New cliffs are created that replace the original hillside scenery (Photo 5.6). Where quarries are sunk into high ground or plateaux, the cliffs may be concealed from surrounding area; and yet the pit can provide an awesome surprise on any path that approaches the rim (Photo 5.7). Deeper resources, with no obvious outcrops, may be detected by geological deduction. Excavation may then be downward into a hole in the flatter or more gently undulating ground, with no visual impact on distant landscape (Photo 5.8).

Cliffs that are created may be features as significant in the landscape as those that are natural. Among Cézanne's many (late 19th century) paintings of Mont Sainte-Victoire in Provence, at least one shows Bibémus quarry as the principal foreground. The mountain's summit cliffs and those of the quarry are contrasted without implying any greater significance for one than for the other. The scale of some modern quarries makes such agreeable compromise unlikely; but high quarry cliffs and deep pits become outstanding geological features in their own right. Bingham Canyon open pit copper mine in Utah, believed to be the world's largest excavation (1.2km deep and 4km across), was declared a US National Historic Landmark in 1966.

Hard mother rock is stable at vertical or near vertical slopes. It is therefore quarried at steep faces between horizontal benches from which holes for blasting are drilled down to what will become the next working level (Figure 5.1).

Quarry faces can be relatively smooth displays of local rock strata and whatever flora and fauna they can support (Photo 5.9), or they may give more rugged impressions of mountain scenery (Photo 5.10).

Hillside quarrying produces flat areas at the bottom of the receding rock face. This space can be used for stockpiles and processing equipment;

Photo 5.6 A complete hill is being gradually excavated by this quarry that produces crushed rock in North Wales.

Photo 5.7 The sight of Torr Quarry that may appear suddenly to a Somerset rambler.

Photo 5.8 Delabole slate quarry, Cornwall, can also only be seen from its perimeter.

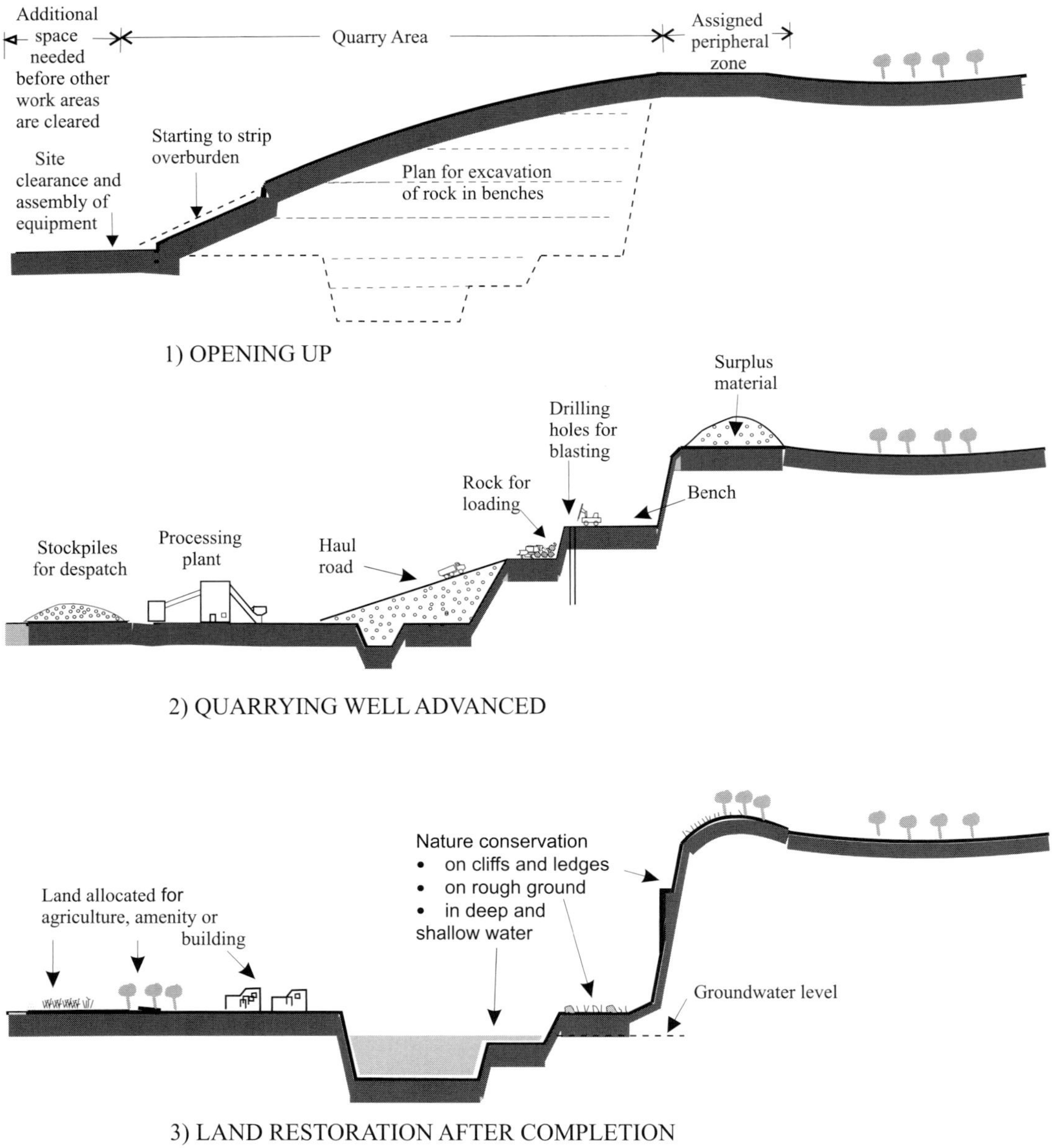

Figure 5.1 Rock quarrying

later it may become available for conservation or building. But, before the face is opened up and this space created, additional land space will be required for machinery and stockpiles.

Eventually quarry excavation often extends below the water table. Water has to be pumped out from the lowest working level. Lowering of the water table in this way may prevent natural discharge into nearby streams. Whether or not this affects flow or threatens aquatic habitat depends on how the pumped water is disposed. At Aggregate Industries UK Ltd's Torr Quarry[8] in the Mendip Hills, a reservoir was constructed for pumped water from which an outlet controls flow through a pipe into an affected stream in an adjoining valley.

Rock quarrying may continue for decades and even centuries at the same place. However,

Photo 5.9 Smooth quarry face, Torr Quarry, Somerset (by kind permission of Aggregate Industries UK Ltd).

Photo 5.10 Mountain ruggedness under snow, Torr Quarry (by kind permission of Aggregate Industries UK Ltd).

whenever excavation eventually ceases, rock exposures become scenery as well as geological evidence. If cliffs, slopes, ponds and water margins are left in comparative seclusion then wildlife and vegetation can take over, perhaps influenced by a degree of deliberate planting or drainage adjustments and subsequent occasional management. Alternatively the flat ground may be invaluable for buildings, in which case layouts should take full advantage of the steep background landscape.

5.4 Gravel, sand and clay extraction

Whereas rock is hard and massive, soil is loose and granular. This section is concerned with non-organic soils – the engineering uses for each constituent, the places where they are found and the consequences of their extraction.

Soil constituents and their engineering uses

Boulders, cobbles and *pebbles* may be used as coarse aggregate in mass concrete, for rapid drainage layers in earthworks, or to fill gabions. Crushed rock, in sharper shapes, can be produced in similar sizes. *Gravel* comprises accumulated unconsolidated rock fragments larger than sand (2mm) and smaller than cobbles (about 75mm). For civil engineering, gravel can provide:

- coarse aggregate for concrete or highway blacktop, or loose road surfacing; and
- permeable (free-draining) material for fill in embankments and other structural earthwork.

In practice, deposits described as gravel may contain significant proportions of finer material, especially sand or silt. For most uses, gravel requires washing, sieving and sorting into particle sizes that can be recombined to meet particular specifications.

Sand (0.053–2.0mm) occurs within gravels or, in a purer often uniformly graded form, in river and coastal deposits. Provided the proportion of fine silt or clay is low, sand can be used as fine aggregate for concrete, as a drainage medium such as in backfill to retaining walls, or as road sub-base.

Silt (0.002–0.053mm) is too fine to permit free drainage but has little of the plasticity of clay. Silt is therefore seldom in demand for engineering material. Where its incorporation cannot be avoided, as in land reclamation in estuaries, a means must be devised for containing silt fill securely and establishing a satisfactory hydraulic equilibrium – to prevent it liquefying like quicksand under any imposed load.

Clays (less than 0.002mm) are cohesive soils; their properties and performance under loads differ widely according to whether they are dry or saturated and to the history of their consolidation beneath other formations in earlier geological time. Clay is the raw material for bricks and an ingredient in cement manufacture. In its more natural state it may be incorporated in impermeable cores for rock or earth fill dams or lining for canals and reservoirs.

Sources of gravel, sand and clay

Sand or clay can sometimes be found in homogeneous deposits. As often, they are found with coarser materials in gravel mixtures. Sources of gravel are:

- glacial drift, found at high latitudes or altitude where there has been glaciation;
- alluvial terraces; and
- river bed and flood plain alluvium in the central reaches of river valleys where seasonally varying stream velocity is sufficient to displace, carry and deposit riverside and river bed material.

Glacial drift, sometimes described as till or boulder clay, is essentially unsorted material of greatly varying composition and thickness. Accordingly extraction of gravel, sand or clay is necessarily selective but any heaps of waste that result do not necessarily alter what may already be naturally rough topography (see Photo 3.1).

Alluvial terraces, ancient valley bottoms and, in arid climates, alluvial fans may be less random in character than glacial drift deposits and yet in drier conditions than can be found on floodplains. The deposits may have formed where river or lake levels subsequently dropped.

River and flood plain gravel pits

River beds and flood plains are, in many well-inhabited regions, the major source of sand and gravel for construction. River valley gravel extraction takes place directly in what are, or have been, active channels. Taken in modest quantities from broad braided river beds and replenished during floods, gravel deposits may be considered a renewable resource. However, excessive extraction might affect a river's sediment transport regime and thus alter its channel hydraulics. Nutrient-carrying sediment transport to farmland may be reduced, scour at bridges and flood risks increased. On the Paute River in Peru gravel extraction was eroding the river bed level to the extent that it had to be discontinued as part of a recovery plan for that valley (Abril and Knight, 2003).[9]

On flood plains there may be fewer restrictions in extracting gravel from wider areas adjacent to rivers, where the water table is high but the ground surface is normally dry. However, after a substantial thickness and extent of gravel has been removed, the resulting landscape is one where lakes replace a major proportion which had been dry land. In the lower Trent, Tame and Thames valleys in England, sand and gravel deposits are being extensively exploited and the disused pits, filled with water, are used as water amenities or become habitat for waterside wildlife or migrating wildfowl.

In the actual gravel extraction process, the wet conditions are overcome by:

- wet digging, in England formerly by suction dredger to a floating barge, now usually by dragline (a bucket thrown out and hauled in from a crane boom) raising material through the water from the pit bottom or sides and discharging it to trucks or heaps on adjacent land; and
- dry excavation in enclosed areas kept dewatered by pumping at perimeter wells.

Typically, excavated gravel is taken by conveyor or dump truck to a washing and grading plant. 'The wastings (tailings) from the plant are run into the silt settlement lagoon. Once the settled silt reaches the water surface, it begins to dry out and consolidate, allowing colonization by plants. Silt lagoons often develop into excellent conservation habitats. Once the deposit is worked out, overburden and topsoil may be bulldozed back into the dry quarry to landscape the banks. When pumping ceases, the pit normally fills up with water' (ARC, 1992).[10]

On any land space acquired for gravel extraction it is likely that the whole area will be excavated, at one time or another, and that space for land operations will be reallocated when necessary. At the end, there will be some reconstituted dry land where:

- settlement lagoons have filled up with effluent silt from gravel washing; or
- excavated pits have been backfilled with overburden.

The remainder of the worked area will remain under water. Its basic pattern can be predicted in initial and interim mineral extraction plans and the ultimate shape and depth of lakes can be finally adapted to suit the type of conservation or development that is intended. Strategies for gravel extraction should take into account the characteristics of the final land areas, the waterside strips and the lakes themselves, and the options for shaping these during excavation or adapting them later.

Figure 5.2 shows the sort of strategy that is adopted, allowing valley bottom lands to be progressively exploited and restored to a former or new land use.

5.5 Restoration or reconstruction of worked out quarries

Quarry features and their adaptation

The way in which quarried land may be restored depends on the use for which it is consequently intended – agriculture, forestry, settlement or wildlife habitat – and what sort of scenery will result. The features of quarries that can be incorporated in their reconstruction include steep and less steep slopes, flatland, open water, watercourses and surplus earth.

Steep slopes, as cliffs blasted for rock, are successively cut back until the limit of a quarry is reached. Any upstanding cliffs then remaining can be conserved as bold landscape; any steep hole may remain as an intriguing abyss or a lake. Alternatively, the local authority may choose to forego any scenic or ecological opportunity so as to backfill any void with surplus material such as municipal waste. *Less steep and undulating land* results from quarrying in softer mineral formations or overburden. It can be reshaped suitably for agriculture or some locally more natural landscape. In arid lands, where many remote mines are located, the more sterile soil may be of no immediate use and can be tidied up in any appropriate pattern.

Flat land may be the hard surface of the floor of a rock quarry or the unexcavated or already refilled ground at alluvial gravel sites. During the operational phase, the land provides space for process machinery, stockpiles, access routes and parking areas. After working ceases, it is available for a variety of uses including built development. Where flat surfaces have been artificially created, it is important to contrive a satisfactory surface drainage system to suit the new topography. Erosion protection may also be necessary if run-off and stream flow are not to wear away soft, newly created banks and ditches.

Open water is an attractive focal feature for an industrial or residential estate (Photo 5.11); or it may be assigned to amenity or nature conservation (Photo 5.12). Wide oblong or round lakes may provide sites for aquatic sport or people may walk around them. But, especially in river floodplains, it is waterside land strips and adjoining shallow water that provide the most valuable ecological habitat.

Watercourses are natural or created drainage routes. In any final landform that results from reconstruction of quarried land, drainage channels have to carry the quantity and quality of water that runs off from rainfall in the catchment or seeps out of the ground or old mine workings. The courses and banks of streams are also important elements in valley ecology and scenery and serve as corridors for wildlife.

Surplus earth, sometimes supplemented by construction or municipal waste, is the main material available for final modification of land shapes and water depths. The quarried surplus can be overburden or extraneous material that has been washed out or otherwise extracted from the main product. Thus, surplus material may comprise:

- topsoil, with organic content and nutrients on which vegetation flourishes; typically 30cm of uncompacted loam, with accompanying earthworms, creates good growing conditions;

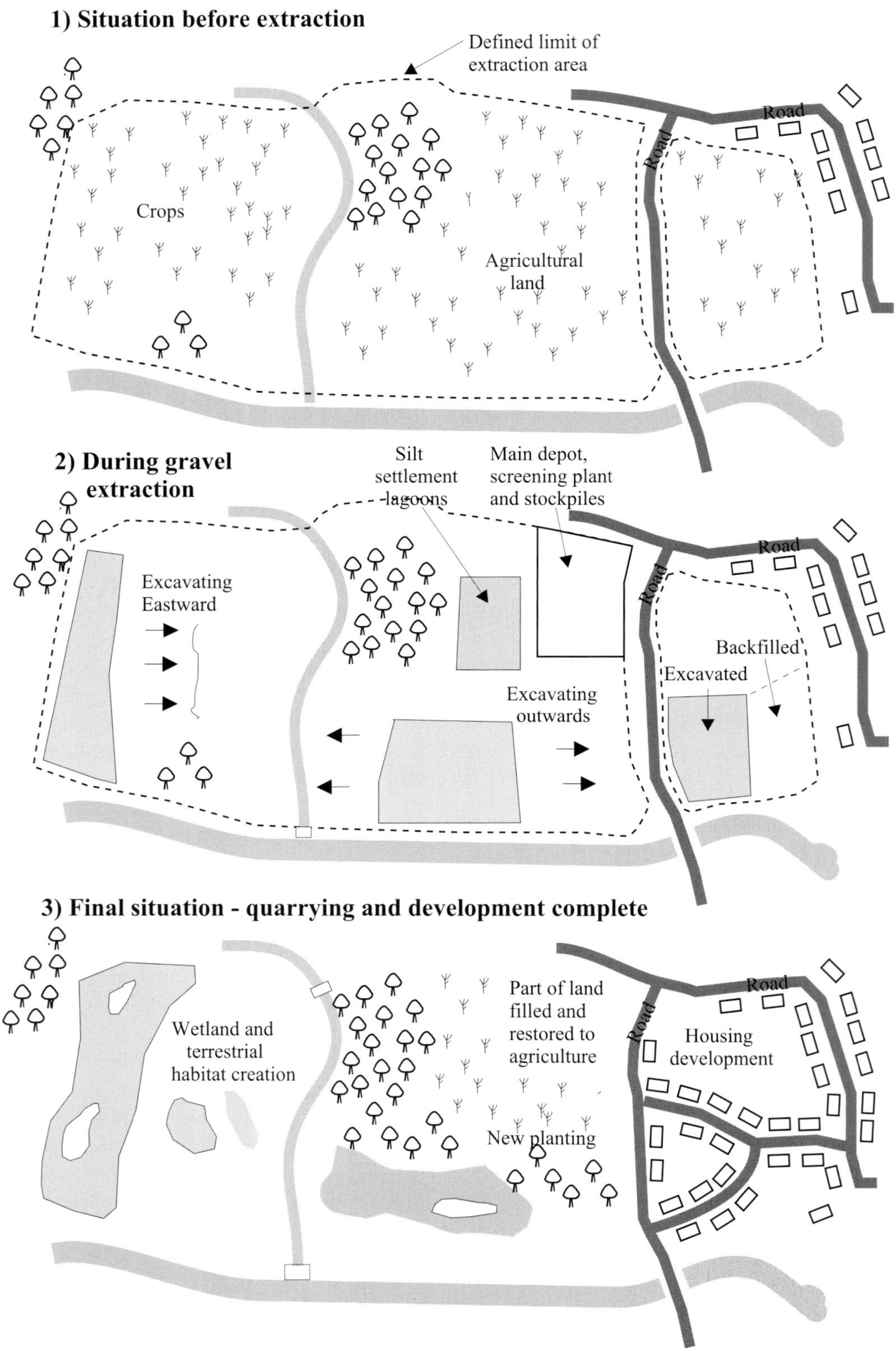

Figure 5.2 Alluvial gravel extraction

Photo 5.11 Residential development by flooded gravel pits, Gloucestershire.

Photo 5.12 The Norfolk Broads, now devoted to boating and wildlife, are the consequence of ancient peat excavation.

- inorganic soil, which has low nutrient content but, placed as a thin topsoil, can support rough grassland flora without risk of dominance by more vigorous plants;
- silt and clay which can emerge as tailings – fine material washed out of gravel or mud settling in streams or ponds; if subject to periodic inundation and drying-out this mud may provide valuable plant nutrients; and
- other inorganic rough material or inert garbage, useful bulk within massive fill.

Placing of these surplus earth materials can occasionally be arranged so that they are removed and placed directly as fill in their new position by one operation of a single machine. More often the material has to be transferred over some distance by dump truck or conveyor belt to the intended destination or to an intermediate stockpile. Occasionally complete chunks of soil and low vegetation can be excavated and transported from one area to be reinstated in another.

Restoration needs and opportunities

Land restoration approaches have developed gradually:

- Deliberate restoration was unusual in 19th and early 20th century industrial and mining practice. During prosperous epochs it was probably never contemplated that closure or abandonment could occur; when hard times arrived there were no funds for making landscape attractive even if there was a consensus about what should be done. Nevertheless, quarrying has sometimes resulted in gradual emergence of a diverse, interesting and productive landscape.
- Worked out land has been restored retrospectively. This has commonly occurred as part of a regional economic recovery plan. Rather than beautifying derelict scenery as an end in itself, the landscape has been made attractive for people to live in, work in or visit. Special measures may be necessary to prevent build up of methane or leachates that could pollute watercourses.
- In Britain and some other populated countries, the later 20th century saw a great need for disposal of unsorted municipal garbage – opportunities to fill up worked out quarries and pits. Bedfordshire clay pits – excavated since the 1930s – have, when exhausted, been allocated mainly for this type of landfill as have several sand pits in southeast England (Photo 5.13).
- Land restoration has now become a statutory requirement at the end of land transformation operations such as opencast mining. Often this has taken place, as in the Durham coalfield, by turning sites 'back into farming land, but the tendency is for the original character of the landscape to be replaced by a featureless

Photo 5.13 Landfill: A worked-out sandpit being prepared to receive layers of municipal waste.

engineered appearance' (Bryson et al, 2000).[11] However land restoration is one of the many concerns of environmental impact assessment, with broader scope than merely reverting to a former land use.

Three situations illustrate the issues in different types of quarry restoration:

1 Steep rock landscapes.
2 Softer landform restoration and that appropriate in remote places.
3 Nature conservation in worked out gravel pits.

Rock landscapes

When rock quarrying ceases, it is too late to think of modifying the appearance of cliffs. Faces may mature, with weathering and vegetation, to more strongly resemble natural cliffs or rock outcrops. The resemblance may partly depend on the sequence and processes of excavation and the shape and rugosity of the final face. But at least the gentler slopes at the top of quarry faces and any wetland at the bottom may be ripe for planting and subsequent management in accordance with specific biological objectives.

The natural habitat of rock quarries typically includes:

- woodland or wilderness on the periphery of quarries or in forgotten or restored remnants;
- loose rocks and the crevices between them – refuges for small mammals that roam the quarry floor;
- cliffs and their ledges – the only inland nesting places for birds like ravens or peregrines; and
- more diverse wild flower and insect habitat on the less steep slopes, such as where the overburden has been reshaped, which may then be replanted or managed.

Boag (1999)[12] described in detail the flora and fauna of Torr Quarry and concluded: 'The effect of the quarry in this part of Somerset has improved the diversity and quantity of wildlife. Once it is restored and given over to nature, as a reserve with country walks etc. no one will ever dispute its benefit to wildlife.'

Soft landform restoration

A great many open pits are mined in comparatively soft rock in formations which, unlike rock outcrops or quarry cliffs, are neither visibly interesting in their natural form nor attractive after the minerals have been extracted; and many such pits are located in remote parts of the world where the climate or soils discourage human settlement.

Mining waste may be tipped from conveyor belts onto prominent cones; or overburden may be dumped in long less striking ridges. In *well-inhabited regions*, there may be a choice between retaining the high heaps – as scenery or viewpoints – or reshaping all the surplus material in a reconstructed landform. Planting trees or managing emerging vegetation helps to stabilize the steeper slopes and gradually improves the initial sterility of newly created agricultural or built landscapes. This may happen naturally, taking 30–100 years in temperate climates, through natural grass/shrub/tree progressions.

In the *remote regions* of the world – deserts or Arctic emptiness – the landscape may be utterly flat and featureless; or there may be striking rock formations, such as long outcrops of hard strata emerging from dusty monotonous surroundings. These, for example, in the eastern Sahara, are seen by people only from aeroplanes or in aerial or satellite photographs. In any circumstances, mining debris can surely be arranged in patterns that are geometrically aesthetically attractive, suit the mining processes and machinery, yet leave the land in a form that might be adapted for secondary mineral recovery or some unforeseen use in the remote future.

Nature conservation in worked-out gravel pits

In Britain, flooded pits attract substantial numbers of breeding and overwintering wildfowl as well as other wetland flora and fauna. To accommodate and encourage wildlife, both the routine working and final renovation of gravel pits should take into account the nature of the dry land, waterside strips and open water that will result.

On *dry land areas* some space, probably near the site entrance, may be allocated for human amenities such as car parks, small buildings and picnic areas. Grassed and accessible open space may merge into wilder growth, a large proportion of which may be dedicated to wildlife. The public can be encouraged to observe the main nature reserve from designated pathways or from well-placed hides or observation towers. Birds are attracted by the waterside features or wetland but also require open grassland, scrub or woodland – for feeding, roosting, nesting or 'loafing'. Each habitat opportunity can be borne in mind in deciding, for example, how filled settlement lagoons are to be used and the pattern in which other fill is placed.

Extra security may be needed for landside habitats. Fences may be necessary to preclude mammals such as foxes. Woodland, shrubs and hedges provide comparatively secure roosts for certain birds and make windbreaks for resting migratory or breeding wildfowl. Islands are particularly safe and, ideally, 'should combine a complex shoreline with downwind bays for shelter' (ARC, 1992).[13]

Some *waterside strip* may be allocated to human amenity such as a beach or boat launching area. A straight stretch of the bank of one of the deeper lakes may be appropriate. But, for nature conservation, waterside habitat is the most valuable asset in restored gravel pits. Accordingly, opportunity can be taken to maximize the length of waterside, preferably vegetated.

Between the dry land and the deep water there is a succession of botanical communities – from terrestrial plants where the soil is dry in summer, through seasonally wet bog and perpetually wet marsh plants, to emergent and then submerged plants in the photic (adequate light) and aphotic (dark) depths of water (ARC, 1992).[14] Earth shapes for waterside habitat can be created as:

- open gently sloping shoreline: necessary for some animals to enter or leave the water; the surface may be harsh gravel or sand, or mud and vegetation created by spreading more fertile material;
- steep banks – habitat for burrowing creatures;
- wide, shallow (less than 1m depth) inlets or bays, extensive enough (at least 2ha) to create reed beds which attract various perennial species as well as autumn roosters; or
- winter water meadows, a comparatively rare habitat in Britain, providing 'feeding grounds for snipe, curlew, redshank, greenshank, godwits and ruff' (ARC, 1992).[15]

Some wetland or waterside habitat depends on seasonal flooding or other timely changes in water level. If gated sluices are constructed where a river is connected to the pit area, a powerful tool for habitat management is added. Water flow control and water level variation can be exercised for a variety of purposes from:

- flooding marginal meadows in winter or raising water level in spring to encourage ducks to nest, to
- drawing down the lakes to expose rich mud with invertebrates to attract waders, to exclude any fish species that predate excessively on other fauna, or for periodic exercises like the rotational burning of reeds to clear out dead stems.

Open water that results from methodical removal of gravel or sand strata is mostly of uniform depth, probably within steep banks and in the shape of wide, rectangular lakes separated by bands of land that were retained for access or built up of surplus material. The ideal situation in a nature reserve, on the other hand, is one of irregular shoreline where much of the water is shallow and only gently sloping. Such a configuration maximizes both feeding grounds and means of access between land and water for a variety of water birds, mammals and amphibians. Deeper water, about 1.5m or more, can be devoted to human water sports or to fish in so far as such activities are compatible with other conservation objectives; or floating rafts can be moored in deep water to serve as perches or nest sites for birds.

The transformation from geometrically regular ponds to a diverse pattern of sinuous shorelines, shallow inlets, peninsulas and islands can be achieved by a combination of:

- adaptations to the commercial abstraction plan: bulk excavation and fill should leave a shape approximating roughly to the finally defined land surface area and water depth, and complying with an overall concept of an ultimate drainage pattern; and

- specific adjustments at the completion of general excavation in each major area.

In 2003, Richard Mabey[16] wrote concerning the creation of a riverine forest near Milton Keynes: 'Gravel is being dug in the river valley to a predetermined pattern of mounds and dips, planted up with alder, willow, black poplar and the like and will eventually be flooded and allowed to get on with its own development plan.'

5.6 Striving to integrate mineral extraction with land resource conservation

The value of minerals and our enthusiasm to exploit them must somehow be balanced against the consequences for future generations and for other land resources. Our descendants will soon suffer from our current depletion of easily extracted fuels such as petroleum and natural gas. Copper is being rapidly being extracted in the remaining rich deposits. Coal and most ores may last for a few centuries. Meanwhile, mining has spoilt some fine scenic resources and many less spectacular landscapes have been at least modified.

Rock is a relatively abundant material and one required in large quantities for construction; but rocky geology tends to be found in the more scenic regions. So the location of quarries has to be sensitively allocated. Glensanda quarry on the Morvern peninsular of western Scotland was located where it is because of its remoteness and distance from the more well-trodden mountains and moors of the rest of that highland region. The quarry is largely concealed, except from the air. It is served only by sea. Excavated granite and other hard rock is dropped into a shaft and conveyed through a tunnel to a deep-water berth (Photos 5.14 and 5.15).

Photo 5.14 Glensanda Quarry from which rock is conveyed through a tunnel ...

Photo 5.15 ... for processing and shipment (both photos by kind permission of Aggregate Industries UK Ltd).

Limestone, which forms the Peak District of northern England, is less universally ideal as construction material but an important feedstock for the chemical industry. Quarries are located on the upland moors and are therefore plainly visible (Photo 5.16); whereas the main beauty spots in limestone country are found in gorges and dales. Controversy continues as to where continued extraction should take place in a region most of which is classified as National Park.

Softer ground where metal ores, thicker coal seams and solid hydrocarbons (such as oil or tar 'sands') are found is generally less naturally scenic and often much more remote areas. In the colder, sparsely inhabited northern forest and tundra regions these minerals are excavated in huge quantities in opencast strip mining with little attempt to avoid spoiling the landscape.

Gold is a highly valued metal because it is decorative. Diamonds are even more decorative. Mining of both is tightly controlled by the concessionary landowners in southern Africa. But in other parts of that continent, diamonds have become the currency of theft, oppression and violence – a source of great wealth to some and great hardship to many. Elsewhere mining for gold and precious stones has combined with unauthorized logging or excessive clearance of ranch land to destroy rainforest or cultivable settlement land. The largest gold mine in the world is probably the Grasberg Mine in Irian Jaya (Indonesian New Guinea).[17] This great pit lies next to the highest snow-capped peaks of southeast Asia, a region that by any ancient or modern standard should be judged as sacrosanct natural heritage of the highest order. Yet it is mined by a foreign company, perhaps mainly for the rich deposits of copper that are mixed with the gold and silver.

Globally, however, there seems to be growing awareness, not only of the impact of large-scale mineral extraction, but also of how controls can be exercised in countries where land-use

Photo 5.16 Limestone excavation in Derbyshire.

planning is effective. Even in those countries, planning permission procedures might be better complemented by national or even international surcharges (taxes) to raise the costs of mining scarce resources to near the level that future generations will have to pay.

In the more remote regions, such as where strip mining of solid hydrocarbons or the scarcer ores takes place, the market value of the minerals might be raised somewhat by imposing extra obligations on mining companies. They could be obliged, within their concessionary contracts, both to restore the land and to set up settlements and develop other land resources for the best advantage of long-term sustainable development. Meanwhile, in many countries, no such sophisticated or even basic restraints are yet practicable. Serious devastation caused by small and large-scale illegal or officially sanctioned spoliation of land resources is difficult to contain until a responsible authority has effective political control of the region concerned.

This need for authority and for fair land use rights and obligations extends throughout settlement or land-use planning. There are similar political issues in the fair distribution of water resources where, as the next chapter shows, construction can play a much more positive role.

Notes and references

1 On 21 October 1966, at Aberfan in South Wales, more than 150,000m^3 of water-saturated colliery waste suddenly slipped, killing 144 people, many of them children in the village school (http://en.wikipedia.org/wiki/Aberfan_disaster, accessed 5 May 2010).

2 Hackett, B. (1971) *Landscape Planning: An introduction to Theory and Practice*, Oriel Press, Newcastle upon Tyne, p57.

3 Ridley, M. (2000) 'The southeast Northumberland coast' in Bryson, B. et al, *The English Landscape*, Profile Books, London, p425.

4 RWE Power Lignite Mines (2009) 'Rhineland lignite mining', www.mining-technology.com/projects/rhineland, accessed 16 August 2009.

5 Swann, P. (1995) 'Planning for the future', *Landscape Design*, March 1995.

6 Cole, D. (2003) 'Brave new world', *Landscape Design*, April 2003.

7 Young, R. (2000), in Bryson, B. et al, *The English Landscape*, Profile Books, London, pp177–178.

8 Foster Yeoman Ltd, who previously owned Torr and Glensanda Quarries and who provided the original photographs 5.6, 5.7, 5.12 and 5.13, were bought by Aggregate Industries UK Ltd in 2006.

9 Abril, B. and Knight, D. (2003) 'Stabilising Paute River in Ecuador', *Civil Engineering*, February, pp32–38.

10 ARC (Amalgamated Roadstone Corporation) and the Game Conservancy (1992) 'Methods of excavation', *Wildlife After Gravel*, Game Conservancy, Fordingbridge, Hants.

11 Bryson, B. et al (2000) *The English Landscape*, Profile Books, London, unspecified author, p390.

12 Boag, D. (1999) *Living Stones*, Foster Yeoman Ltd, Frome, Somerset, p128.

13 ARC (Amalgamated Roadstone Corporation) and the Game Conservancy (1992) as Note 10, section on 'Islands and shallows'.

14 ARC (Amalgamated Roadstone Corporation) and the Game Conservancy (1992) as Note 10, figure showing 'Ecological succession of the botanical community'.

15 ARC (Amalgamated Roadstone Corporation) and the Game Conservancy (1992) as Note 10, section on 'Wet meadows'.

16 Mabey, R. (2003) 'Countryside futures', *Countryside Voice*, summer 2003, p22.

17 http://en.wikipedia.org/wiki/Grasberg_mine, accessed 11 August 2009; see also 'Irian Jaya' in *National Geographic*, February 1996.

6

Dams and Other Hydraulic Structures

River control is vital to store and distribute surface water resources. But the effects of structures like dams on valley lands can be profound and sometimes irreversible.

The first section of this chapter explains the unequal global occurrence of water and describes the development of control works constructed on the world's major river systems. The main impact of these works on global land resources is to bring fertile but otherwise arid or semi-arid land under irrigation. This has increased world agricultural land by a sixth and agricultural produce by a third. Besides meeting the more direct needs of growing human population and industry and enabling the survival of wildlife communities, other benefits of well-planned river control structures may sometimes include flood alleviation.

Section 6.2 describes the land resources of river valleys, from catchment highlands to lowland flood plains, and the opportunities that these provide for water storage and diversion structures. Section 6.3 describes how the location and design of dams and the provision of reservoirs affect the useful land and habitat that they damage or flood; and Section 6.4 identifies any additional issues in hydroelectric schemes.

Section 6.5 is concerned with the construction of channels to distribute or drain water so as to raise the value of land resources. Section 6.6 then summarizes the issues in determining the most equitable and sustainable ways of relating hydraulic engineering to sustainable land development.

6.1 Water resources – occurrence and control

Water occurrence

Water is both essential for all life forms and a powerful agent in the formation of landscape. Occurrence of rainfall and river flow determines the pattern of human settlement and wildlife habitat as well as that of erosion in valleys and sediment deposition on floodplains. Annual patterns of precipitation and snowmelt produce wet season excess and dry season shortage. Wet season storms or wide flood inundation threaten agriculture, settlements and built infrastructure. Normal dry periods can be anticipated but severe droughts result in thirsty populations, crop failure and wind erosion.

About 115,000km^3 of rain or snow falls annually on the land areas of the Earth. Perhaps about:

- 70,000km^3 of this amount is evaporated either directly, from water surfaces, bare or paved ground, or transpired through vegetation;
- 40,000km^3 runs off into streams and rivers; and
- 5000km^3 reaches the sea through the ground or in icebergs.

These are order-of-magnitude global indicators only, based on various estimates.[1] There are wide geographical differences. Regional water resources assessment also has to take account of seasonal and year-to-year variations in river flow, local groundwater conditions and human intervention in the hydrological cycle.

Of all the river flow across the continents, the greater proportion crosses regions where it is impracticable to abstract water or there is no need to do so. Thus most of the huge flow that crosses the Siberian plains (2000km^3 per year) and the Amazon basin (6500km^3 per year) runs freely to the sea. About 580km^3 per year flows through the Mississippi basin but, although there are vast areas of fertile soil, the flatness of the riverine terrain makes flood protection a higher priority than river water abstraction. However diversion works on the Indus River and its Punjab tributaries mean that a large proportion of the annual flow of 210km^3 per year is withdrawn for irrigation projects in Punjab, Rajasthan and Sind in both India and Pakistan. In the Aral Sea basin in Central Asia so much water was abstracted from the Syr Darya and Amu Darya rivers that the flow of 100–120km^3 per year that emerges from the mountains was reduced in some years to no flow at all into the diminished inland sea.

Water use

The total amount of organized water abstraction in the world can be roughly estimated[2] at about 4000km^3 per year. In southern Asia more than 90 per cent of abstraction is used for irrigation, in Europe most is for industrial or domestic use. Perhaps two thirds of total demand is met from rivers and the remaining third from groundwater.[3] Where water resources are meagre, there is often contention as to how to allocate water for competing demands in the different parts of each river basin. A large proportion of the surface water abstraction total has been achieved by storing seasonal high flows in reservoirs and by constructing water transfer aqueducts. Groundwater is usually pumped out.

Dams are constructed to impound reservoirs. Cross-river weirs or barrages divert water into canal, pipeline or tunnel aqueducts; and irrigation or water supply systems convey the water to wherever the engineering works can be justified by the benefits of water provision. The value of water (per cubic metre) is usually high for municipal and industrial use, much lower for the relatively modest cash returns in irrigated agriculture. So irrigation schemes can justify large engineering works only by economies of scale. Energy, from fuel or as electricity where these have been cheaply available, has greatly accelerated some forms of water abstraction, notably by pumping from rivers, lakes or tubewells.

Development of river water diversion and control

Early civilization flowered in arid or semi-arid regions along major rivers, for example in Mesopotamia, Egypt or northern China. The ingenuity and effort of the Babylonians could not tame the Tigris or Euphrates directly but, even 6000 years ago, people undertook earthworks to protect their cities from seasonal floods and diverted the waters onto fertile croplands. Egyptians took advantage of flood water for basin irrigation in the Nile delta. In China, for thousands of years, flow has been diverted from rivers such as the Yangtze. Other early achievements, not related primarily to agriculture, include long distance ship canals, especially in China,[4] and large municipal water supply schemes such as those supplied by Roman aqueducts (Photos 6.1 and 6.2).

Photo 6.1 Roman aqueduct for water supply across a rural area in northern Spain ...

Photo 6.2 ... and in an urban setting (photo by Robin Carpenter).

The major river basins of the northern Indian subcontinent are comparable in total flow to the Chinese rivers (1000–2000km^3 per year). At the western end, the Indus and the Punjab rivers are perhaps more predictable as to the volume of annual high flow, and in some stretches more topographically suited to diversion of water. The Indus Valley Civilization flourished on basin irrigation 4000 years ago. By Moghul times, a system of 'inundation' canals for diversion of seasonal floods was in use in Punjab and Sind. Some of these eventually fell into disuse but others were renovated in the 19th century. Most were replaced by cross-river barrages and headworks diverting water into formal canal, irrigation and drainage networks (Photo 6.3).

From the Ganga (Ganges) river, huge irrigation projects have also been established since the 1850s. Diversions have reduced the river's flow in the dry season to perhaps a tenth of its former level but have had little effect on the much greater peak flow from July–October. Because of the width of the main stem river, diversion on the plains has been associated with weirs rather than gated barrages. However the Farakka barrage was built across the Ganga in West Bengal in 1971 with a capacity for passing floods of 70,510m^3 per second, greater than any barrage on the Indus; but its purpose is primarily to feed water into a canal which flushes out the bed of the Hooghly river to maintain shipping depths in the port of Calcutta.[5]

The circumstances in crowded east and south Asia contrast sharply with those in most of the American continent. America covers nearly one third of the world's non-polar land surface, has about 45 per cent of global water run-off but supports only 14 per cent of the population. However, there is still contrast, for example, between conditions in the Amazon basin – which accounts for 38 per cent of America's total river flow but with little necessity to consume it – and the US state of California, where water is in such

Photo 6.3 River barrage and canal headworks, Punjab, Pakistan.

demand that the state's surface water flow is mostly harnessed, its groundwater is over-exploited and much of its supply is by long aqueducts bringing water from afar. Between these extremes of water abundance and scarcity is the basin of the Mississippi. Much of the river's vast basin is fertile but, although many dams have been built on tributary rivers, no substantial storage or control is practicable on the main stem. Farmers on fertile land not readily served by river diversion draw supplementary water by pumping from the underground aquifers; this has to be licensed to conserve the resource.

In the world as a whole, the 20th century saw a proliferation of storage dams and river diversion schemes at all scales. Where an underground aquifer is available, 'conjunctive use' balances shortages of seasonal surface water flow with timely (hopefully sustainable) drawing of groundwater. There was also very wide development of hydroelectric energy potential, usually involving dams, even in regions where storage of water for consumptive use was not the prime necessity.

There are comparatively few examples yet of long distance transfer of water from wet to arid climates or from remote to populous regions. Among notable mid-20th century achievements were the Colorado River Aqueduct and the All-American Canal, which both transfer high value water from the Colorado River mainly to the cities and farms of southern California.

Two ambitious but problematic schemes were completed in Soviet Central Asia. The first was a $75m^3$ per second water transfer scheme, including 22 pumping stations (total 350MW) to raise water 418m and carry it from the Irtysh River in Siberia to the Karaganda region of Kazakhstan, completed in 1975 (Tanton, 2002).[6] The scheme, a key element of one to transfer water ultimately over a distance of 1300km, supplied an average of $0.883km^3$ per year (out of a possible continuous flow of 2.3 km^3 per year) for ten years, operating costs being greatly subsidized by the USSR government and irrigation water being supplied free. With the end of the Soviet Union, governments of the newly independent states were unwilling to maintain subsidies for the system's operation. Requirements that water users should pay resulted in greatly reduced demand; and by 2002 there was danger of abandonment, threatening the regional economy and discouraging investment in such transfer schemes.

The second scheme was the Karakum Canal, which runs for 1100km, carrying water by gravity from the Amu Darya (the ancient Oxus) to the arid but fertile lands of southern Turkmenistan. Because of the distance involved, the nature of the terrain and the economic philosophy of irrigated agriculture in the former Soviet system, the canal is unlined. In places, the route is along existing or ancient river beds, and there are a number of large balancing reservoirs. The result is a profusion of wetland situations adjacent to the canal with heavy losses of water that might have better continued to feed settlements further downstream along the Amu Darya towards the Aral Sea.

Work is now in hand for China's South-North Water Transfer Project. This is to transfer water from the Yangtze basin to that of the Yellow River (Huang He) and northeast China.[7] The labour force available and total number of people who could potentially benefit are much greater than in thinly-populated Central Asia.

Still conjectured but economically unproven are schemes to connect up rivers flowing to the eastern coast of India into the Bay of Bengal. Elsewhere there are prolific rivers, which may be harnessed to generate electricity but in which most of the flow reaches the ocean. The Columbia River, which rises in western Canada and flows through the mountainous northwest US, has an annual discharge ($260km^3$ per year) that is an order of magnitude greater than the fully used surface resources of the Colorado basin. It has been proposed that a transfer scheme could be of huge benefit to California. But the amount of tunnelling needed to transfer Columbia River water to California would be prodigious; and the energy needed to pump thousands of cubic metres of water per second over the intervening watersheds is probably beyond the hydroelectric capacity of that river or any energy source likely to be practicable for many years.

However, the majority of the world's water demands are met within more modest river or groundwater basins. In temperate, relatively wet countries, abstraction is less in demand for agriculture and more for municipal and industrial water supply. The higher value of water for these applications justifies higher cost supply and greater use of energy – for pumping, treatment or even desalination. There remains scope for more frugal use of water of each appropriate quality.

Meanwhile, in many densely populated countries, the main role of water is still to help grow more food. Optimum operation of existing and new storage reservoirs will have to be combined with maximum retention of flood water and strict allocation and measurement of irrigation supplies.

Trying to cope with floods

Protecting people and property from exceptional flows has proved much more difficult than diverting more modest river flow. In China, confining the river channels within naturally deposited or man-made levees has resulted, with sedimentation, in their beds rising considerably above the surrounding land. As a result, maintenance and raising of flood containment banks is a continuing necessity; failures in this respect or any exceptional flows can result in disastrous flooding. Examples[8] of such catastrophes have included:

- An 1887 flood in which the Yellow River (Huang He) inundated 130,000km^2 of land, causing the direct deaths of about a million people, followed by more through the ensuing famine.
- Floods on various Chinese rivers in 1931 killing 2–4 million people.
- At least three occasions when Yangtze river floods killed more than 100,000 people.

As recently as June 2005, flooding in China's southern provinces resulted in more than 500 deaths and the forced evacuation of 1.4 million people. Indications for the northern Chinese rivers are that difficulties in integrating any management of the rate of river bed siltation with the increased diversion of water on the plains result in higher stage (river) levels and therefore greater damage than formerly resulted from the same magnitude of flow (Wang and Plate, 2002).[9] The Three Gorges Dam Project on the Yangtze, with a reservoir capacity of 39km^3, was implemented with flood protection as a prime objective as well as incorporating the largest hydroelectric power facility in the world. Whether that reservoir capacity can be used to effectively tame a total annual flow approaching 1000km^3 per year remains to be seen.

Downstream of India's Farakka Barrage, the Ganga flows into the vast delta shared with the Brahmaputra and Meghna rivers with a combined annual flow that is the third highest in the world (after the Amazon and Congo rivers). On the flood plains of Bangladesh, people have struggled to cope with floods from the rivers and surges from the sea, to protect their homes and livestock against the water flow while using it as effectively as possible to grow crops for the ever-increasing population. Much of the flood plain morphology is so dynamic as to surpass man's capability to fashion a stable landscape; and forest clearance has reduced the land's natural flood retention capability and exposed the soils to deposition of coarse infertile sediment. Flood protection embankments are practicable only at certain strategic locations; and, unless there are adequate release mechanisms for water impounded on the protected side, they can impede the run-off of monsoon rainfall.

While total flood protection may always be impracticable on the plains of Bengal (82,000km^2 flooded in 1988), sustained effort on those of the longer, narrower Mississippi has culminated in perhaps the world's greatest river flood protection scheme. It involves thousands of kilometres of flood protection banks with locks for release of tributary water. The Mississippi is now channelled between banks that are designed to contain discharges of 80,000m^3 per second. Nevertheless, there are occasional serious breaches such as occurred in 1993. At the delta, flood defences are now thought capable to have resisted Hurricane Katrina in 2005 but risks of exceptional events remain.

River basin management

Most people live in river basins where there are needs both to maximize sustainable abstraction of water and to share it equitably among competing consumers. Even in regions of high rainfall, strategies have to be devised for periods of comparative drought, as well as accommodating storm run-off and reducing flood damage. Examples of attempts at very large-scale river basin management can be cited for central and southern Asia.[10] Sophisticated irrigation systems are believed to have been in operation from the Amu Darya river in the Aral Sea basin from more than 2000 years ago until they were destroyed or abandoned at the time of Mongol invasions. On the other side of the central Asian mountain core, the Indus Valley civilization irrigated Punjab and Sind from even earlier. By the beginning of the 20th century, the Indus Valley was well populated, but the Aral Sea basin was much less developed.

To the Soviet regime in the 1950s, recovering from World War II, the Aral Sea basin seemed a region of vast untapped water, soil and mineral resources ideal for colonization and economic development. A huge programme of construction and settlement included and hydroelectric power schemes in the Tajik and Kyrgyz highlands as well as storage reservoirs (the world's highest dam) throughout the basin. Development of cotton-growing irrigation schemes on 70,000km^2 on the Uzbek and Turkmen plains incurred the Water transfer schemes described on p87. Long-term consequences were new settled populations and huge agricultural schemes, but also excessive water diversion for irrigation (and evaporation in reservoirs) which caused waterlogging and salinity in some areas[11] and serious water shortages downstream, disastrous for people and wildlife near the Aral Sea. Electricity is produced for the whole region (except when the mountain catchments are frozen up in winter).

Management of the basin became particularly difficult after the Soviet Union was dissolved in 1991. The newly independent states of Tajikistan and the Kyrgyz Republic controlled most of the hydroelectric power production while the downstream nations of Uzbekistan and Turkmenistan used that power and consumed most of the water. There was no longer a strong central government for direction and financial management of the operation, maintenance and renewal of the hydraulic systems. But concerted international cooperation has started to rescue what is left of the Aral Sea.

The development of the water resources of the Punjab rivers and the Indus followed a steadier course, although the perils of rising water tables and salinity were becoming evident on some of the irrigated lands by the 1940s. Then fierce contention, even a threat of war, arose as a consequence of the partition of India in 1947. India controlled most or all of the headwaters of the five Punjab rivers (with a total flow of nearly 100km^3 per year) and wished to construct dams whereby they could use most of the flow of at least the easternmost rivers for Indian agriculture in East Punjab and for new schemes to be served by long canals into arid areas as far as Rajasthan. This might leave Pakistan's West Punjab – itself a huge irrigated area – bereft of supplies. Pakistan had control of the Indus River itself (110km^3 per year) but that river had little command of fertile land except further downstream in Sind. Solutions to these serious problems of basin management were:

- to combat waterlogging and salinity by improving drainage, leaching salt out of the soils, pumping out excess water from the ground at tubewells and mixing saline groundwater with controlled fresh surface water supplies in more effective irrigation;
- to resolve the regional water distribution dispute, an extensive system of link canals – supplementing those that already existed – was built to transfer part of the water from the western rivers (the Indus itself and the Jhelum and Chenab) to the eastern lands within Pakistani West Punjab, leaving India full scope to build dams and divert water upstream of the frontier on the rivers Ravi, Beas and Sutlej; and
- dams to regulate river flow were also built in Pakistan on the Jhelum river at Mangla and on the Indus at Tarbela.

The solution for equitable water distribution (under the 1960 Indus Basin Waters Treaty) was essentially constructional, that to cure waterlogging and salinity mainly operational.

Water resources management of basins as large as that of the Amu and Syr Darya rivers and of the Indus and its tributaries has been significant in the economic development of the nations concerned, particularly as to the consequent problems in ex-Soviet Central Asia and to solutions reached in Pakistan and India.

For dam and diversion projects, water control strategies concern release of water to meet often-conflicting demands of various users. Besides non-consumptive needs for hydropower demands for release may concern:

- municipal and industrial supplies, usually as a comparatively constant demand if local service reservoirs cope with hourly variations;
- irrigated agriculture for which water demands are much more seasonal, depending on cropping patterns and any contributions made by rainfall;
- compensation flow for wetland or other wildlife habitat if timely supplies would otherwise be cut off by abstractions upstream;
- empty capacity in reservoirs reserved for flood control storage, requiring lowering reservoirs in anticipation of exceptional flows and therefore possibly reducing the amount stored for subsequent releases;
- periodic flushing of sedimentation from reservoirs which may require reservoir levels to be drawn down even further; and
- navigation or fish migration, necessitating maintenance of adequate channel depths and flow as well as construction of locks or fish ladders at dams.

For total river basin planning, account has to be taken of the amount of diverted water that is ultimately returned to the river as drainage, and of abstraction, outflow and recharge in groundwater aquifers.

6.2 River landscapes and engineering opportunities

River basins encompass headwater catchments, valleys, gorges and alluvial plains. Their land resources comprise:

- the scenery of rock and vegetation, as well as water itself in the form of prominent glaciers, tumbling rapids or mature river channels;
- the moorland or forest that collects precipitation on higher ground and attenuates the flow of run-off into streams;
- fertile land suitable for agriculture; and
- sites suitable for dams, weirs, bridge crossings or navigation works, and space for human settlement and industry.

Headwater landscape

Headwaters are streams flowing from their sources. The sources may be glaciers, springs or run-off from mountainsides; they may be in moorland which soaks up rainfall, provides rough pasture for sheep and is attractive to leisured man mainly for solitude, or in terraces and plateaux more conducive to settlement; or main river sources may be large lakes which have their own catchment areas further upstream. Thus the Blue Nile emerges from Lake Tana in Ethiopia, the White Nile from the swamps of the Sudd in Sudan or the higher branches of that river from Lake Victoria in Uganda or the Rift Valley lakes in Congo and Rwanda.

In the mountains, the beauty of rock, snowfields, forests and waterfalls should deter most forms of development. But, where there is hillside agriculture, then painstaking construction and maintenance of terraces ensure conservation of the soil and a special type of heritage landscape. Less assiduous land clearance, cultivation and building, without adequate drains and soakaways, increases run-off and aggravates erosion.

Valleys

Where river bed gradients are less steep and valley bottoms wider, there may be opportunities for reservoir storage if there is, just downstream, a constriction in the valley suitable for dam construction (Photo 6.4). Suitability of sites for reservoir creation depends on geological conditions, reservoir depth/volume/area characteristics, lifetime

Photo 6.4 A dam was built here to impound a large reservoir in the wide valleys of two rivers that join, upstream of this constriction, to form the Sefid Rud river, Iran.

of the site for useful storage, and anticipated impact of inundation on life and landscape in the valley. Justification for the costs of dam construction then depends on the value of timely water released to downstream users.

Valley landscape comprises far background such as mountain peaks, then rocky, forested, cultivated or inhabited slopes and, at the bottom, any foreground space such as meadows or braided (multi-channel) river bed. A reservoir turns that bottom space into a lake; a dam probably blocks some sort of gorge.

Figure 6.1 shows the profile of three river valleys in different geological conditions. The highlands of Sri Lanka comprise highly metamorphosed Archaean rocks. The scenery is forested except where it has been cleared, totally for tea plantations or more spasmodically by shifting cultivation. Rivers

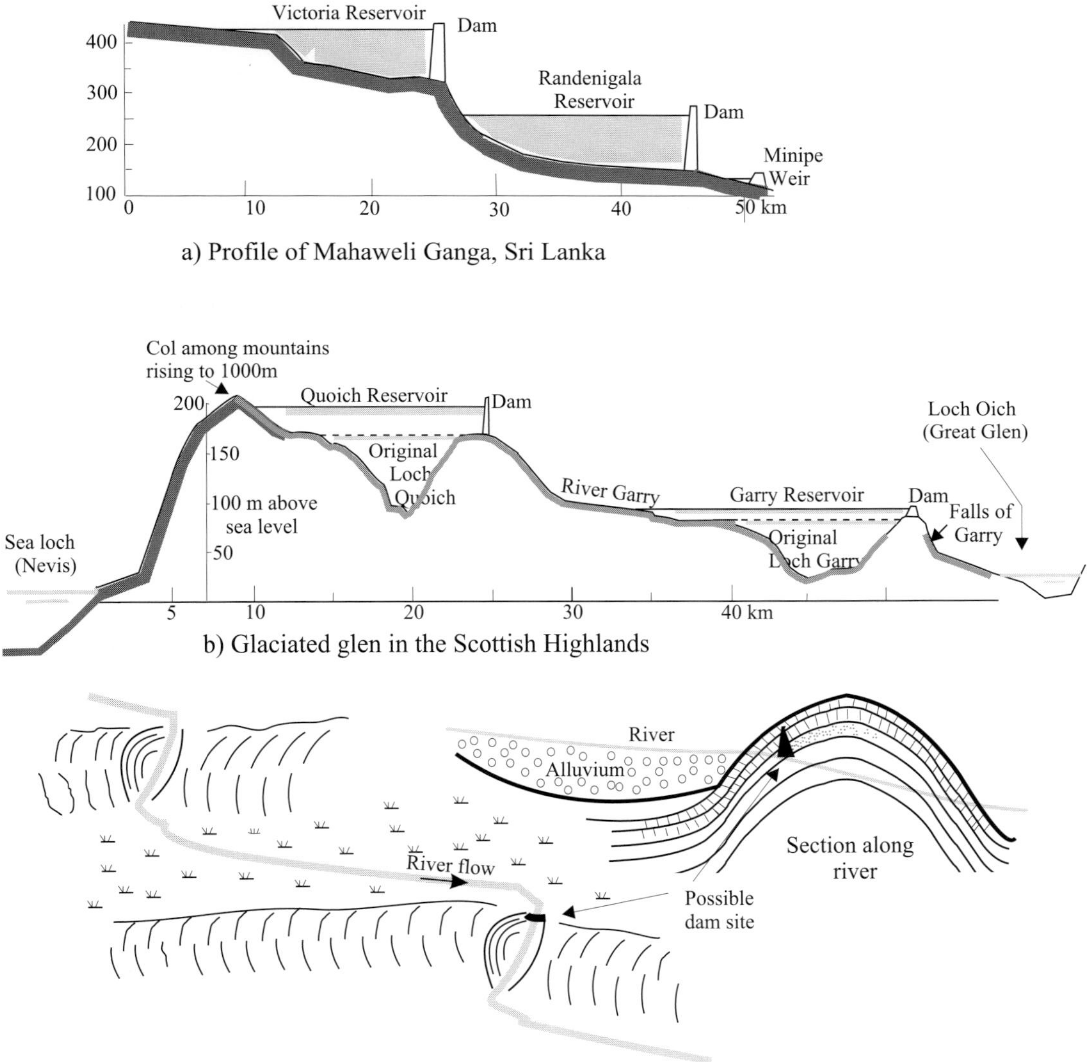

c) River course along synclines and through anticlines in Zagros Mountains, Iran
(figures are not to scale)

Figure 6.1 River valley profiles and their suitability for dams and reservoirs

Photo 6.5 Possible dam site at entrance to an anticline gorge, Karkheh River, Iran.

such as the Mahaweli Ganga erode the ancient hard rocks very slowly and, with few weaknesses, there are variations in a general gradient rather than either abrupt falls or wide mature open valley sections. However, such variations as exist give opportunity to dam rivers at the head of the steeper sections for reservoirs in the less steep stretches, and to generate power across the total drop of the valley through a cascade of dams and power stations.

In the highlands of Scotland, the effects of glaciation have resulted in U-shaped valleys interspersed along rivers by much steeper sections or rapids where harder rocks were more resistant to the ice. Dams at the top of steeper sections have created reservoirs in the broader valleys which may have been already occupied by deep lakes. But tunnels or pipelines divert water to lower level power stations, reducing the flow through often fine woodland-surrounded rapids to whatever compensation release can be negotiated.

Some parts of the Zagros Mountains in southwestern Iran exhibit a remarkably uniform topography of folded sedimentary rock strata. Wide valleys with slopes bedecked in scrub or oak trees lie in the synclines and harder rock within the adjacent whaleback anticlines. River flow deposits alluvium along the synclines and occasionally finds faults where it can break through gorges in the anticlines. Dam sites exist in the harder outer rocks at the entrance to those anticlines (Photo 6.5) and their reservoirs flood the synclinal valleys behind. In the Zagros Mountains, the most robust rock stratum is limestone. Where the occurrence of any karst conditions in limestone can be overcome by extensive grouting, dam construction has been successful.

In any formation of relatively recent geological age, such as in the Himalayan, Central Asian or Alpine mountain ranges, the stability of valley slopes must be seriously investigated in considering opportunities for dams and reservoirs.[12]

Gorges

Gorges as fine scenery are the downward equivalent of mountain peaks. The Grand Canyon is probably the most popular scenic attraction in the world. Through it runs the Colorado river – indeed most of that river runs through canyons cut down by differential erosion into a rich variety of coloured rock strata. The river was first brought under control by the completion of the Hoover Dam in 1936. The dam is 176m high; its reservoir, Lake Mead, extends 185km upstream towards the Grand Canyon and had an initial capacity approaching double that of the river's average annual flow of $20km^3$. The reservoir is not expected to fill with sediment in less than 400 years. More dams have since been built both upstream of the Grand Canyon and downstream of the Hoover Dam. A large proportion of the river's energy is now converted into electricity, although a significant part is still dissipated in rapids of those scenic sections not inundated by reservoirs. In most years, all downstream flow is eventually diverted for irrigation or municipal water supply.

Gorges in harder rock, with walls rising in places to 1000m above river level, contain the huge discharge ($995km^3$ per year) of China's Yangtze Kiang. The Three Gorges Dam Project, already mentioned, impounds a reservoir raising the low season water level by about 100m where pre-dam seasonal levels varied by 30–60m in one of the gorges. The sight of flow in the bottom of the gorge or of the more spectacular seasonal changes in level are no longer to be seen, but navigation in the reservoir is easier and the tourist attraction of at least the gorge walls has been preserved.

Numerous dams have totally altered the character of, usually shorter, gorges worldwide. However, many steep-sided valleys are less suitable for dams; their beauty has to compete only with less dramatic structures such as weirs, bridges, navigation locks or riverside transport routes. The Rhine, one of Europe's main rivers, carries about a tenth as much flow as the Yangtze and with less seasonal variation (between $800–2000m^3$ per second at Mainz). There is a comparatively swift current where its course passes through sections of gorge between Bingen and Bonn. The modestly high valley sides are wooded with rock outcrops. The landscape would be ill-served by any impoundment of the river, which is in any case unnecessary. Engineered structures are mainly river walls, which, for much of their length, support railways or roads on both sides of the gorge (see Photos 4.13 and 4.14 in Chapter 4). With alignments unsuited to wide motorways or very high speed trains, these routes give leisurely views to passengers comparable with those seen afloat on the river itself.

South of the Alps, the railway from Maribor to Ljubljana follows a more tortuous scenic course through the Slovenian river gorges. Other routes, as in parts of southeast France, find easier access along wide-bottomed gorges, often in limestone with comparatively modest or underground watercourses. Occasionally roads are built, with some rock blasting, along the side of narrower gorges, an infringement of the wild situation but allowing views of places otherwise inaccessible except to intrepid exploration.

Particularly attractive features of gorges are cataracts, within the gorge itself or spilling into one over the rim of a plateau. A large example of the first is Niagara Falls, of the second Victoria Falls. Niagara Falls, 50m high and currently divided by an island, pours over the edge of a layer of hard dolomite that recedes as softer shales and sandstone beneath are eroded. Thus the Falls are moving upstream at a rate that is rapid in geological terms. Because of the clear-cut line of the Falls and uniform flow over them, their appearance is not unlike that of a massive man-made overflow spillway in action; indeed the Falls are a spillway in that the channels for hydroelectric schemes could take most of the flow and thus bypass the Falls. But, as tourist revenue far exceeds that for electricity, upstream abstraction for power is strictly limited so as not to markedly reduce the spectacle of the falls or of rapids in other bypassed sections of the gorge.

At Victoria Falls, the Zambezi, with a similar annual flow volume to the Niagara River, drops from a high plateau into a fissure that is much narrower, more jagged and deeper (128m) than the Niagara gorge. The wildness of the Zambezi gorge is as much part of the awesome scenery as are the various falls and their foam and spray. Only a 200m single span railway bridge spans the cleft nearby.

Alluvial plains

Alluvial plains were created by waterborne deposition of sediment. Flood plains are those where flooding and sediment transfer still occurs today. The ever-varying patterns of deposition and scouring result in fundamental instability in the course of the main channels. The Yellow River (Huang He), the Indus and the Amu Darya have changed their downstream courses in historic times. Other rivers that convey heavy sediment loads do the same within whatever measures are built to constrain them.

Vegetation cannot easily gain a foothold in the coarse material of braided channels, where gravels are often a ready source of aggregate for construction. Botanic growth is much more prolific in the siltier deposits and may form a sponge to attenuate the passage of floods, a windbreak against dry season soil erosion or a defensive barrier against the force of waves across lakes or estuaries.

Opportunities to control water on alluvial plains include:

- storage only where there is still sufficient variation in ground relief to impound comparatively shallow wide reservoirs;
- ponding up river flow sufficiently, by cross-river weirs or gated barrages, to divert water to flow by gravity through headworks and along canals commanding fertile land; and
- flood protection measures, river training works and limits on built development to better define what areas of land should be subject to flooding and what should not.

High dams are not practicable on plains except at the very upstream limits where the rivers debouch from hillier terrain. At Tarbela, where the Indus leaves the Himalayan foothills, a dam 140m high was built across a 2.5km-wide gap in these hills. However, the river bed in the wide gap was of mainly coarse permeable alluvium to a maximum depth of nearly 200m. Leakage beneath the dam is inevitable in these circumstances but has been kept within acceptable limits by installing and maintaining a blanket of impermeable material on the reservoir bed for 1740m upstream of the dam.

On some of the wider, but still gently sloping, river plains there may be occasional emergence of hard geological features such as form the rapids on the Blue Nile at Roseires in Sudan. Although the high ground near the rapids is barely 20m above the river bed, a dam was built to impound a reservoir to about 40m above that bed level by constructing long earth embankment wings of the dam on either bank extending to a total length of 10.7km. The central 1km long concrete structure is more than 50m high, making provision for a future extension of the flanking embankments and their heightening by a further 9m, increasing the storage capacity from about 2–7km^3. Roseires Dam also serves as the headworks for two major irrigation canals and thus combines the functions of a storage dam with that of a barrage of the type built across even flatter alluvial plains to abstract unstored 'run-of-river' flow by gravity.

Most of the Punjab is a flat plain. Its five rivers rise in the Himalayan mountain range or its foothills in India or Kashmir. Their courses across the plains were capricious in the days when only inundation irrigation was practiced and before the main river training works were built. However, the Punjab landscape and economy have been transformed since the late 19th century by the construction of diversion barrages and irrigation canal systems. The topography has been further stabilized by the construction of flood protection embankments ('marginal bunds') upstream of the barrages and bridges and by drainage systems inherent in canal and transportation route construction. The average gradient in south-western Punjab in the general direction of the rivers is about 1 in 5000. The Indus Basin Project link canals that cross the natural drainage were constructed in the 1960s at a gradient of about 1 in 10,000 on careful alignments with barrages at river crossings and intermediate facilities for cross-drainage under or into the canals.

On the other side of the Indian subcontinent, river channels across Bengal flow in shifting channels at gradients even less than 1 in 10,000. More disciplined channels of the Rhine and Maas flow between the flat polders of the Netherlands created by centuries of sustained effort and huge twentieth century engineering works.

6.3 Dams and reservoirs

The effects of dams on land resources and scenery are strikingly different from those of the reservoirs they impound. Yet it is impracticable to consider them separately because the location and height of a dam directly determine the depth, area of land covered and volume of a reservoir. Furthermore, the geology of the wide and narrower parts of the valley has to be assessed both as to the potential watertightness of the reservoir and the structural strength and stability of the rock formations at the dam site.

The impacts on local land features of reservoirs are the inundation of valley bottoms and their replacement by lakes, while dams lead to the destruction of the wild nature of the site and visual intrusion in the valleys or gorges where they sit.

Inundation of valley bottoms

To maximize storage volume, the optimum sites for reservoirs are those in the widest parts of valleys. Thus, impoundment often submerges the least steep, most fertile and most economically significant land areas found in upland regions. Valley bottom land that is lost may include pasture, cropland and settlements. Gentle riverine topography cannot be replicated on steeper slopes above reservoirs; nor can some of the wildlife on the banks and beds of free flowing streams flourish beside a lake.

A common objection to large dams is that many people lose their homes and livelihood when the reservoirs are impounded. Of very large reservoirs (exceeding 1000km^3 in area) in inhabited regions of Africa, the lakes at Volta and Kariba displaced rather less than ten people per square kilometre, the High Aswan lake 25–30 people per square kilometre. Local communities do not usually benefit directly, as do city water or electricity consumers or downstream farming communities. However, in most cases, it is possible to demonstrate that the net national or regional benefit is positive. The 1 million people displaced by the Three Gorges Project in China is an exceptional figure but has to be compared with the many millions who have lost their homes or lives in catastrophic floods of which avoidance is a major project objective. Equitable solutions depend on ample planning, funding and implementation of adequate resettlement and compensation for those who lose their homes.

When settlements are threatened by plans for a large reservoir, it is sometimes suggested that a number of smaller schemes might achieve the objective with less loss of land. In most circumstances, however, it is a myth that a number of lower dams can be as effective as a single high one. The storage capacity of a reservoir increases roughly in proportion to the third power of the dam height, the area of land submerged only to the square.

Effects of lake creation

The impacts on the landscape of a new reservoir include views over the lake, popular amenities on or beside the lake, and different riparian habitat and altered agricultural practices around the lake, with limitations depending on the way the reservoir is operated.

Lakes are visually attractive, not so much as water surfaces but as foreground with reflection of islands, waterside fringes, forests and steep scenery beyond the far shore. The latter may have been visible before the reservoir was created if there was any intervening open ground; but any new land clearance may provide opportunity for new viewpoints along or above the lake.

Popular amenities can be provided in conjunction with the attraction of new views, although, where reservoir water quality is paramount, it is necessary to regulate waterside activity and to prevent discharge of polluted wastewater into the lake. Besides lakeside recreation, there can be water sports; but swimming, paddling or launching of boats may have to be undertaken at varying water levels and precautions applied if there needs to be rapid filling or drawdown of the reservoir.

Reservoir operations comprise storing up much of the river inflow into the reservoir during the high flow season and release of water through outlet structures – to water channels or pressure pipes – or over spillways, regulating outflow to best meet the downstream demands. The consequence is a rising and falling water reservoir level, producing conditions at the margin quite different from those beside a natural lake.

The variations in level in a *small reservoir*, operated for example in a pumped storage hydroelectric scheme to meet peak demand, may be rapid and on a daily or even twice daily cycle. In these circumstances there is little possibility or even desirability of attracting riparian wildlife. However, in *large reservoirs* there may be relatively little variation in level; and that which does take place may be seasonal and similar, if in a different pattern, from that which occurs at riversides during natural floods. Towards the upstream perimeter of an extensive reservoir, the adjacent land may be only gently sloping, suitable as nature conservation land or for flood recession agriculture. Riparian agriculture has long been practiced on the banks of rivers such as the Nile, Niger and Senegal and around Lake Chad. Seeds are sown on the wet perimeters as soon as the seasonal floods have receded and the resulting crops are harvested before the onset of the next season's flood. Known in Sudan as *gerouf* cultivation, Chapman (2001)[13] has described how such agriculture is successfully undertaken on extensive areas of the Roseires reservoir on the Blue Nile. Elsewhere, it may be possible to contrive embayments suitable for growing of crops or providing habitat suited to the changes in reservoir water level, for example in polders isolated from these changes by embankments.

Around reservoirs, the practicability of hillside cultivation is unaffected. However there may be socio-economic changes and new building as a result of dam-related activity. There will be more direct access to the outside world and perhaps a road around the reservoir.

Destruction of wild landscape

The impact on natural conditions at dam sites appears most blatant after initial clearance for construction. Vegetation is stripped and hillsides near the construction site are laid waste as access routes are cut along them and the excavated debris spills onto the slopes below. Dry space at the bottom of the valley is soon covered in materials storage yards, processing plants, vehicle parks and temporary buildings.

Wide slots are cut into the valley sides, deep enough to expose firm rock abutments for the dam walls; and cofferdams are constructed on the river bed, within which excavation for foundations can be undertaken. Then, as the dam wall begins to rise, so the new character of the blocked valley landscape starts to emerge. On completion of dam construction, clear-up of the surrounding land space ensues. Ground surfaces are shaped, surfaced or sown and trees or shrubs planted in accordance with visual concepts previously planned for the project site as a whole. Certainly the wild character of that stretch of the river will have been permanently altered. The new scenery may involve:

- for modest dams, a main wall – possibly a grassed embankment – and spillway, perhaps a power station building, sometimes a fish ladder, certainly an access road, all within original verdant or rockier surroundings; or
- for uncompromisingly large structures, a monumental feature in the valley whose character is thereby changed – hopefully to a dam landscape that is a worthy example of its type and a major attraction in its own right.

Any permanent road across steep ground or wild landscape undoubtedly interrupts the natural scenery. But access is invariably needed to any dam and its associated power station. In steep gorges, access roads may be best partly underground – for practical reasons as well as to avoid further damage to the hillside. Furthermore, tunnelled roads can lead to the crest of the dam or to more discreet viewpoints – with sights of the dam wall, the spillway or wilder downstream stretches of the gorge not previously visual to visitors.

Visual impact of dams

The visible manifestations of a dam are the structure itself and any associated or separate water intake/abstraction structure or spillway over which surplus reservoir water can flow. Dams can be concrete or masonry walls or earth or rock embankments. No doubt landscape architect Dame Sylvia Crowe (1958)[14] had concrete dams in mind when she described them as 'perhaps the most magnificent of all organic structures' that 'grow out of the landscape with the impersonal grandeur of a glacier or outcrop of rock'.

The scale of large dams can be marvellous (or dismally obtrusive if you dislike engineering); the purity of a clear steep wall seen from downstream can be elegant (or the clutter of gates, power penstocks or abrupt changes in profile can be clumsy, even ugly); and the achievement of holding back millions of tons of water can be awesome (or alarming if you doubt the structure's competence). Any dam has to be:

- of a shape and mass adequate to resist overturning by the pressure of reservoir water; this is no problem with broad embankment dams but dictates the conceptual design of a more compact concrete structure; and
- sufficiently impermeable to prevent water leaking through it and washing it away, this is no problem with well-constructed concrete dams, but embankments are made up largely of permeable soils and pieces of rock, so they must incorporate the sort of seepage resisting measures suggested for water-retaining structures on p54.

Concrete or masonry dams whose weight is the sole resistance to applied loads are known as 'gravity' dams. There is no tension in such dams if the resultant of all forces acts through the middle third of any horizontal section of the structure. An arch dam, on the other hand, transmits most of the thrust of the reservoir on to the abutments. This solution, which can be applied only in a narrow valley or gorge of sound rock, enables substantial saving to be made in the volume of concrete compared with a gravity dam. Buttress dams are gravity structures in which the volume of concrete is reduced by taking the hydraulic pressure on a relatively slim (reinforced) concrete sloping slab, itself supported by a number of buttresses carrying the load to their foundations. A multiple arch dam is similar except that tensile loads on the upstream face may be eliminated. Common types of concrete dam are shown on Figure 6.2.

Gravity dams of uniform cross-section have been the frequent solution in situations where rock or aggregates are available and where the local rock conditions are not suited to any arch solution at the abutments or to heavy buttress loads on the foundations. Gravity structures remain common for low dams, especially those incorporating gated spillways or low head power stations; but they have also been built to great heights, including the highest concrete dam in the world at Grande Dixence (285m) in Switzerland. The downstream face of a concrete gravity dam is plain and steep. Its starkness is best left uncompromised by any protruding ancillaries, discordant colours or irregular texture. If possible, any concrete surface should remain unmarked by stains from spills or cement seeping through joints.

Many early dams were built of rock block masonry rather than concrete. The confined interior of any gravity dam can be of various massive materials from compacted rubble to mass concrete with large stone fillers, or leaner concrete using pulverized fuel ash or blast furnace slag as the cementing binder. A compromise between the gravity concrete and earth embankment dams is today found in rolled concrete construction; the main mass of such a dam comprises weak concrete that is placed and compacted in the same manner as earthwork.

Concrete *arch dams* range from gravity-arch structures, in which only a proportion of the reservoir's load is transferred to the abutments, to double curvature overhanging 'cupola' (upstream facing dome-shape) structures possible in very narrow gorges where almost all the load is transferred sideways by thin arches which seem to approach the ultimate in daring construction. Such dams are beautiful in their slenderness and curvature whether seen from above in the arc of the dam's crest, from downstream looking up into the concave side of the cupola, or from upstream when the even more striking convex side can be seen during construction or when the reservoir is drawn down (Photos 6.6–6.8). Perhaps opportunities to build new arch dams are now rare; many of the suitable sites have been dammed already; which makes it all the more important to take the best advantage of any remaining chances to store water in such an inspiring fashion.

Buttress dams (Photo 6.9) are much more linear, comprising a sequence of geometric shapes. Accepting the stark obtrusiveness of any such artificial structure across a valley, there is a military neatness in the repetitive buttresses along what are often comparatively long dams on wide river beds.

Embankment dams of earth or rock fill are commonly chosen at wide sites or those lacking

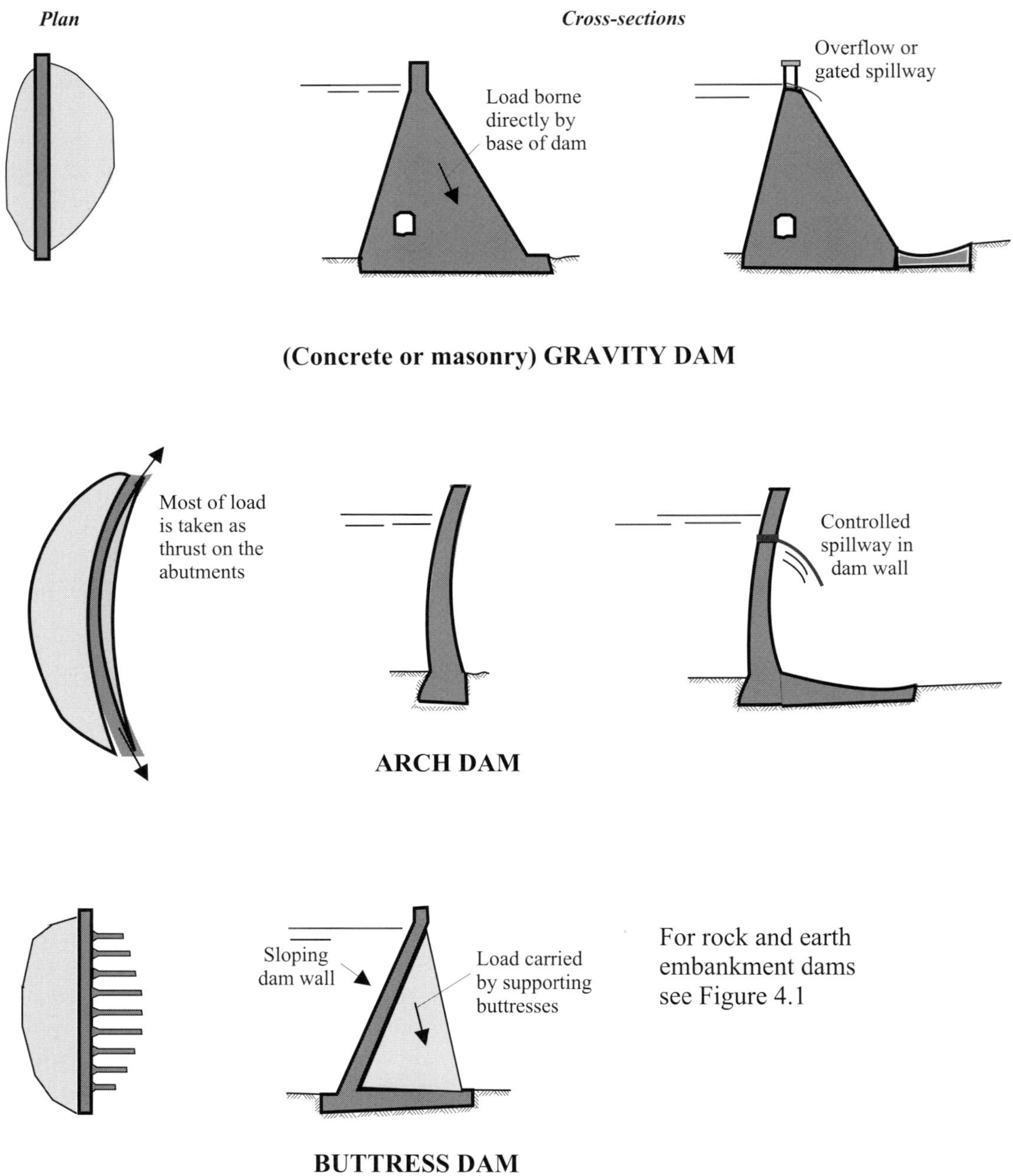

Figure 6.2 Types of concrete dam

sound rock foundations. Essentially embankments are more massive, wider and less stark structures than concrete dams; their slopes are less steep than 1 in 2, making it possible to face them in rock or plant them in vegetation to blend into flanking hillsides. The success of such blending, at least in low embankments, can be judged from the difficulty in recognizing at a distance whether some mountain lakes are natural features or man-made reservoirs. Failure to suit the surroundings easily results from clumsy shaping of benches or ends of embankments or scruffy arrangement of dumped material on the visible downstream slopes.

Photo 6.6 El Atazar concrete arch dam, Spain (photo by Robin Carpenter).

Photo 6.7 Karaj arch dam, Iran, seen from downstream during construction.

Photo 6.8 An arch dam in northern Spain seen from upstream during construction and with the reservoir already flooding some of the agricultural land.

Water intakes control the release of reservoir water into diversion channels or hydroelectric power conduits. The intake structures may be integral with the dam wall or may be constructed as

Photo 6.9 Stwlan concrete buttress dam impounding the upper reservoir of Ffestiniog pumped storage scheme, North Wales.

freestanding towers in the water – near the dam (as in Photo 6.9) or as lakeside architectural features elsewhere along the reservoir boundary.

Spillways, in operation, are among the most spectacular of civil engineering structures. The spectacle is the discharge of great volumes of water from top reservoir level and the dissipation of the huge energy generated by the water's fall to the toe level of the dam. Types of spillway in use include:

- overflow sections with or without gates at the top of a concrete or masonry dam or openings high on the dam wall from which the water drops, the energy being dissipated on the face of the dam or in a stilling basin below (Photos 6.10 and 6.11);
- separate concrete chute structures in which the energy is mainly dissipated in steps or a stilling basin at the bottom (Photo 6.12);
- by a tunnel around the dam through its rock abutments, discharging the water into the gorge downstream;
- through tunnels constructed for power generation, but from which excess flow energy can be dissipated into the air through outlet valves brought into use in addition to, or instead of, discharge through turbogenerators; and
- low level sluices suitable for rapid reservoir drawdown or transport of silt loaded flow.

However, the awesome display of a spillway in action only occurs when water has to be openly discharged, when it is in excess of that normally carried by concealed (pressure) pipes or high level outlet channels. Thus for much of the year – and at some sites almost always – there is no spill. In such circumstances, the elegance or inelegance of the spillways depends on the same factors as the appearance of the downstream face of a concrete dam. The smooth curve of the dry spillway may contrast with the rugged pattern of impact blocks in the stilling basin.

Options for type of dam

The most practicable form for any dam and its spillway is largely dependent upon the shape of the valley and the nature of the proposed foundations and abutments. Generally, the likely location, height and alignment for a dam is first conceived in the light of topographical and geological survey. Then there may be a number of structural and geotechnical options, depending on detailed investigations and design.

At comparatively wide dam sites, the choice often lay in the past between a concrete buttress dam and a rockfill embankment, the latter gradually becoming the more common preference in the later 20th century. Even non-site-related factors could influence the choice. In the 1960s, Latiyan Dam in northern Iran was built as a buttress dam rather than rockfill because the appointed construction contractor could then make use of the steel formwork available from a dam of the same height recently completed on another river (Sefid Rud).

The final choice of design can be also be made in the light of aesthetic appearance. Scott and Molyneux (2001)[15] note that Wimbleball Dam in Cornwall was completed in 1978 as a buttress dam

Photo 6.10 Free flow spill over Caban Coch gravity dam, Wales. Some of the energy of the falling water is dissipated by the intentionally rough surface of the dam face (photo by the late Don Brooks).

Photo 6.11 Crest spillway at Maentwrog Dam, North Wales.

because that type 'was preferred over a rockfill for aesthetic reasons'. The Claerwen Dam, completed in 1952, was the last of a series built in the Elan Valley in Wales. 'Although the dam is a mass concrete structure it was decided to have masonry facing so that the new dam would blend with the existing dams in the cascade.'[16]

Photo 6.12 Stepped spillway, at far side of dam, Central Wales.

6.4 Hydroelectric power

Opportunities for harnessing hydraulic energy

Hydroelectricity is still, in 2010, the world's major form of renewable energy. Projects concerned include:

- the conventional majority, which intercept river flow and capture its energy in steep sections of passage, anywhere between the highlands and the sea; and
- pumped storage schemes in which the water is used as a medium for short-term storage of electrical energy.

Conventional hydroelectricity can be generated using uncontrolled river flow; or that flow can be withheld in reservoirs to provide electricity on a timely basis to suit demand. Low head *run-of-river* water power has been utilized for centuries at mills where machinery was driven by a water wheel. Greater advantage of large rivers is now taken by installing turbogenerators in high flow barrages such as are intended primarily to head up water for abstraction into aqueducts for irrigation or urban water supply. At large capacity run-of-river (uncontrolled flow) installations, excess electricity can power discontinuous processes. At high head sites on smaller streams, mini-hydroelectric plants are now used in remote locations for local electricity supply if low season flows are sufficient for modest power demands.

Substantial *control of flow* requires storage dams, particularly if dry season flow is to be increased by retaining some of the flood season excess. However, the height of the dam provides not only reservoir storage volume but also raises the head of water that directly determines the output of electrical energy. Indeed many dams have been built solely to maximize that head. For example, in the middle

Photo 6.13 Ffestiniog pumped storage hydroelectric scheme, North Wales. The lower reservoir and power station are in the foreground. The upper reservoir is behind Stwlan Dam (see also Photo 6.9) near the top of the picture.

reaches of the Naryn (Syr Darya) river, where it descends from the high basin of the Kyrgyz Republic towards the wider plains of Uzbekistan, a large upper reservoir at Toktogul regulates flow for a power station at the dam and for a downstream cascade of five other power stations in the steep valley, each fed from its own dam providing maximum head but only nominal reservoir capacity.

Some important hydroelectric dams have been constructed at steep narrow sites on major rivers, creating very large reservoirs where the upstream valleys are wide and gently graded. The huge lakes above Kariba and Cabora Bassa dams on the Zambezi River are notable examples of land flooded behind dams built high enough to create power sources capable of meeting national and international energy demand.

Pumped storage schemes store surplus (off-peak) electrical energy by pumping water uphill from low to high level reservoirs; from the latter, water is released a few hours later to run down again, generating energy to meet peak period demand. The greater the difference in elevation between the two reservoirs, the greater is the energy storage capacity; whereas reservoir water storage capacity need only be modest (say for 12 hours pumping or generation). For these requirements, small high-level reservoirs behind relatively high dams can be built nearer to the tops of mountains than are conventional reservoirs (Photo 6.13). Alternatively, upper reservoirs may be lakes in valley bottoms or on plateaux, and lower reservoirs may be located in disused mines or below adjacent escarpments.

Land-take

Run-of-river hydroelectric projects take up little more land than the weirs, diversion structures, pipelines and power stations involved. Nor need much of the land space occupied by a reservoir be ascribed to power production if the reservoir stores water primarily for some downstream consumptive use. However, for dams constructed mainly for power, the benefits must be offset against that loss of land resource.

According to Meyer (1996),[17] Swiss hydroelectric projects have a capacity in excess of 10,000MW based on storage reservoirs covering only 123km^2; whereas 'the Tucuruí dam on the Tocantins River (Amazon basin), for example, has flooded more than 2000km^2 of rainforest to produce 4000MW of power' and other dams built since have flooded even greater areas in proportion to power output. The natural value of inundated land and the significance of the area-flooded/power-generated ratio may be more relevant in the Amazon basin than in the comparative wastes flooded by similar schemes in Labrador or Siberia.

Effects on landscape

In addition to impacts associated with dams and reservoirs, effects of hydroelectric power schemes on landscape relate to:

- high level intakes and aqueducts, and downhill penstocks (pressure pipelines or tunnels) for delivering water to the low level turbogenerators;
- power stations that house the generating equipment;
- switchyards where the electricity is transformed;
- high voltage long distance transmission lines; and
- variations in stream discharge consequent upon diversion and control of flow.

High level intake works, such as inlets to power tunnels, are usually unobtrusive and minor elements in the landscape compared with the dams and reservoirs with which they are associated. Sometimes long *headrace channels* are constructed to take advantage of high river flows over long distances with substantial cumulative energy generation opportunity. An example is the 1600m^3 per second Ghazi Barota canal in Pakistan which extends for 52km from downstream of Tarbela Dam to gain a head at its tail of 72m above the Indus river bed, sufficient for a power station capacity of 1450MW. The landscape impacts are similar to those of other large water transfer canals, reviewed in the next section.

Penstocks are often tunnelled through rock, putting them out of sight. However surge shafts – necessary to contain the energy of water flow when valves are being closed – may protrude above ground level; or pressure pipelines laid on the surface create a straight line down the mountainside (Photo 6.14). At some high flow but medium head installations a line of parallel penstocks may be prominently connected to a power station at their feet, the whole complex occupying all the side of a gorge from its rim down to the riverside.

Power stations in similar settings may be exemplified by the two fed by canals on either side of the Niagara River downstream of the Falls. Except where vegetation is now flourishing, much of the scenery is artificial. The slopes of the gorge and the river channel itself have been geometrically fashioned to secure stable slopes

Photo 6.14 Steel pipe penstock, constructed in 1928 and still visible amid woodland at Maentwrog, North Wales.

and banks. The power station buildings are overshadowed as structures by an impressive steel arch bridge spanning over of the gorge.

If space for an outside power station is not available, for example within a narrow rocky gorge, it may be practicable to build it underground in a cavern excavated within the rock. At modest-sized surface stations in more natural more intimate surroundings, such as in the Scottish highlands, sympathetic architecture is the key to preventing serious visual intrusion.

Power stations have to be located at a level low enough to gain the maximum effective head of water, but not so low as to prevent discharge being obstructed by appreciable back pressure in the tail race. Actual station configuration depends on the type of turbogenerator sets (with vertical or horizontal shafts), the operation of cranes needed to lift the equipment, the means by which the largest elements can be transported to the station, and the routes for cables from power station to switchyard.

Switchyards are outdoor areas where electricity is converted into the form required for long distance transmission (high voltage three-phase, perhaps double or multiple circuit). Flat or terraced land is required for the adequately spaced layout of steel frameworks supporting cables, transformers, switches/circuit breakers and insulators. In secluded or unremarkable landscapes, slender, complex equipment can be fascinating rather than brutally intrusive, often seated on a gravel or visually neutral ground surface.

High voltage electricity transmission lines are considered further in Chapter 11. They are noticeable mainly because of the towers (pylons) from which the overhead cables are suspended. Lines from hydroelectric stations often traverse scenic hill country and may be considered sharply intrusive where swathes are cut for them through forests or where they cross the skylines of established and esteemed views. Their route should therefore be planned in close consultation with whatever agency is guardian of the landscape; normally it is too expensive to transmit high voltage electricity in cables buried in the ground.

Variations in stream flow occur when valves controlling water flow through pipelines are opened or closed affecting, for example, the flow

Photo 6.15 Steep natural watercourse in which uncontrolled flow descends almost geometrical but natural steps across hard rock strata and a cataract before meandering across softer material below, Ordesa National Park, Spain.

downstream of hydroelectric power stations using stored water but operating only during peak electricity demand periods. More significant is the complete bypassing of flow for much or all of the year along the stretch of river from a dam to a power station tailrace. Compare the vibrant features of steep natural rapids (Photo 6.15) with those of a dry gorge (Photo 6.16). The adverse impact can be partially mitigated by compensation flow in the old course or occasionally offset by the sight of a spillway discharging excess water from a full reservoir. Shaw (2003)[18] assesses the visual features of waterfalls and the impact of various degrees of abstraction of flow for hydropower. Appropriate thresholds of permissible abstraction can limit any ill effect.

6.5 Canals, irrigation and drainage

Canals are man-made watercourses for conveyance of water or boats. Drains are pipes, open ditches or hydraulically efficient constructed channels that carry away excess water. There is great variation in the size and type of canals and drains and the extent to which they resemble natural watercourses or stark concrete conduits. Navigation canals are

Photo 6.16 Deep gorge bereft of flow diverted to generate electricity (Maentwrog, North Wales). The gorge is in a Site of Special Scientific Interest (SSSI).

transport routes, considered in Chapter 8 (p138). Here we are concerned with channels allowing water to flow gently downhill. They are designed for a stable cross-section and gradient best suited to economic water transfer from a controlled source – at headworks fed by a river or springs – to outlets such as for an irrigation distribution system or city water supply. Drains remove excess water where natural drainage cannot function satisfactorily, especially in an artificial landscape.

Water transfer canals

Any open waterway cuts across the land surface, creating a barrier to terrestrial movement of people or animals. It may complement or obstruct natural drainage patterns; it probably combines with roads, railways and land enclosure boundaries in defining patterns of flat landscape; and it may introduce waterside features that are welcome in that landscape. Medium-sized canals comprise:

- the water channel itself, lined or unlined and with the bed generally below natural ground surface level;
- canal banks, formed partly of material excavated from the channel; and
- any control structures and over- or under-crossings.

In a typical *lined channel*, water flows on smooth-faced concrete or brickwork incorporating a bitumen or other impermeable layer. If the lining and its surface are well maintained, leakage losses are low and stream flow is laminar. But if people or animals fall into the water, they find it difficult to escape. An *unlined channel* is cut through, or built up in, the earth or rock that exists along the route. It resembles a natural stream in that it will scour its banks, remove or deposit sediment or meander according to whether flow is greater or less than that for which the cross-section is designed or that for which a natural 'regime' can be established. There may be some turbulent flow around irregularities on the banks or bed or at any imposed obstacles such as bridge piers. Grazing animals may regard the canal as a drinking trough; vegetation and riparian life may arise beside it.

Depending on the local geology, unlined canals are liable to leak. Their permeability can be reduced if suitable fine-grained soils are available for placement on the canal bed or are deposited there by flow. It may be more practicable to construct an earth canal than a rigidly lined one if the surrounding water level is already high or if there are likely to be varying back pressures on a fragile lining when the canal is empty. On long canals, overall conveyance efficiency depends on how much water evaporates, especially at any wide, shallow ponds en route, as well as on leakage.

Canal banks often carry a path or roadway. The riparian strip can also support vegetation, drawing water from canal seepage on the wet side or from drainage ditches on the other. Trees – part of hydraulic construction's contribution to a greener landscape – should be planted where their roots will not destabilize the banks.

Structures on canals include flow regulators and headworks for subsidiary canals; excess flow escape facilities; cross-drainage inlets, culverts or siphons; and bridges. Most earth channels, and the footings or abutments of structures within them, need frequent routine maintenance and occasional reconstruction. Canal lining is more robust and should fail only infrequently; but, when it does, early remedial action is necessary. Hydraulic structures ensure that land space is better managed and water conveyance is more efficient than in natural watercourses. The structures are also points of pleasant interest along the way.

Canals for water movement only are not usually such formidable barriers as ship canals because they do not require headroom. Open canals can be bridged above freeboard (emergency) water level, mountain sections can be tunnelled and uphill (pumped) sections have to be in pipes. For security and to prevent pollution, open municipal water supply aqueducts are not usually adjacent to public roads. Some existing and possible future long distance water transfer schemes have been mentioned on p87. The major obstacles to their success are that, for uphill sections of aqueducts, the cost of pumping water is difficult to recoup in the benefits, for example, of irrigated agriculture. Also, on long gravity flow sections, there may be considerable losses due to evaporation in open channels and intermediate reservoirs. However, the great majority of water transfer canals take water from rivers and deliver it effectively in the same basin at no great distance. The delivery points of main canals are generally water treatment plants or service reservoirs for municipal or industrial water supply systems, or headworks of irrigation distributary canals.

Water treatment plants are partly or completely covered and often associated with more prominent administration buildings. Service reservoirs may also be covered tanks and water distribution pipes are invariably concealed. However many village and small town water supply systems utilize elevated tanks (water towers) which can be architecturally attractive features of the landscape. Irrigation distribution systems are much more extensive and visible and their influence on land resources justifies separate consideration.

Irrigation systems

Mountain valley irrigation is undertaken by constructing seasonal (brushwood and stone) or permanent (masonry) weirs across streams, heading up the water. The diverted water feeds channels running along the valley sides at a gentler gradient than that of the stream bed. Temporary cuts enable water to flow from these canals on to the fields running down to the river. Visually, there is sharp contrast on hillsides between higher open brown unwatered land and the lush green cultivated fields within the command of the irrigation channels.

Where river courses are deeply incised into the valley bottom alluvium, simple weirs are not capable of heading up water sufficiently to command adjacent fertile land; and that land may not be extensive enough to justify construction of a higher permanent dam. In these circumstances, farmers may be able to afford to pump directly from the river where their forebears raised water by means of animal-driven Persian wheels or man-operated shadoofs.

Hill-foot irrigation is often fed from springs, for example at the foot of limestone mountains or alluvial fans. Again there is clear distinction, seen from above, between the watered and unwatered land. Springs may discharge into ponds or more formerly fashioned pools. Springs can also be contrived by construction of subterranean channels – known as *qanats*, *fallaj* or *karez* in the Middle East and Central Asia – which tap the sloping groundwater level within hill foot fans. The tops of shafts from which qanats are excavated have been very distinctive landscape features seen from aircraft approaching, for example, Tehran Airport.

Irrigation schemes on plains are more extensive and yet, because of the flat terrain, can be viewed as systems only from the air or on a map. Main, distributary and minor irrigation canals, drainage ditches, roads and settlements are distinctive, usually linear, features. At ground level, it is trees that are most visible, in distant lines along canals or around villages. Pleasant environments can be created, beside canals and at headworks, by trees and their shade and by a slightly cooler temperature than further away as a result of evaporation off the water surface.

While surface water distribution channels are most commonly earth canals or ditches there was a tendency in the Middle East in the second half of the 20th century to install raised precast concrete semi-circular half-pipe aqueducts (Photo 14.1 in Chapter 14). These are not elegant, although the junctions are ingeniously geometric; they may ensure accurate water allocation, but can be obstacles to traffic.

Effective surface irrigation may require considerable land levelling on irregular surfaces; and adequate drainage is paramount where there is likely to be excess water, either as a result of overwatering or as run-off from rainfall.

Furrow and basin irrigation – surface supplies flowing on to the fields – are the most common but often the least efficient approaches to meeting optimum crop water requirements. Sprinkler or drip ('localized') water application systems are more concentrated, less wasteful and particularly suitable for row and tree crops. Pipes are necessary in these switch-on/switch-off systems, perhaps fed from local head reservoirs. Alternatively, water can be pumped; groundwater, if available in adequate quality and quantity, is ideal for localized irrigation application. This gives rise to a landscape in which there is little evidence of actual water supply systems, only the greenery of production. An extreme example, seen from the air, is the dark growth within the circumference of long radius centre-pivoted spray irrigators, fed from desert aquifers and contrasting sharply with the empty aridity around them. However, over-pumping of any aquifer can eventually lead to failure of supplies.

Land, stormwater and foul drainage

Drainage channels or ditches on agricultural land have the reverse function to that of irrigation supply canals. Their capacity increases as they unite and lead back to the main watercourses. Because the water is excess and because it may be contaminated by fertilizers or animal-derived nitrates, it may be desirable to slow the drainage flow by absorption and transpiration in vegetation adjacent to the main ditches. Possibly the greenery may create wildlife habitat as well as an absorbing sponge.

However, stormwater drains in or around built-up areas are often large concrete-lined or masonry-walled channels. The greater their hydraulic capacity the more rapidly flood waters can pass – which may or may not cause back-up problems at downstream constrictions. At most times these channels are dry or the flow is insufficient to do more than deposit debris or sediment – a problem that can be resolved by constructing a deeper, narrower open conduit for low flows along the centre of the main channel bed. Alternatively, occasional deliberate diversion of stream flow into certain channels might enable cultivation of a more attractive low form of vegetation that could attenuate the impacts of floods.

Planning for stormwater in cities is considered in Chapter 15. A major issue is whether or not stormwater and foul sewers should be combined. Waste (foul, used) water treatment requires space, especially for open water (pond) processes such as sedimentation and biological treatment. Some such ponds even provide an attraction to birds or riparian wildlife. If sufficient land space is available and if climatic conditions are suitable, it may be possible to treat waste water in artificial wetlands or semi-natural marshes such as in beds of phragmites reeds (Sendich, 2006).[19]

Watercourses and wetland for nature conservation

Aquatic, riparian and wetland habitats exist in the beds and banks of rivers and streams, in marshes and swamps, and in grassland or forest that is periodically flooded. Yet the needs of civilized development have necessitated the formal channelization of waterways, for example, for navigation, and the extensive draining of marshes to eliminate disease vectors and to provide prime agricultural land. The processes have long been known as river training and land 'improvement' or 'reclamation'.

Some large wetlands remain in remote regions – such as the Okavango Swamp in Botswana and much of the Amazon basin. In many of the others, drainage and reclamation have not been ruthlessly complete. Valuable remnants survive in the deltas of the Danube, Rhône (Camargue) and Guadalquivir (Doñana) and in assigned polders in the Netherlands and even some Californian wetlands.

Much of the historic drainage of wetland has been major civil engineering. So it remains the responsibility of engineers to make the best of opportunities (to create zones of viable habitat and to manage them and their water requirements) within land in which drainage ditches and waterways are constructed.

There is little scope for attracting wildlife to concrete-lined channels; but, even if most of a water channel is lined, it may be practicable to incorporate unlined, naturally banked and even meandering sections of a cross-section, gradient and (reduced) conveyance efficiency that can provide riverine conditions encouraging locally appropriate nature conservation. For example, where the gradient is so steep as to require drop structures on a lined canal it may be possible to choose a steeper unlined profile where hydraulic energy is dissipated in riffles on the bed or niches in the banks.

There are further considerations in making hydraulic engineering sensitive to natural ecology:

- Hard lining is not always necessary to achieve moderately efficient water transfer. The cost of such lining may not be justified for channels through relatively impermeable ground or if a layer of local silt or clay can be a firm effective seal.
- Where a waterway's prime function is to effect rapid discharge of stormwater run-off, then seepage of water into the ground or into riverside vegetation and backwaters can be more effective than a smooth flood relief channel which might rapidly and disastrously deliver peak flow downstream.
- Some already straightened rivers courses can be adapted ('naturalized') to restore some of the informal elements that typify 'unchannelized' streams. Guideline principles have been devised for river channel restoration, for example, by Brookes and Shields (1996).[20] Hydraulic research continues on conditions that satisfy flood defence needs and yet remain environmentally attractive.
- Measured performance of unlined channels can indicate relationships between such factors as the roughness of vegetated channels and river flow pattern at bends; or it can identify the effects of introduced complications such as low stone weirs or two-stage channels.
- River restoration involves both bed and riparian works and must balance ecological needs against the realities of river flow and sediment transport.
- Advantage can be taken of poorly drained but adaptable areas cut off by such engineering works as irrigation systems or railway embankments.

Plants in the beds and banks of watercourses provide habitat for a variety of fauna. Plant stems also bend with the water flow absorbing some of the stream's energy. Their roots entrap sediment and help to keep the river profile stable; but riparian trees absorb water through their roots and, especially in arid climates, may take substantial proportions of stream flow.

Actual measures undertaken to incorporate natural habitat into watercourse engineering depend on identifying appropriate ecosystems or species whose proliferation or recovery is regionally desirable. The systems should be suited to the soils and nature of the waterway and adjacent land and to the anticipated water levels and flows. Subsequent management measures may involve the selective cropping or replanting of vegetation and adjustments to water flow control strategies.

Seasonal water supply to wetland habitat can be provided:

- by runoff, arranged in planning all artificial and natural drainage routes within an area;
- in planning releases of excess water through escape structures on canal systems;
- in releases ('compensation water') to ensure adequate flow in pools and rapids where minimum or periodic flows are necessary to maintain fish or natural habitat, or to flush out mosquito larvae from stagnant pools;
- by restricting maximum releases from dams; the controlled flows down the Colorado River now permit beaver dams to be constructed where they would have been washed away by floods before the man-made dams were constructed; and
- in planned release of 'artificial floods' for downstream wetlands, dependent on seasonal

inundation, where such floods have otherwise been eliminated by upstream diversion; part of such releases might be combined with other operational requirements, for example with seasonal sluicing of sediment in reservoirs.

6.6 Hydraulic engineering for sustainable land use

Water's role in determining land capability

Man needs land on which to live (habitation) or grow food (agriculture). People cannot live, nor can food grow, without water. People's most basic need is enough water to drink (2–4 litres per day). The extent to which their quality of life is then improved by using water for cleansing and cooking depends on how much more is available or can be brought from more difficult or distant sources by means of hydraulic engineering.

Ground food crops grow as a consequence of planting seeds on fertile land where there is seasonal rainfall. Yields harvested are then strongly dependent on that rainfall being adequate; when it is only marginally sufficient, yields can be greatly improved by timely irrigation – supplementary water from another source. There are inhabited areas that are short of fertile land. There are many more that are short of water. Water may be copious – and fertile land short – in the headwaters of large river basins; but further downstream the climate is often arid and the land spacious. Some of that arid land can be brought under irrigation by canals fed by gravity from the rivers; but there will seldom be sufficient water, even in a river dammed upstream, to serve more than a proportion of the potentially irrigable land.

Engineering to optimize use of available water

Water available in each river basin or other hydrogeographical unit may comprise:

- rainfall which falls directly on cropland or may be collected locally for agricultural or domestic use;
- streams that can be diverted, rivers that can be dammed and lakes from which water may be drawn;
- groundwater, which can be over-exploited by pumping too much but can sometimes be artificially recharged; or
- sea or other saline water; but this can be converted to fresh water only where there is abundant cheap energy.

In practice, the amount and quality of water that is supplied to a community depends on the extent to which individual landowners can or are allowed to pump out groundwater; or what engineering works are practicable to increase surface water supplies; and whether the users or their representing authority have the capability to organize these works and to pay for the supplies.

Domestic water, organized by municipalities, is delivered in pipes to public standpipes or individual houses according to what can be afforded. Poor villagers and some urban fringe dwellers have no such municipality and may have to carry supplies from a distant source. Assistance in gaining safe water supplies has been promised to the poorer nations by the richer ones; some help has been forthcoming, but a billion people are still at risk. *Industrial water use*, on the other hand, is dictated by functional needs rather than living standards and demand need not necessarily rise as fast as that for domestic supply. Economies can be affected by less wasteful production and by use or reuse of water of adequate rather than potable quality.

Irrigation dominates water demand in semi-arid countries. According to Meyer (1996),[21] the amount of water abstracted for irrigation rose 25-fold in the last three centuries while the world's population increased only about sevenfold. Yet, as the world's population rises in the many areas where water resources are scarce, so the proportional amount that needs to be abstracted from watercourses or groundwater also rises. Almost everywhere, the supply of water is only as satisfactory as it is because of the storage dams, aqueducts, pumping and distribution systems, and treatment plants that have been built. Groundwater is often over-exploited and should somehow be controlled so that it is fairly and sustainably allocated.

So, wherever fertile land lacks water, people look to engineering works to provide water from some distant source. Even more dams, aqueducts and pumping will be called for, although these can only be provided where additional supply is practicable. Schemes have been mooted for very long distance water transfer; but experience to date indicates that these will be uneconomic for agricultural schemes until a vast source of cheap energy is available – currently an unlikely prospect. The remaining available strategies for maximizing crop production with available water concern more efficient water application in the fields and choice of food crops that meet adequate human diets without excessive consumptive use of water.

To make best use of any supplementary surface or ground water that can be provided, regional long-term weather forecasts may one day be good enough to assist river engineers, hydrologists and hydrogeologists to predict what quantities of irrigation water are likely to be available in the coming season. Farmers could then plan the extent of land that can be irrigated, or – if they rely mainly on rainfall – whether, when drier than normal growing seasons are predicted, some land should be planted with particularly drought-resistant crops or left fallow to regain fertility.

Risks and safeguards in planning water control structures and land development

Risks in constructing water control works lie in any uncertain performance of dams and reservoirs. If a dam collapses there are dire consequences for people downstream. If the slopes above a reservoir slip into the water in a landslide, a wave may endanger the dam structure. If a reservoir leaks too much it cannot store water effectively; nor can it do so if it fills with sediment. If the dam outlet controls fail to release water as planned, or the reservoir's storage capacity is seriously depleted, then the site that they occupy is a valuable land resource lost.

Safeguards in river control must protect the quality of the catchment in which river flow originates; others must be exercised in the siting and design of dams and reservoirs. Safeguards in headwater land management include the conservation of forests and cropland and arrangements that runoff from cleared ground or paved areas is not too rapidly discharged into watercourses – all to minimize erosion and limit the amount of sediment carried downstream. Siting of reservoirs must consider their watertightness and the stability of hillsides in the various conditions imposed by varying reservoir water levels. Design of dams must ensure that the foundations and abutments will be enduringly robust, that the dam can be raised in the future – if that is practicable – on the same foundations, and that arrangements are included to ensure that sediment can be flushed or sluiced out of the reservoir. Until adequate safeguards to ensure the perpetual or very long-term survival of a proposed dam and reservoir can be guaranteed, then investment in dam construction and realization of storage benefits may have to be foregone.

Risks in land development related to water engineering concern the flow regimes of rivers or water flow networks. Badly managed irrigation can lead to soil salinity; or excessively robust flood protection can worsen rather than alleviate the consequences of floods.

Safeguards in irrigated agriculture lie in control and measurement of timely application of water, adequate drainage, allowing time and means for recovery of soil nutrients, and measures to reclaim degraded land. In *flood plain development*, sustainable flood alleviation needs to consider more flexible land management and, in some circumstances, less total protection. Actual construction of buildings and infrastructure should take into account what land is to be seasonally flooded, what occasionally or exceptionally flooded, and what never flooded.

The *final exhortation* for hydraulic engineers is to ensure the continuing function of their structures, by appropriate design and operational and maintenance strategies throughout future centuries. Only thus can valuable land resources such as fertile irrigable land, reservoir sites or wetland habitat be sustained by timely allocation of the finite quantities of water available.

Notes and references

1 The sources vary as to different or common official data used and possibly as to its interpretation. Published sources considered include:

- Gleick, P. H. (ed) (1993) *Water in Crisis: A Guide to the World's Fresh Water Resources*, Oxford University Press, Oxford.
- Gleick, P. H. (ed) (2006) *The World's Water: Biennial Report on Freshwater Resources*, Island Press, Washington, DC.
- Goudie, A. (2001) *The Nature of the Environment*, 4th edition, Blackwell, Oxford.
- Newsom, M. (1992) *Land, Water and Development*, Routledge, London.
- Veltrop, J. A. (1991) 'Water, dams and hydropower in the coming decades', *Waterpower and Dam Construction*, June.

2 Gleick (1993) indicates that world abstraction in 1975–1977 was about 3000km^3 per year, Veltrop (1991) that it was 3528km^3 per year in 1980 and, expanding with population increase, might exceed 4600km^3 per year by 2000. Gleick (2006) gives data on water abstraction reported for each country, but evidently does not think this data consistent enough to assess a total (it actually adds up to rather less than 4000km^3 per year).

3 Global data is scarce because groundwater/surface water use ratios vary greatly according to geographical conditions. Newsom's 1992 estimate regarding annual water *availability* is that 41.5km^3 per year of total water flow comprises 27.0km^3 per year as surface (river) flow, 12.0km^3 per year groundwater and 2.5km^3 per year as glacial melt. It is conceivable that the ratio of surface to groundwater *use* might be similar.

4 China's Grand Canal, construction of which took place between the 5th century BC and the 14th century AD, is a ship canal stretching for 1776km from Hangshui (south of Shanghai) to Beijing, crossing both the Yangtze and Yellow rivers and climbing (with locks) up to 42m en route (http://en.wikipedia.org/wiki/Grand_Canal_(China), accessed 9 February 2010).

5 http://en.wikipedia.org/wiki/Farakka_Barrage, accessed 9 February 2010. See also Stephens, J. H. (1976) *The Guinness Book of Structures*, Guinness Superlatives, Enfield, London, p195.

6 Tanton, T. (2002) in *Water and Maritime Engineering*, vol 148, no 4.

7 The eastern route of China's South-North Water Transfer Project will make use of a 400m^3 per second pumping station already built by the Yangtze river, the route of the ancient Grand Canal, further pumping stations and a tunnel under the Yellow River (http://en.wikipedia.org/wiki/Grand_Canal_(China) and http://en.wikipedia.org/wiki/South_North_Transfer_Project). A central route is to connect a reservoir on the Han River (a northern tributary of the Yangtze) with the Yellow River and thence by gravity to Beijing. A western route nearer the headwaters of both rivers may follow if a feasible scheme is resolved.

8 The examples quoted are from numerous sources plus a brief review in http://en.wikipedia.org.wiki/Natural_disasters_in_China, accessed 8 April 2010.

9 Wang, Z-Y. and Plate, E. J. (2002) *Proceedings of the Institution of Civil Engineers Water and Marine Engineering*, vol 154, no 3.

10 For more information about the Indus and Aral Sea basins see:

- Duder, J. N., 'Large reservoir storage in Pakistan'; and
- Carpenter, T. G. and Halcro-Johnston, J. F. 'Land use and water transfer in Central Asia';

both in Carpenter, T. G. (ed) (2001) *Environment, Construction and Sustainable Development*, Wiley, Chichester, pp223–243 and 259–277.

11 If too much irrigation water is applied and it does not drain away, then the water table rises until the ground is waterlogged. Salts are drawn up by capillary action and what were fertile soils become saline and barren.

12 For references to disastrous blockage of rivers by landslips see Carpenter (2001) as Note 10, p253 (Indus river in 1840/1841), p274 (Lake Sarez in Tajikistan) and p465 (overtopping of the Vajont dam in Italy in 1963).

13 Chapman, R. J. (2001) 'Reservoir storage on the Blue Nile in Sudan', in Carpenter (2001) as Note 10, p254.

14 Crowe, S. (1958) *The Landscape of Power*, Architectural Press, London.

15 Scott, C. W. and Molyneux, J. D. (2001) 'Britain's concrete dams:The final 50 years', *Civil Engineering*, November, pp171–172.

16 Scott and Molyneux (2001) as Note 15, p175.

17 Meyer, W. B. (1996) *Human Impact on the Earth*, Cambridge University Press, Cambridge, p137.

18 Shaw, T. (2003) 'Over the edge', *Landscape Design*, November.

19 Sendich, E. and the American Planning Association (2006) *Planning and Urban Design Standards*, Wiley, Hoboken NJ, p334.

20 Brookes, A. and Shields, F. D. (1996) *River Channel Restoration*, Wiley, Chichester.

21 Meyer (1996) as Note 17, p121.

7

Coastal and Estuarial Construction

Seen from a human standpoint, the sea:

- has destructive power that often makes coastal scenery more spectacular than that of adjacent inland country;
- accommodates ships, still a major form of freight transport, needing sheltered harbours along the coast; and
- supports fish – a limited food resource – and is associated with oil and gas, wind, wave and tidal energy resources.

Section 7.1 describes coastal landscape features as they constitute scenery or relate to human activity and settlement. Section 7.2 presents the options in defending or managing coastlines. The sections that follow describe the role of construction:

- in protecting particular sections or features of coastline (7.3);
- in making use of, or stabilizing, river estuaries (7.4); and
- in providing ports for shipping (7.5).

The final section reviews options for conservation or development of coastal hinterland, a particularly significant land resource.

7.1 Coastal landscape features

Globally there is a great variation in seaside landforms according to the local geology, climate, tidal range, nature of river entrances, and sediment transport in these rivers or along the coast itself. Because of this variation, it seems practical to introduce coastal landscape by taking examples from the diverse coastline of the main island group of northwest Europe, but mentioning further forms or striking examples from other parts of the world.

British coastal landscape

The British Isles comprise the two islands of Great Britain and Ireland, together with a large number of smaller adjacent islands. The most spectacular coastal regions and those least suitable for settlement are found along the rugged and irregular western coasts. Softer or lower-lying topography is featured along the east coasts, and mixtures of the two on coasts facing south or north.

England and Wales alone have a coastal frontage of 2741 miles of which three quarters was substantially undeveloped in 1970. The Countryside Commission (1970),[1] making proposals for designation of heritage coastline, selected 34 areas with a frontage of 730 miles (26.6 per cent) in various categories of high quality scenery. This included 19 miles of settled frontage, so recognized scenery is not necessarily remote from habitation.

The total coastline of mainland Scotland and of Ireland are each of comparable length to that of England and Wales, the undeveloped proportion being greater and the proportion of high quality scenery perhaps greater too. The following review of mainly British coastal landscape commences with the more spectacular hard rock geology of the Atlantic coasts and continues through softer formations to dynamic landforms such as beaches, dunes and estuaries.

Hard rock coasts

Hard coastal geology provides a sharp contrast with smooth inland slopes or plateaux. In hard rock a range of scenic situations is typified by:

- substantial monolithic extents of cliff face such as at Moher in County Clare, Ireland, Croaghan in County Mayo and Scottish islands such as Orkney and St Kilda;
- rougher landscapes of headlands and coves with offshore stacks, inshore caves, rock pools and cliff-top paths attracting intrepid exploration; these arise from differential erosion in more complex geology of folded strata and igneous intrusions, for example, in Cornwall and Devon;
- isolated bosses of hard rock where surrounding softer material has been eroded, such as Great Orme's Head in North Wales, St Michael's Mount in Cornwall or the natural bastion on which Bamburgh Castle is built in Northumberland; offshore examples include Bass Rock and other nearshore islands which provide isolated seabird habitat;
- exceptional natural architecture such as the columnar basalt at Staffa, Scotland, and the related Giant's Causeway in Northern Ireland;
- wave-cut rock platforms or hard tidal beaches, as in South Wales (sedimentary) or Mull (basalt); these arise where there is insufficient sand for beach material and may exhibit sharp dykes or ribs of hard strata that sometimes even resemble rough types of man-made training works;
- drowned valleys where sea level has risen at the end of the last ice age as in southwestern Britain, Ireland and northwestern Spain; these 'rias' may fill with silt or may remain deepwater havens; above the water they may be bedecked with woodland down to high tide level or, in the fiords of Norway or New Zealand, may rise as sheer rock faces; and
- coastal features of volcanic origin such as the Pacific atolls and some natural harbours like Aden; islands like Iceland or the Canaries are scenic entirely in volcanic terms.

Steep eroding coastline

The process of coastal erosion has been introduced in Chapter 3 (pp39–40). The predominant geological formations of southern and eastern England are soft rocks and river alluvium. Alluvium accumulates mainly in lowland and is associated particularly with estuaries. The higher cliffs of relatively soft formations comprise Mesozoic and Tertiary formations such as of chalk, sandstone, mudstone and clay. Inland features of chalk scenery are escarpments; sandstone outcrops hard enough to survive are relatively rare; and clays or loose soils have little obvious scenic effect other than in the vegetation they encourage. A sea cliff, however, is an eroding hill. Its scenic character depends on the nature of the rock, the eroding forces and the currents carrying away the eroded material. These factors can result in a cliff partly buried in scree (as at Portland Bill), a vertical cliff face with little remaining loose material (as at Shakespeare Cliff, Dover) or the complexity of the Jurassic coast in Dorset and East Devon, the only natural World Heritage Site in England.

Chalk is a white limestone, occurring in England and northern France as a relatively soft but remarkably pure deposit of calcium carbonate. It is laid in thicknesses of hundreds of metres with occasional thin bands of hard flint stones. Because of their texture, homogeneity and porosity, chalk cliffs are eroded by wave action at the base and collapse in vertical slices. Thus the cliffs at Beachy Head maintain their height, steepness and stark beauty, contrasting with the receding grassland that provides a natural pleasure ground at the top.

Landslides are a much rougher formation occurring in less homogeneous material. Marine erosion at the cliff toe may be supplemented by seepage from above which builds up pressure in porous layers and lubricates planes on which sliding may occur. Wide strips of confused unstable ground have arisen as a result along the Devon/Dorset and North Yorkshire coasts and at Folkestone Warren (Photo 7.1). On the south side of the Isle of Wight, in the area known as the Undercliff, porous Greensand rock has slid down towards the sea on a layer of Gault clay. The landscape value of such undercliff lies in its very wildness. Amid the confusion of heaps and

Photo 7.1 Folkestone Warren, Kent, an area of continuing landslip affecting both alignment of the railway and the coastal defence strategies.

ponds, and without much human interference, a profusion of plants and associated fauna may arise which, even in such a dynamic situation, can create special ecosystems and intriguing scenery.

Between the clean facade of chalk cliffs and the chaotic wilderness of landslides is the middle course of erosion that gradually eats away straight coastlines. On the coast of Suffolk, the receding coastline has continued to foster a sparsely inhabited area of attractive sandy heath on the heights, or marshy lagoons behind receding sand dunes of the lower coastline. The North Sea coast of Yorkshire is also receding by slippage into the sea, at Holderness by about 2m per year.

Beaches

Shingle and sand on beaches – the littoral zone between high and low water – are components of eroded cliff debris; some lie on wave-worn rock shelves or other hard forms of shallowly submerged land. The finer material is usually washed out to beyond the low water level whilst the shingle and sand may be carried further along the coast (littoral drift). Beaches provide:

- protection for the dry land, especially if the beach is wide or robust enough to dissipate much of the energy of storm waves;
- habitat for marine and riparian wildlife, suited to the abrupt changes in dry and wet conditions, especially in rock pools but also in sandier situations (where on some coasts creatures such as turtles may bury their eggs) or in sand, silt or mud that provide marine foods for wading birds; and
- sandy strands which are the delight of human visitors and provide splendid viewpoints of any fine scenery rising in the hinterland.

While unprotected cliff coastlines can only abrade, beach systems can erode, accrete or remain stable according to any changes to windward landforms or structures.

Sand dunes are accumulations of sand shaped by the wind. By the sea, their origin includes beach material. Sand dunes develop from 'an abundant supply of light, small grains produced by erosion of cliffs, onshore wind of considerable strength, and a low-lying coast' (Burrows, 1971).[2] Dunes, with the vegetation that grows on or stabilizes them, are significant landscape features:

- in protecting their hinterland from flooding by high tide surges (which they did well, in comparison with man-made protection, in the 1953 North Sea onslaught);
- as hills up to 80m high at Penhale, Cornwall, striking scenery in their own right, or ridges affording views of the rest of coastal strips and hinterland; high sand dunes also characterize the Lithuanian and Polish coastline, that of the North Sea in Denmark and the Netherlands, the Atlantic shores of Les Landes in southwest France and the Namib Desert in southern Africa; and
- as nature reserves because of the special habitat that arises in the moving or plant-stabilized dunes themselves, and in the dune slacks of thickly vegetated, even marshy, space between successive lines of dunes running parallel to the beach (Photo 7.2).

Accumulating land arises from deposit of littoral drift as spits of sand or shingle, *recurrent* where their extremities turn inward along the coast, or in broader pointed *cuspate* forelands resulting from drift in opposing directions. Dungeness in southeast England is a prominent cuspate foreland which has emerged in a few centuries while Orford

Photo 7.2 Morfa Harlech, North Wales. Drifting sediment first created river alluvial, then seaside 'links' terrain. This area has been allotted for built development, a premier golf course, a dunes nature reserve and a long sandy beach.

Ness in Suffolk is a long recurrent spit. Because of its coarse size and salty situation, shingle is infertile and less attractive to people than sandy shores or coastal heath; but shingle spits may provide rare forms of habitat.

Estuaries and deltas are where the regime of the sea meets that of freshwater outlets from the land. Rivers drop their coarser sediment in their rapid upstream sections but deposit the finer muds where the channels widen out on flatter land. Where rivers diverge into a number of relatively narrow courses across flat coastal alluvium the fine sediment is deposited in the sea itself forming a delta.[3]

In wide estuaries the inter-tidal zone may comprise mudflats, flooded at every tide. Around high tide level, there may be salt marsh where salt-tolerant vegetation thrives on trapped sediment. The salt marsh is interspersed by muddy tidal streams that convey the seaward and landward flows at mid and lower tide levels (Photo 7.3). Providing wilderness rather than scenery, much remaining salt marsh can be visibly appreciated from adjacent embankments where these have been built to reclaim former marsh for agriculture or settlement. Meanwhile, in north Norfolk, the beaches, dunes and marshes are in a constant state of evolution and 'changes of some magnitude are not infrequent' (Countryside Commission, 1970).[4]

Many deltas continue to grow; a few have receded since sediment flow has been reduced by damming upstream. The deltas of the Rhine, Maas and Schelde rivers entering the North Sea have been completely stabilized by massive embankments and barrages, coupled with continuous dredging to maintain harbour basins and access channels.

Lagoons are found where freshwater drainage from the hinterland is enclosed behind spits formed by littoral drift (Photos 7.4 and 7.5). Drainage from lagoons into the sea can occur only by seepage or through any remaining narrow openings. While primarily of ecological value, lagoons can be attractive sections of water flanked by palm trees, admired from boats plying on them or from shoreline paths.

Photo 7.3 Salt marsh, Porthmadog, North Wales.

Photo 7.4 Minsmere, a Ramsar Wetland site. Parts of this coastal lagoon have been modified to optimize conditions for wading birds. During construction of Sizewell B nuclear power station in the background, special precautions were taken to prevent drawing down the water level in the wetland.

Mangrove trees or shrubs are important features in shallow tropical or subtropical waters fed by fine river-borne sediment. Their growth traps further sediment and, undisturbed, may gradually extend the land area seaward. A valuable ecological resource, mangroves are also a significant natural defence against the ravages of sea storms. In warm conditions, where there is insufficient sediment to encourage mangroves in shallow water, coral may grow on reefs. These are the protective limestone skeletons of living polyps. Not part of normally visible scenery, corals are important biologically diverse natural resources of great beauty and high tourist interest.

7.2 Coastal defence or management options

Coastal defence implies measures to defend lengths of vulnerable coastline against sea storms. Coastal management accepts that it may be impracticable to provide total defence and settles for measures to influence natural forces for the best net benefit. Defence is resisting attack whereas protection is keeping safe from harm. The two terms are often used synonymously, but here defence is for options (7.2) and protection for structures (7.3).

Where cliffs erode or tidal surges inundate coastal lowland, should robust defences be provided? Or should nature take its course? Perhaps such modifications should be made as will establish a new equilibrium balancing changing conditions against any more critical human needs. Answers must be sought through:

- experience of natural balances that have arisen along studied coastlines and how these have been upset by historic interventions, coupled with modelling or prediction of the consequences of proposed works, based on established hydraulic and geomorphological principles;
- comparison of the effort and results in construction, maintenance and renewal of engineering works with the consequences of not undertaking these works; and
- the economics of saving land and property – by complete or partial protection – and who pays for this protection.

Situations vary from those in which the very existence of cities depends crucially on secure defences to those where limited agriculture or isolated settlement can no longer justify the expense of resisting coastal erosion or occasional inundation.

Complete coastal defence

In the estuaries and coastal flats of the Netherlands, reclamation of land by building protective embankments and creating polders has been in progress in stages starting with the district of Holland in the 12th century.[5] Polders are areas of land surrounded by dykes and lying below high tide level, and in many cases below mean sea level or even low tide level. Polders are drained by opening sluices at low tide or by pumping. In conjunction with continuous pumping out of

Photo 7.5 Marazion Marsh, a coastal lagoon, was divided by 19th century railway and road construction into three parts, all now elements of a nature reserve.

water, the protected area of the Netherlands has extended to all that within the outer coastline and includes wide areas formerly below the sea. The result is contrived landscape achieving a new balance between agricultural land, wildlife habitat, human settlement and road and drainage systems.

In low-lying Bangladesh, no such solution is possible. Not only is the annual river flow through that country 25 times greater than that of the Rhine, but the cyclonic storms of the Bay of Bengal are much more serious in impact than North Sea storm surges because of the dense population living in some unprotected areas. Rather than build higher coastal embankments that would further impede the escape of river floods, reliance has to continue to be placed on conservation or restoration of natural defences such as the mangroves and other vegetation of the Sundarbans coastal zone.

Natural coastline defences

Where there are not hard rock cliffs, natural buffers against storms include:

- high natural barriers such as sand dunes which, for example, lie along much of the Netherlands coastline north of The Hague;
- wetland – wide tidal flats, lagoons, saltings or marshes;
- coastal beaches and shingle spits against which the force of waves is dissipated as the water flows in and out between the stones;
- the surface roughness of vegetation and the build-up of sediment within it; and
- shallow water features, particularly coral reefs along many tropical coasts.

Sand dunes are robust and potentially dynamic. Coastal wetlands are much more sensitive to earthworks and drainage. Seaside vegetation is reduced by clearance for built development or fish farming. Shingle beaches are directly related to the drift of sediment along the coast and their protective value may be threatened if coastal works upset that drift.

Sections of secure defence

Along the south coast of England there is no 'low country' comparable in extent with sub-sea-level lands of the Netherlands or the East Anglian Fens. There are estuarial flat lands and considerable lengths of soft cliff landscape. Among these are sited various seaside towns – in Sussex mostly conurbations – which demand protection along their low-lying fronts or where settlements lie near fragile cliff-tops. The construction of hard protective walls along these sections eliminates most of the sources of sand or shingle which would feed the littoral sediment drift and replenish beaches further along. It is therefore at undefended sections of coastline where the consequences of additional erosion must be accepted. Accordingly, strategies for management of long coastlines are best devised for lengths that are self-contained in terms of the movement of beach sand and shingle (Payne, 2003).[6]

Partial defence at particular places

Defence works may protect some aspects of conservation interest but imperil others. At Hengistbury Head on England's south coast, important geological exposures would have become overgrown had full sea defences been implemented to prevent cliff erosion; but, without any such defence, rare cliff-top plants were already endangered (Coker, 1992).[7] In this case, limited hard defences preserved enough cliff-top without halting all erosion. Nature conservation interests often equate to scenic ones but cliff erosion prevention measures always alter the appearance of at least the bottom of a cliff face.

Accepting coastal attack while adapting key infrastructure

Some balance between security and scenic wildness can be achieved by flexible approaches to the upkeep of roads, promenades or footpaths where these are undermined by wave action or inundated by landslides. Temporary closure, repairs, rerouting and occasional property abandonment are all measures by which coastal amenities can be managed, depending on the funds that are available, the effects on nature conservation and people's acceptance of limitation and change. Along the coast of Suffolk, defences have been maintained in recent centuries, allowing only occasional damage at towns such as Southwold. Nearby, however, there has been considerable land lost to erosion over a longer period, including the complete city of Dunwich. Land and property are currently being lost rapidly on parts of the coasts of Norfolk and Lincolnshire.

Managed retreat from previously defended coastline

Some coast defence works that once enabled wetland reclamation, usually for agriculture, may be inappropriate to modern economic conditions. The productive value of reclaimed land may have decreased while the costs of maintaining defences may not. Furthermore, the old lines of defence may not be well suited to new strategies for longer sections of coastline. Cooper (2003)[8] defines 'managed retreat' as the technique of deliberately breaking, entirely removing or relocating landward an existing tidal flood defence or coastal protection structure. Cooper further defines five sets of circumstances where managed retreat is appropriate:

1. Flood defence and coastal protection can be improved by using natural buffering and innate protection properties of natural landforms.
2. Accommodation space for natural change, such as rising sea level, or to avoid 'coastal squeeze' can be found by moving natural or other defences landward.
3. Wetland can be created or re-created, for example, as intertidal flats and salt marsh.
4. In economic solutions, the cost of maintaining existing defence works or constructing new ones might not be justified compared with the value of assets at risk.
5. For public safety, the risks and costs of protection of assets remaining in precarious locations may have become unacceptably high.

Managed retreat can become contentious when a decision has to be made as to which currently protected land or property does or does not justify the expense of maintaining robust defence. Mabey

(2003)[9] notes that while 'managed retreat … has worked wonderfully' in Essex salt marshes, it 'would be a heartless solution in areas like north Norfolk, with its string of anciently inhabited coastal villages'. Sometimes adoption of ill-judged retreat policy might be politically expedient in avoiding or deferring necessary investment in coastal protection.

Predictors of substantial sea level rises during the 21st century see the Fenland of England as one of the more extensive regions under threat. Some form of retreat may have to be contemplated to new lines of strategic protection. On either side of these, natural equilibria will eventually be re-established. Such changes have profound consequences for coastal and estuarial landscape, possibly with a net gain to the wilder land and wetland resources.

7.3 Coastal protection structures

Coastal features that may need protection include:

- cliffs that could be undermined by heavy seas or subject to landslips;
- transport routes or promenades, which run along coastlines just above sea level or are constructed at the foot of cliffs; and
- low-lying land that may be worn away or flooded beyond planned lines of maximum retreat.

A first measure is to conserve the beaches that already absorb much of the wave energy attacking coastlines. Historically, the engineering solution was often a solid sea wall at the top of the beach, perhaps combining its defensive function with provision of a road, rail route or promenade. However, the effects of such sea walls on the coastline is now well understood; they reflect wave energy rather than absorb it, leading to increased levels of wave energy immediately in front of the sea wall and in turn to increased risk of scour of wall foundations. A less elegant but often more effective protection against storms is revetment, typically a slope of large rough-shaped rocks or loose fitting concrete blocks; or an artificial beach of sand, shingle and cobbles may be effective if its be constructed to remain stable. In conjunction with any of these measures, the need remains to conserve the best features of coastal scenery, such as the cliff faces, and to enable people to enjoy that scenery on seaside walkways with access to beaches.

Ensuring protective beach material remains in place

Beach shingle and gently sloping sand absorb much of the energy of waves. Short beaches, typically crescent-shaped and bounded by rocky headlands, can be stable; but littoral drift causes long beaches to lose material. So, firstly, there must be a source of sufficient top-up beach material that is allowed to continue to erode from other points on the coast. Then, if natural littoral drift exceeds the necessary supply of top-up material, the rate of drift can be artificially reduced to trap the beach material. Groynes built transversely across beaches can be at least partially effective in stabilizing sediment drift along stretches of popular public beach (Photo 7.6). These weather-beaten wooden walls anchored by timber piles are convenient beach furniture, serving also as windbreaks and frames to sit or climb on.

In some conditions, a series of detached groynes, constructed parallel to the shore (Photo 7.7), can be more effective in stabilizing beaches than groynes built across the beach perpendicular to the coastline. Offshore breakwaters – founded beyond the low tide line or even fully submerged – refract waves and reduce littoral drift, sometimes leading to the formation of tombolos (spits) between groynes or connecting them to the shore. The consequences depend on the water depth at the new breakwaters, their length, spacing and distance offshore and the pattern of sediment drift. Usually the planned purpose is to reduce erosion and sediment transport, especially at high tide; but shore-parallel breakwaters have also been constructed simply to increase the area of beach in popular seaside resorts (notably Monte Carlo). Submerged breakwaters (reefs) have even been constructed to increase the height of waves on surfing beaches in weather conditions otherwise too calm for surfers.

Photo 7.6 Seawall, railway and stream outlet at Dawlish, south Devon. The foremost groyne across the beach is of cemented rock and others are of timber construction.

Photo 7.7 Seawall and railway near Dawlish. At this point, short lengths of rock groyne were also constructed parallel to the shoreline.

Sea walls

Smooth masonry or concrete-faced walls are traditional defences against wave damage to cliffs or incursion of high tide storm surges into seaside settlements. The stability of sea walls depends on their depth, mass and shape. If a wall is solid and vertical it bears the full brunt of storms, sending spray high into the air. Perhaps a considerable proportion of that spray, blown by the wind, will shower down on the inland space – a road, railway or promenade – although with considerably less effect than the deluge that would have resulted if there had been no protection. Some reduction in the spray effect can be achieved by constructing the wall face as a concave curve with an overhanging lip; but the full impact of the waves is still taken by the wall.

The spectacle of waves striking a sea wall is an awesome one hinting at the destructive forces that will seek out any weakness in the wall's structural joints or erode its foundations. This is why walks on sea walls or breakwaters have a hazardous fascination during storms. However, where there is no substantial beach material to protect them, solid sea walls sometimes fail as scour attacks their foundations.

Rock or block revetment

Revetments are a more flexible form of construction. Large quarried rocks or preformed concrete blocks are dumped or placed so as to absorb the energy of waves in the voids between them and to settle into any cavities caused by erosion. Experiments have shown that irregularly shaped rocks are more effective in dissipating energy than closer-fitting or rectangular blocks of greater size and weight. The first step in planning rock protection is to determine the necessary extent of protection and the weight and disposition of the pieces needed to absorb the energy of destructive storms. Engineers must then identify a local or distant source of suitable material, the means of access for rocks and machinery to the site, and the equipment needed to place the blocks in position.

The scarcity and expense of large blocks of hard rock at most softer rock coastlines led to alternative artificial armouring solutions in which precast concrete shapes are substituted for natural rock. Patented types of precast concrete blocks such as *Accropodes* or *Tetrapods* can be custom-built to suit particular conditions and to interlock with each other in such a way as to inhibit subsequent movement yet still maximize their role in wave energy dissipation. Such block defences can be built in an variety of forms, textures and colours to contrast or blend in with adjacent features; however, at close quarters, they are uncompromising in appearance. Rock is sometimes provided as additional protection on the seaward side of conventional sea walls.

Conserving the scenery and amenity of protected coastlines

Aesthetically satisfactory protection plans may need to incorporate measures to conserve natural features like rock pools or cliff faces and, access to coastal walks and beaches. Solution need some affinity with the local geology and established man-made structures.

Rock pools are found on intertidal hard strata free of sediment and are generally stable in natural conditions. In softer formations cliffs may need protection both by sea walls at their base and by measures to prevent rock falls or landslips higher up. Wild landslip territory is valuable scenic heritage (see p115), but where transport routes or settlements are affected, stabilization may be needed.

Ensuring the durability of the upper parts of cliffs involves adequate internal drainage, bolting or containing in mesh any potentially loose rock surfaces, or cutting back the slope of the cliff face to increase its stability. Such measures can spoil the appearance of the cliffs, especially if modifications and repairs are frequently required. If a period of several years can elapse without adjustments, then weathering, establishment of vegetation and even minor slips can result in a more mature aspect. This is unlikely to happen if cliff faces are sprayed with mortar or massive concrete or masonry buttresses are constructed to enforce stability. But there can, if necessary, be less massive, still structurally effective patterns of anchored beams and columns of more aesthetic design, especially if their appearance harmonizes with any sea wall, paving or buildings at the bottom of the cliff.

Promenades, with ramps or steps down to the beach, are common features on hard sea walls. Indeed, a feature of sea walls and even some harbour breakwaters is their use as an amenity. Where rock revetment is needed as an extra or more effective form of storm protection, it may be a barrier to access or an intrusion on traditional seascape. Meanwhile, the expense of revetment normally only justifies its adoption along a limited number of particular stretches of shoreline. Longer coastlines should be planned and managed as a whole, in terms of what parts should be conserved, modestly adapted or stoutly defended according to engineering, ecological, social and economic needs.

7.4 Estuarial construction

Estuary landscapes

An estuary is the wide tidal mouth of a river. Estuaries occur in the flat terrain that results from centuries of deposition of river alluvium, modified by tidal ebb and flow and by littoral drift of material along the coast. Salt marshes and tidal mud flats occur more at river estuaries than on straight coastline; and, at estuaries, natural coastal formations (such as dunes) merge with flood deposited river levees. The interaction between tidal salt water inflow and outflow and the one-way but seasonally variable outputs of fresh water creates areas of special characteristics in which wildlife finds habitat. The worms, crustaceans and shellfish that burrow in mudflats are a great attraction to wading birds. Wildfowl may feed on salt marsh plants, and the same birds need nearby dunes, shingle banks or fields to roost on in large numbers at high tide. Many estuary lands are also wintering or transit grounds for migrating species.

Human activity in these conditions traditionally comprised fishing, hunting wildfowl and sometimes cropping of wetland vegetation. The landscape, to a wildfowler or boatman, is a flat one. Particularly when it is misty – and fogs are common in these situations – there is no scenery. Instead the sensual environment comprises feelings of climate, a tang of salt or salt vegetation, and the sound of birds, rustling reeds and lapping water.

Constructional modifications to estuary landforms have comprised:

- land reclamation – by constructing or extending defence works and draining the land thereby isolated from tides and floods;
- making road or railway crossings of the estuaries on bridges or causeways and of adjacent land on embankments, some doubling as coastal or river flood protection banks; and
- creating channels for navigation of ships in the river.

In planning construction, consideration has to be given to conserving, creating and maintaining viable areas of wildlife habitat as well as to human amenity. Man-made embankments or bridges afford views – which would not otherwise be seen – of the flat estuarial land, whether wild or developed.

Flood protection or wetland?

Reclamation for agriculture or settlement has been achieved mainly by drainage within embankments created to protect land against tidal or river flood inundation. This has been at the expense of tidal or wetland flats. For example, more than a quarter of the US eastern seaboard wetlands has been lost and a much higher proportion has gone in southeast England. Conservation interests welcome anything that can be done to redress the balance in favour of dwindling wildlife habitat. To this effect, managed retreat is now being practised in some of the estuarial lands of Essex, England.

Wildfowl and some forms of farming make use of land on both sides of the flood embankments, for example:

- many types of birds that feed on exposed estuarial flats need adjacent dry land – which often happens to be farmed – as a roosting area during high tides; and
- sheep and cattle can graze land in tidal and semi-tidal conditions.

Estuarial crossings

Long bridges and causeways over extensive waters are addressed in Chapter 9. Embankments carrying transport routes across low-lying areas may be combined with coastal or river flood protection banks; and similar banks form polders for agriculture or settlements. All these earthworks form regular patterns of fields, drainage channels and more random intentional or incidental wetland.

Barrages are gated weirs constructed across rivers, for example to head up water and divert some of it for irrigation or other consumptive use. In tidal waters, barrages or movable barriers have been constructed:

- to protect cities such as London, Rotterdam and St Petersburg from excessively high sea levels caused by storm surges;
- to manage water levels at city waterfronts – for flood management and to encourage waterfront development and use, for example, in Cardiff Bay;
- to generate hydroelectric power from tidal currents, as at La Rance in Brittany; and
- to facilitate navigation in tidal rivers.

These operations alter hydraulic and morphological regimes both upstream and downstream.

Estuarial ports

For historical and geographical reasons, many ports are located in estuaries or tidal sections of rivers. Particular needs in estuary ports include:

- dredging to achieve adequate depth for ships at berths, along river channels and through bars (shoals) obstructing access to the sea; and
- assignment or reclamation of land areas for quays, involving either drainage of existing land or creation of new land intruding into the tidal water.

Both these are issues in harbour construction.

Many cities were located at the upstream limits of navigation for sea-going ships. However, as ship sizes have grown over the centuries, so port facilities have had to be constructed further downstream in deeper water. In London, starting in the 18th century, riverside wharves close to the City were supplemented by enclosed docks, allowing ships to remain afloat at low tide (and providing secure space for cargo handling and storage). Even larger docks were constructed during the 19th and 20th centuries, each being further downstream and providing deeper water for larger ships. At the same time, deep wharves were constructed in the river, providing sufficient depth by dredging instead of by impounded docks. This process continues with construction of the London Gateway container terminal close to the downstream limit of the estuary. This involves channel dredging, land reclamation and brownfield (disused oil refinery) remediation.

7.5 Harbours and ports

A *harbour* is an area of water that is deep enough and sufficiently protected from winds and waves, either naturally or by construction, to constitute a haven for ships. *Port facilities* are those provided for loading and unloading ships. There can be berths around natural harbours or along shores protected by breakwaters; or there can be on-shore installations connected by jetties, submarine pipelines or boat access to deep water moorings. On the land side, there must be facilities for storage and transfer of cargo to land transport or to other ships. Today, some ports concentrate on construction and servicing of offshore structures like oil and gas platforms and wind turbines. The significance to land resources of harbours is the impact of maritime activity on land use in the hinterland; and port facilities, harbour walls and seagoing vessels are all elements of coastal landscape.

Natural and artificial harbours

Natural harbours occur worldwide, for example in bays at Sydney, Rio de Janeiro or San Francisco or behind islands such as Hong Kong. Artificial harbours have been created along exposed coastlines by constructing breakwaters extending into the open sea as, for example, at Colombo, Dunkerque, Beirut or Dover (Photos 7.8 and 7.9), or small harbours can be created behind protective seaward walls at inlets such as that at Mevagissey (Photo 7.10).

Many new ports have been created, for example in the Middle East, in conditions where

Photo 7.8 Dover harbour, enclosed by breakwaters on three sides.

Photo 7.9 Dover port facilities – from the customs and immigration facilities and roll-on roll-off terminal in the foreground to the town beach and the original rail passenger terminal on the far side of the harbour.

Photo 7.10 Mevagissey harbour, Cornwall.

the sea is relatively shallow and the land is mainly flat and often undeveloped. In these circumstances, substantial earthworks and dredging are required to form a satisfactory interface between deep-water berths and landside service space. Three construction approaches are possible:

1 reclaiming land from the sea out as far as deep water; but this is more practical in steep topography;
2 building causeways or piers out to berths constructed in naturally deep water – with little dredging but with long breakwaters if necessary (a strategy adopted at Dammam and Jubail in Saudi Arabia); and
3 dredging basins inland to provide deep water berths behind the existing shore line (as was done in the early 19th century for London's docks and in the 1970s at Jebel Ali in Dubai).

In harbour landscapes, the role of nature conservation is one easily disregarded among all the human endeavour. Construction of quays and breakwaters, reclamation of land and dredging of channels can all disturb or destroy fragile ecosystems that cling precariously to the special conditions of the coastal zone or river estuaries. Fragile features may be in the water and on the seabed – marine breeding and feeding grounds – or among land and waterside habitat.

Sensitive planning for harbour construction should therefore strive:

- to exclude from development any seabed and waterside areas of exceptional ecological value; and
- where habitat has to be disturbed, to replace it with something similar but viable, by creating and managing new conditions at a suitable location unaffected by dredged channels or landside construction.

The capability of a port is limited by the deepest draught of ship that can be accommodated at berths and can negotiate the access channels and by the capacity of port cargo handling equipment, storage space and onward transport facilities.

Development of new port facilities can entail construction of:

- breakwaters;
- moorings, jetties or landside quays;
- dredging for access channels and berths; and
- reclamation for quays and landslide storage areas and – in a broad interpretation of the role of ports – for other city development.

Breakwaters

The primary function of harbour walls built from the shoreline out into deeper water is to create a haven for shipping. They are designed to absorb or reflect waves; and their layout is contrived to maximize calm conditions on the sheltered side. The configuration of breakwaters may take advantage of natural coastline features such as islands or headlands. Thus the Phoenicians built harbours on either side of the peninsula at Tyre. More often, stable coastlines are comparatively straight. Then, the further that breakwaters extend out to sea, the more they interfere with littoral currents and littoral drift. The consequences have to be taken into account in determining the configuration of breakwaters and harbour entrances.

As with shoreline sea walls, wave energy is most effectively dissipated by heavy, loosely interlocking, irregularly-shaped rocks or precast concrete blocks. However, before machinery was widely available to place very heavy rocks in rough conditions and before the benefits of randomly placed artificial concrete armour blocks had been recognized, many breakwaters were constructed as vertical masonry walls in more precisely geometrical forms. Some such walls could perform two roles, one side as protection against the sea, the other as berths for ships. Accurate positioning in terms of line and level is vital if breakwater berths have to serve the same purpose as landside quays.

Formal blockwork walls serving both as breakwaters and quays enclose many picturesque older and smaller harbours (Photo 7.11). In fine

Photo 7.11 Breakwater protecting Watchet Harbour against heavy seas.

weather, these elegant and robust features are now often regarded as much as stimulating amenities as functional structures. It should not be too difficult to provide some sort of paved public access along the top of modern rougher breakwaters.

Moorings and jetties

Solid or liquid cargo can be transferred from ships at deep anchored mooring (by barges or flexible pipelines) or at jetties (by fixed pipes or conveyor belts or on roll-on roll-off vehicles). Moorings need not interfere with seabed sediment flow; nor need piled jetties. The main visible impact of jetties is in their structures and any tall equipment which they carry; that of moorings is only the sight of the ships and lighters that tie up there.

Natural deepwater inlets such as Bantry Bay or Milford Haven, make ideal pipeline transfer points for tankers – from a mooring or at jetties. But perhaps neither of these examples has achieved its potential in this respect, being oil or gas terminals only for Ireland and Britain respectively. Tankers carrying oil to continental Europe could more conveniently discharge at Rotterdam or Le Havre, once those ports were able to accommodate deep draught vessels, than on the remote southwest coasts of Ireland and Wales.

Landside quays

Discharge or loading of boxes and machinery cannot easily take place on narrow jetties or at offshore moorings. Such 'general cargo' must be transferred directly to firm land surfaces by ship's derricks or quay cranes. Today most small items of cargo are packed into standard (20 or 40ft) containers. Containers have to be offloaded from the ship by very large cranes and then transferred by special carriers to be stacked up in extensive storage yards or loaded directly onto trains, road vehicles or other ships.

General cargo, containers and dry bulk goods are normally transferred at shoreline quays. These require wide handling space well above high water level and must be immediately adjacent to a depth of water sufficient for any ship to remain afloat at low water. The engineering problem is to provide structures offering this abrupt difference in level from dry land surface to sea bottom on a long straight line. In some situations, and particularly for smaller craft, vessels unload onto floating pontoons, connected by hinged rising and falling bridges to the fixed waterfront. More commonly the quay walls are fixed and their construction involves dredging on the waterside or reclamation of land up to the quay front.

Quay wall construction can be:

- of concrete or stone masonry blockwork built up on a secure and level foundation, preferably on rock but otherwise in firm ground excavated or dredged deeply enough to preclude subsidence due to scour;
- of hollow caissons floated into position and sunk precisely on firm foundations and then filled with concrete or rubble to add weight; or
- on piled foundations with pile caps incorporated into flat slab deck construction.

Most quay structures are too massive for removal. So the fixing of a new line of quays is decisive in determining the permanent shape of the adjacent landform.

Dredging

Dredging for berths or navigation channels may have considerable impact in shallow water habitat. But it is relevant to land resources mainly when the dredged material is used for land reclamation.

Reclamation of land from the sea

In steep topography, such as surrounds Hong Kong harbour, there has always been a demand for land space for long-term city construction. This involved the flattening of hills, in turn providing a substantial quantity of material for creating land in shallow water. Dredged sand or gravel can also be used for fill; so, with precautions, can finer soils if they are then drained, consolidated or strengthened. New land can be formed by placing any of these materials within banks of more robust material in shallow water. The reclaimed land can then be used for extra municipal space or even airports as well as for port facilities. Japan and the

Netherlands have made major increases to flat land areas by means of extensive reclamation.

However, the ecological value of shallow seabed, intertidal zones and riparian wetland is now widely recognized. Where this habitat is significant, land space creation is unlikely to be acceptable unless it can be shown that the seabed or riparian conditions can be replicated nearby.

The first stage of land reclamation in coastal water, after thorough site investigations, is construction of a perimeter embankment that will act as a breakwater and cofferdam during construction and as a permanent sea wall thereafter. Within the area enclosed, stages of work may then include:

- removal of any material likely to subside or to pollute subsequent fill; this may involve dredging or dewatering of suspect layers;
- placement, as fill, of hard or coarse material excavated elsewhere;
- fill of softer but more abundant materials like dredged silt in conjunction with adequate measures to extract or release excess water; and
- artificial consolidation of fill or periods of waiting for natural settlement between stages of fill.

In some materials, settlement may continue for years. The ideal form of pavements, buildings or informal space and of any drainage systems will depend primarily on the nature and long-term stability of the resulting landform.

Structural foundations on reclaimed land can be:

- a firm bedding layer for flat pavements such as airport runways or container storage yards;
- raft foundations for structures wide enough or flexible enough to permit modest differential settlement; or
- piles, either load bearing on their toes on rock or dense sand, or held by friction in cohesive soils.

Complete cities, some now of outstanding cultural and scenic repute, are largely the result of reclamation from the sea, for example Venice or St Petersburg.

At Al Khairan in southern Kuwait on the Persian Gulf a scheme is being undertaken to create a new residential seaside city. 'Creating it involves carrying out an interconnecting network of channels, each around 100m wide, from the salt marsh and desert, to make 340km of artificial coast' (Mylius, 2003).[10] A seemingly empty landscape has the fortunate attributes of some low variation in micro-topography including rock outcrops up to 15m high. The 'frond-like' canals and landforms are being 'developed on the basis of the existing topography', excavating from the low parts and building up around high ground. Excavated rock is used to construct groynes protecting the channels against tidal scour, other spoil is used for the high fingers of land, and sandy beaches are created 'by blowing out fines from excavated material'. Areas of relatively sheltered seabed are expected to become a haven for rich plant and fish life. Similar major residential land reclamation was taking place in other Gulf ports, such as Dubai, Abu Dhabi and Bahrain while early 21st century regional economic optimism remained. All these schemes demand careful planning to provide the long lengths of beach that are the intrinsic attractions as well as sufficient water circulation to maintain good quality sea water.

7.6 Coastal and hinterland development

Historically, people settled close to the sea because of the livelihoods offered at harbours by fishing and maritime trade. More recently, after the arrival of mechanized transport systems, people have tended to visit or settle in coastal areas because amenities, such as beaches and scenery, may be more attractive than are found further inland. The ratio of coastline to land area is a high one in Britain, making its island population relatively maritime in outlook. In Australia, most people live in towns near the sea even if much of its very long coastline is deserted. In US, nearly 30 per cent of the population live within ten miles of the coast (Berz, 2000).[11]

For centuries, inaccessibility or infertility discouraged settlement or over-exploitation of steep coasts, salt flats, lagoons or estuarial wetland, thus conserving some fine scenery or wildlife habitat. However, apart from occasional response to the changing needs of ocean-going ships, the concern of coastline planners has, for many decades now, been an attempt to control residential, commercial and infrastructure development that otherwise threatens to destroy the very amenities that are attractive. Conservation policies must be implemented for recognized heritage coastline; new ports must be planned in existing harbours or with great care if new sites are essential; seaside cities must develop with conservation of their best waterfront features; and new settlements must be kept back from the unspoilt coastal features that remain.

Heritage coastline

A high proportion of people live in coastal regions, perhaps partly attracted by fine coastal scenery. We must first protect from further spoliation the area defined as heritage coastline. The best way to do this is to keep all forms of built development and access, especially vehicle parking areas, at a sufficient distance from that coastline. This is usually no nearer than is already provided. Sometimes it should be further away.

British designated heritage coastline includes some settlements, such as harbour villages along the Cornish coast. Many Mediterranean harbours are similarly attractive. If the layout of buildings, streets and quays follows the traditional patterns, then these are the equivalent of the 'old city' quarters that are such a significant part of urban heritage (pp294–296). More remote, grander stretches of coastline are still natural landscape where there can seldom be any justification for built intrusion.

Meanwhile, in less rich countries, most people living in coastal provinces do not have the leisure or wealth to live other than on productive land or near city workplaces. However, some of them may soon be able to afford greater mobility. So exceptional scenic resources of wooded cliffs, coves and islands – for example, along part of the coasts of Vietnam or southeast China – may be at risk from speculative developers or trunk road planners unless official control is introduced before it is too late.

New port development

A common justification for new ports at new sites is that deeper draught ships must be accommodated. In historically developed regions this should leave vacant older port facilities land that are no longer adequate.

There are indeed many ports where a profusion of quays, railway sidings and warehouses long ago became redundant. With the decline of general cargo carriage in crates and sacks, shipping moved elsewhere to be accommodated by different berth formations and quayside cargo handling techniques. The wide land spaces left in redundant ports have been allocated to industrial, commercial or residential use, hopefully preserving the waterfront amenity. Some of that space has been assigned to landside activity associated with current forms of marine cargo transport and transfer. For freight, these activities may include conveyor, pumped and crane transfer of dry bulk (coal, ores, grain), liquids (crude oil, liquefied natural gas and petroleum products) and containers for direct transfer to other ships or modes of transport or for nearby open, shed, silo or tank storage.

There may also be roll-on roll-off ferry operations at the end of short flexible jetties, loading facilities for passenger cruise ships, marinas for smaller leisure craft berthing at a complex pattern of narrow jetties, or on-shore pumping stations at the end of pipes serving tankers moored in deep water. In London, some dockland has been converted for a short take-off and landing aircraft runway.

With specialized handling facilities for very high throughput, large vessels discharge rapidly and with a short berth occupancy. This leaves some waterfront space vacant at some of the older ports. Rather than selling all the land to commercial interests, there may be new opportunities for public amenity or conservation space.

In spite of extensive rationalization, there are still calls for new port locations where:

- there is deep enough water access for large deep draught vessels such as container ships;
- there is ample space for storage and related industrial production; and

- there can be connections to a well-developed inland transport system.

The freedom for spatial planning associated with greenfield sites is always commercially attractive. But opportunities for redevelopment of existing sites should always be considered first.

Seaside city development

Maximization of waterfront amenity value should be a prominent objective in the redevelopment of port areas. Especially where the front is already dedicated to the enjoyment of people, that value should be enhanced through the conservation or improvement of spacious promenades, shady avenues, ornamental gardens and views in and out of the city, the latter's architectural treasures contrasting with the masts and paraphernalia of ships and port equipment. Even strategically secure parts of the harbour need not be completely cut off from public interest. Where high port walls have created an opaque barrier in the past, there may be scope for redevelopment creating new frontage, access openings or vistas. The attraction of port frontage is not so much the sea itself – although many ports, such as Dover, incorporate a section of beach – rather it is the sight and proximity of maritime operations.

Waterfront space provides great opportunity for leisure. Inevitably in large cities it also carries routes of railways, tramways and roads. There are many options for accommodating, limiting or altering these to permit pedestrians the choice of a number of safe corridors including one immediately beside the shore. On city waterfronts, road vehicle parking space is at best a luxury, at worse an abominable intrusion.

General settlement spread along coastal hinterland

The beauty of coastal landscape is appreciated in views along the shoreline, from the top of cliffs or from beaches or boats. Building cliff-top houses or hotels for people to live in and to privately enjoy the views (often of an empty sea) spoils the finer prospects of the cliffs from below. Properties that already encroach on coastal landscape may not deserve costly protection against natural forces such as erosion of their foundations. In some Portuguese resorts, houses have been built on top of or half-way up the modest cliffs that are often the main scenic features of the coast. Elsewhere, the houses themselves have in due course been destroyed by landslides and cliff erosion, or the cliff's natural beauty has been spoilt by concrete defences against erosion which is inevitable there or further along the coast.

Designated heritage coastlines are exceptional perimeter features at the edge of areas of fine scenery; the whole strip is worthy of conservation as far inland as historic development leaves practicable. Port city hinterland can be improved as certain harbour activities are reduced or rationalized. Where new ports are proven necessary, the plans and investment should include expertly conceived mitigation measures increasing the net long-term richness of an even more complex landscape. Seaside cities have, in and around their harbours, assets in which maritime activity, popular leisure and architectural splendour can be blended together.

On coastal strips that are neither striking enough to be specially protected nor practicable for port operations, there remain risks that buildings and roads may take over; such development must be firmly contained behind boundaries to prevent it imposing on the semi-natural coastline or complicating the fragile stability of its geology and vegetation. Where structures already exist in these fortunate situations, let their owners pay for the privilege or share it in some way that could fulfil the opportunity of conserving the total landscape and the enjoyment it engenders.

Thus, for future planning of construction, the case of coastal land typifies how:

- outstanding scenery should be conserved perpetually;
- development on flat land, for example of possible new ports, can be planned as to how and where this might best take place, but can then be deferred until there is an overwhelming need; and
- city environments should take maximum advantage of features such as waterfronts or steep background.

These are the subjects of Chapters 13, 14 and 15 respectively in Part III.

Notes and references

1 Countryside Commission (1970) *The Coastal Heritage*, Her Majesty's Stationery Office (HMSO), London.

2 Burrows, R. (1971) *The Naturalist's Devon and Cornwall*, David & Charles, Newton Abbot.

3 Deltas are substantial only at the mouth of greater rivers than those of the UK. Some deltas may act as filters sifting out river borne toxins and algae arising upstream. Drainage of the Danube Delta destroyed this filter threatening the protected life of the Black Sea – see Martin, J. (2006) *The Meaning of the 21st Century*, Eden Project Books, London, p31.

4 Countryside Commission (1970), as Note 1.

5 Substantial areas of polders also exist in the UK – in the Fens of East Anglia, the Somerset Levels and the Gwent Levels in South Wales. The lowest point is thought to be Holme Fen in Cambridgeshire, nearly 3m below mean sea level.

6 Payne, H. R. (2003) 'Sustainable coastal defence', *Engineering Sustainability*, Issue ES3.

7 Coker, A. (1992) 'A case study of Hengistbury Head', in Coker, A. and Richards, C. (eds) *Valuing the Environment*, Belhaven, London.

8 Cooper, N. J. (2003) 'The use of "managed retreat" in coastal engineering', *Engineering Sustainability*, Issue ES2.

9 Mabey, R. (2003) 'Countryside futures', *Countryside Voice*, summer 2003, p22.

10 Mylius, A. (2003) 'Desert pearl', *New Civil Engineer*, 8 May 2003.

11 Berz, G. (2000) 'Flood disasters: Lessons from the past', *Water and Maritime Engineering*, March 2000.

8

Transport Routes and Infrastructure

Roads, railways and canals promote travel and trade, improving access to places and facilitating economic development. They also take up land space, adapt landforms, create or sever local connections and alter drainage patterns.

Transport affects scenery close at hand, by the appearance of route features and vehicles or by abrupt intrusion in confined but spectacular scenery such as gorges. Distant views can be affected by earthworks and structures. New routes influence the occurrence of flora and fauna, destroying habitat along the alignment or creating it in new boundary reserves.

In urban areas, streets are principal components in a topography originally defined by watercourses and subsequently by property boundaries, access routes and longer distance canals and railways. Port construction has modified waterfronts on estuaries and coasts. More recently, airports have taken wide areas from other land use.

The first section of this chapter reviews briefly the engineering, operational and visual features of different forms of road access and then of each other mode of transport. Section 8.2 deals with the implications on land use of route selection and alignment of linear infrastructure, Section 8.3 with those of junctions and activity at the nodes of routes. Section 8.4 summarizes how the economic benefits of transport and the pleasures of travel should be balanced against impacts on land and human or biotic communities.

8.1 Transport networks

Rural and cross-country road access

Any rural areas of the world still not connected by roads retain a landscape, especially in regard to human settlement, that is reminiscent of a bygone age. Survival is precarious for such remote communities that have to strive increasingly hard to compete economically with those that are better served. Construction of roads gives access to new technical and commercial opportunities, but also to exploitation and potential spoliation of traditional lifestyles and landscape.

This author travelled in the Elburz Mountains of Iran in 1957. Access to the high valley settlements was on foot from the few road heads that then existed. Horses carried baggage and agricultural produce. Eight years later, similar exploration in the Zagros Mountains of the same country was undertaken by Land Rover using tracks that had by then been prepared for lorries to reach anywhere where crops were grown.

In other more arid tracts of the Middle East, rough roads into the hills follow the course of dry river (wadi) beds. The routes have to be cleared of rocks afresh after each occasional spate of flow; and there is little permanent impact on the scenery. The road route from Port Sudan into the Red Sea Hills and thence to inland regions of Sudan followed a long narrow defile until a paved highway was constructed in the 1970s. However, the old routes had very low traffic capacity and high vehicle operating costs compared with the permanent roads that inevitably replaced them.

To negotiate steep territory the need for earthworks and structures – retaining walls, bridges and drainage culverts – becomes paramount. In remote areas, some roads are first built to provide access for heavy equipment to mines or construction projects such as dams. These roads then enable all forms of development – welcome or otherwise – to take place in the regions they serve.

In flat country, motor vehicles can take almost any clear dry routes. In much of semi-arid or savannah Africa, drivers navigate by distant landmarks or a compass or follow previous wheel tracks. However, only permanent low embankments can overcome seasonal flooding; and rivers can be crossed only by bridges or ferries. Mud and rutting can be avoided by a rigid surface of fitted road stone on a permeable sand base. Gravel is a well-drained material that for long formed the predominant surface for rural main roads and still does if traffic is light. But only sealed surfaces can avoid wet weather softness or dry weather dust.

Gravel surfaces can be assimilated into the scenery. They require frequent grading, not necessary for more costly, blacktop sealed roads. But even the latter require periodic renewal; if paved surfaces are allowed to deteriorate under traffic, they can become much more difficult to negotiate or to repair than gravel roads.

City and suburban streets

Town centre space was once primarily for pedestrians and some towns are now reverting to that concept. The open ground of squares, small parks and wide sidewalks is essential foreground in city views. Congestion of streets by vehicles was started by horse-drawn traffic. Today, access for motor vehicles and especially private cars is often restricted in city centres. However some city streets still accommodate trams (Photo 8.1) and occasionally even trains (Photo 8.2).

Suburban areas have to accommodate both through routes and residential streets. Photo 8.3

Photo 8.1 Tramway on city street, Prague.

Photo 8.2 Mixed mode traffic situation, between the pedestrianized city centre and a ring road in the small city of Canterbury.

Photo 8.3 The Kingston Bypass, southwest London, was completed in 1927 and immediately attracted developers who built roads and houses wherever access had been provided. After 1935, such ribbon development was not allowed. Features in this 1996 picture are: the suburban estate and 'service' road to the left, with cars parked in the street or in front garden spaces; limited access between the service road and the highway; the highway itself, still serving as a bypass but subject to heavy peak hour congestion; and a footbridge (from which this picture was taken) connecting the communities on either side.

shows an early example of a scheme which attempts to do both, not entirely unsuccessfully. Modern urban throughways and inner ring roads are usually isolated from their immediate surroundings through which they cut wide swathes of land and sever structural coherence. In some parts of extended cities, such as Los Angeles, through motor routes so dominate the land space as to make pedestrian travel impracticable. Residential streets, shopping areas and industrial estates are greener (with more planted space) where land is cheaper in the suburbs. Buildings, gardens and yards, sidewalks, roads and vehicles are main elements of a man-made landscape in which managed nature is allowed.

Highways

The Romans and the Incas built robust roads. But most medieval and subsequent main roads were unpaved until the 18th and 19th century improvements by Trésaguet in France and then by Telford and McAdam in Britain. Major roads, on substantial earthworks and structures and with gravel or hard surfaces, were widely constructed throughout the 20th century to accommodate motor traffic. The influence on the landscape of early main roads was largely indirect and concerned the construction and activity which took place beside them and at the centres which they served.

Arterial roads, autobahns and motorways built since the 1930s were the out-of-town solution to increases in road traffic. Characteristics of modern major highways are:

- multilane carriageways, permitting speed without risk of head-on collisions;
- restricted access, severing land that is cut through by the highway yet denying its use to adjacent communities not near to interchanges; and
- grade-separated (two-level) interchanges which require extra space for high speed slip roads and bridges.

The width and direct course of high capacity, high-speed roads can result in visual scars or severed landscapes, unless constructive design achieves a blending of new with existing features. In regions such as Western Europe, where substantial motorway networks have already been constructed, recent policy when demand continues to increase has been to widen these roads – if possible within their existing acquired space – rather than to build new ones (Photos 8.4–8.7). This is laudable in terms of limiting land take but can still be costly

Photo 8.4 A four-lane section of the M40 motorway in Buckinghamshire in 1990 ...

Photo 8.5 ... was later upgraded to six lanes with no additional land-take ...

Photo 8.6 ... but the 1990 overbridges ...

Photo 8.7 ... had to be completely replaced.

unless adequate allowance for widening has been made in the original plan.

Inland waterways

Navigable rivers and canals were much more effective for inland delivery of goods than animal-drawn carts had been.

Because of the horizontal course that navigation requires, waterways need locks to raise or lower vessels in their passage through undulating territory. Channels for large vessels can be a considerable barrier to land movements. Roads or railways must cross over them by high bridges or under them through tunnels.

Although railways have played the major role in carrying inland freight since the mid-19th century, some major rivers are also still important transport routes. The Rhine, Elbe and Danube and their connecting canals are European examples (Photos 8.8–8.10) and there are others in the Far East and North America. The landscape of a constructed or controlled waterway typically entails:

- on the larger canalized rivers, stable banks, sometimes doubling as retaining walls for railways, roads or promenades;
- on narrower canals, such as those now devoted to leisure in England, one 'tow-path' bank for access by people and an opposite less accessible bank suited to aquatic and riparian wildlife; and
- often, fine waterside scenery.

Photo 8.8 Container yard by the River Elbe.

Photo 8.9 Container ship on the River Rhine.

Photo 8.10 Waterfront in Friesland, Netherlands.

Main line railways

By 1800 in Britain, canals were the cheapest form of industrial transport, but they were severely limited as to speed and points of delivery. From the 1830s, railways rapidly took over the role. Iron tracks afforded greater speed, could be laid almost everywhere and coal-fed steam locomotives could haul trains up modest gradients. However, unlike contemporary roads, the alignment of railways required earthworks – cuttings or tunnels, which were hidden from sight, and embankments or viaducts, which blocked views or presented structural achievement according to one's perspective. Early railways set the pattern for nearby land development. More recent high-speed lines have required straighter alignments, presenting new routing difficulties. All differ from highways in that train movements are completely controlled and scheduled. The principal purposes of 21st century main line railways are:

- long distance freight, such as US double-stacked container trains or southern hemisphere iron ore traffic;
- high speed railways for passenger trains travelling at speeds upwards of 250km per hour; and
- mixed traffic (freight and passenger) lines, common in Europe, which can achieve high route capacity by four-track (fast and slow) configurations or by running all trains at 120km per hour or faster.

Urban railways

Steam railways and – by the 20th century – electric ones, encouraged commuting to work and promoted residential suburbs, types of landscape in their own right. Most city metro railways are underground, thereby relieving what might otherwise be excessive road traffic and street congestion. Further out, suburban railways are usually at ground level. Some urban railways were originally built, and some remain, elevated; others have more recently been constructed above road level in recent developments such as in the London Docklands (Photo 8.11) or new residential areas of Hong Kong.

Photo 8.11 An elevated metro railway giving good views of redeveloped dockland area, London.

In all these circumstances, it is the high capacity of railway services that is their attraction in passenger transport planning. Two-track metro railways are commonly provided where 15,000 passengers can be expected in either direction in a peak hour; and maximum throughputs of 30,000–60,000 per hour have been exceeded. In contrast, the capacity of each lane of even grade-separated highways seldom exceeds 5000 people per hour in combinations of nearly full buses and cars.

Aqueducts and pipelines

Romans could build inverted syphons – pipes that could contain hydraulic pressure across the bottom of a valley sufficient to compel flow up the other side. However, most Roman water supply channels maintained the slight gradient sufficient for gravity flow by following the land surface, cutting or tunnelling through high points, and constructing impressive structures across intervening valleys (Photos 6.1 and 6.2). These structures took the name 'aqueduct', a word intended originally and still used for the whole length of constructed water channels – open or in pipeline – for example, up to 99km long into Rome and later much further from the Owens or Hoover dams to Los Angeles. The word viaduct was adopted first for multi-arch railway or road bridges of a nature similar to Roman aqueducts of this type.

Pipelines replaced open channels, where necessary, after it became possible to pump water and to maintain high pressure in gravity flow. They have been increasingly used for bulk transfer of oil and gas. Pipelines are usually buried except where conditions – such as rocky ground or permafrost – make surface installation and operation more economic. They can also be buried across ground in sensitive landscape. Routinely buried pipelines, such as for water and gas in Britain, show no superficial evidence of their existence except along vistas where trees have been cut for construction and maintenance corridors.

Other transport modes

Marine transport relates to landscape at ports, described in the previous chapter. Transfer of electric power by transmission lines is mentioned in Chapter 11.

Air traffic can be observed in a variety of situations – from commercial aircraft on established flight paths to military aircraft manoeuvres and private or leisure flying. These are mildly interesting spectacles divorced from serious landscape consideration. Airports, however, take up considerable portions of flat land space. Much has been converted from agricultural or other productive use; some is created by land reclamation in shallow water. The main space, for which sites of a suitable configuration have to be sought, is for runways and aircraft ground movements. The total area is usually much greater than the operational minimum, to provide for emergency space as well as buildings and connecting land transport systems.

8.2 Route selection, alignment and design

This section concerns the planning of linear routes. In a modern context, new routes are often multilane highways or high-speed railways. Where there is a need for these, a broad route band has to be selected, within which precise alignments can be devised, satisfactory for the rapid flow of traffic and as compatible as possible with existing land use and scenery. Relevant to both broad route selection and precise alignment are the effects on drainage, the concept of transport corridors, and strategies for adjusting the impact on the surrounding landscape.

Broad route selection

Organized communities created cities. Mercantile civilizations extended the synergy of organization by establishing communications and trade between cities. Hence, the inhabited parts of the world were soon covered by a network of roads or waterways for the passage of pedestrians, horse-riders and animal-hauled vehicles or boats. After the invention of steam and petrol engines, not only did freight become faster and cheaper but people began to travel further. The extra capacity in land transport infrastructure was provided by railways, sometimes following the traditional road or waterway routes because these already recognized landform opportunities, but usually taking new alignments because of limitations on gradients and curvature and to provide a reserved space for trains.

Fundamental to any new transport link is its intended function. If a highway, is it to be for a specific connection between main centres, or is there to be intermediate access at frequent junctions? If access is limited, what are the beginning and end points and where are any intermediate connections to be made? If access is freely accessible but in congested areas, are any restrictions to be placed on the vehicles permitted to use the road? For a railway, what will be the mixture of fast and stopping passenger and freight trains on how many tracks? For roads, how many lanes and is traffic to be primarily long distance or local? The answers determine width and alignment parameters for road or railway formation and whether or how these can be accommodated in particular topographies.

Across hilly land, routes have to make optimum use of ground contours, following valley courses or climbing over or through intervening ridges. Excavation of cuttings and construction of embankments enable track or pavement to overcome surface variations; valleys may be crossed on bridges or viaducts (Photo 8.12). In steep country, train operating speeds may have to be reduced over certain sections. Or, as in Switzerland, there may be justification for construction of more direct railway routes in long 'base tunnels' deep under whole mountain ranges – for trains carrying passengers, freight and road vehicles.

In flat or gently undulating territory, there may be a choice of routes within broader bands of land between the end points. Specific routes can then be chosen to avoid:

- certain settlements or intermediate towns – in the distant past, this was to spare their inhabitants the depredations of armies living off the land, in modern times it is to provide

Photo 8.12 The A21 trunk road runs on a viaduct and through a tunnel to negotiate the steep chalk downland between Folkestone and Dover.

bypasses for traffic that would congest the built-up areas; and

- areas of unspoilt, tranquil or scenic character, of significant scientific or cultural value, or property of high commercial value or strategic location.

There may have to be compromises, for example by choosing longer or slower route options to preserve conservation, heritage or recreational land, or by creating a bypass through previously unspoilt landscape. At Winchester, the 1930s bypass was the heavily congested slow option before a nearby missing section of the M3 motorway was completed. The former was closed and returned to nature (Photos 8.13 and 8.14) while the latter

Photo 8.13 In the bottom foreground is a strip of open space where there had been the heavily trafficked and congested Winchester bypass that was replaced by the M3 through Twyford Down. In the trees is a long-abandoned railway line and the River Itchen. Beyond are water meadows in a total area that encompasses a wetland SSSI and recreational walking land for citizens of Winchester meadows.

Photo 8.14 Looking in the opposite direction is the now reconnected path up the scenic face of Twyford Down, the motorway cutting being concealed in this view.

made a controversial deep cutting, completed in 1994, through part of Twyford Down (see Colour Plate 11). Some of the cut land was part of a chalkland Site of Special Scientific Interest (SSSI), but much of it was ploughed land. Perhaps the controversy lies near the border between economic justification for new roads and allowing congestion to limit traffic.

For protecting outstanding landscape, it may be necessary to avoid a whole region, such as the Lake District of northern England. The routes over Shap Fell of the 1840s Lancaster and Carlisle Railway (now the West Coast Main Line, Photo 8.15) and the 1960s M6 motorway are through bleak but quietly impressive upland scenery, seen from a train or the Tebay motorway services area, in contrast to the outstanding beauty of lakes, verdant or rocky hillsides and picturesque settlements in the landscape jealously guarded by John Ruskin and the founders of the National Trust.

By contrast, there are other railways that took any available route rather than negotiating steep country. What seemed the most practicable routes at the time were along coasts, lakes or river valleys whose beauty was not publicly recognized until it was seen from trains. The coast of Devon near Dawlish (Photo 7.6) and several points on the Mediterranean Riviera are examples. Meanwhile, in modern times, a new route may still spoil established beauty. A road bridge across the Elbe Valley near Dresden has lost that section of the river its World Heritage status (NCE, 2009).[1]

In the past there was a wide choice of routes between city end points. For example, 100 years ago there were three main railway routes between Liverpool and Manchester and three from Edinburgh to Glasgow. In each case, two of these routes are still operating today, serving much denser populations; and new motorway routes have been constructed through whatever space was still available. For the Channel Tunnel Rail Link between Folkestone and London, four broad route options were identified in the 1980s, of which one was then selected for further investigation in 1989. Eventually yet another option was taken for most of the route on a more northerly course crossing under the Thames Estuary. Although the route was on a very high speed alignment, a major part of it was in fact close to existing railway or more recent

Photo 8.15 The main line railway and, later, the motorway between Lancaster and Carlisle chose a high route over Shap Fell, partly to avoid the Lake District. In this region of open mostly grazed moorland, patterns of wildlife habitat are on the railway cutting slopes as much as in the more accessible semi-natural landscape.

motorway routes, so within already established transport corridors.

Within any broad route option, particular difficulties and opportunities – concerning, for example, river crossings, junctions and geographical features – become evident as specific alignments are examined.

Precise alignment

The alignment of highways and railways has to meet operational requirements as to the width of formation and limiting gradients and curvature. Within the geometrical options satisfying these criteria, the chosen course has then to avoid or minimize damage or disruption to the landscape. Finally, iterative adjustments can be made to find the most satisfactory solution.

Width of formation is determined by the number of road lanes or railway tracks. Typically the ballasted width of two-track and four-track railway 'permanent way' is 12m and 20m respectively, while a six-lane dual carriageway highway pavement extends over 30m (Photo 8.16). To these widths should be added the land area covered by the slopes of embankments or cuttings unless these are eliminated by construction of viaducts or tunnels or reduced by retaining walls. It is also generally desirable to maximize the width of land that can be secured and held in reserve. Then it may be possible to:

- keep houses well back from transport routes (typically 50m for railways);
- make space available for wide verges and slopes – for fenced-off wildlife or accessible public amenity and for better views for passengers; and
- allow space for future widening or adjustment of alignments.

Gradients as steep as 1 per cent (1 in 100) were a formidable if routine undertaking for steam-hauled trains on the British West Coast Main

Photo 8.16 A transport corridor – the Midland Main Line and the M1 motorway built about a century later.

Line; but electrification and substantially increased tractive power made light of these gradients. In more mountainous regions of the world, long climbs with gradients as steep as 3 per cent are tackled by freight trains with multiple locomotives. French Train à Grande Vitesse (TGV) high-speed trains and German Inter-City Express passenger trains can, by their momentum, maintain 250km per hour up short gradients as steep as 4 per cent. Gradients up to 3–5 per cent are routine (albeit with crawler lanes for heavy lorries) for motorways or US interstate highways in hilly country over lengths that would not normally be practicable for fast or heavy trains. Mountain railways can use track-and-pinion (rack) mechanisms for gradients steeper than about 7 per cent, but these are essentially low speed, moderate capacity lines, usually for carrying tourists through splendid scenery.

Horizontal alignment – concerning curvature and changes in direction – is also easier for road vehicles than for trains. Indeed, straight sections of highway are considered to be physically and visually tiresome for drivers. Whereas visually unobstructed 500m radius curves can be taken safely and routinely by motor vehicles without skidding at 112km per hour, trains require an 800m curve radius and a practicable degree of outer track superelevation to remain comfortably stable at the same speed. For higher train speeds, comfortable curve radii are at least 1000m radius at 125km per hour and 4000m at 300km per hour. Trains can run along these curves still safely at somewhat higher speeds and 'tilting' trains are supposed to correct for any discomfort that might then be felt by passengers. Photo 8.17 illustrates the alignment differences between modern motorways and high speed railways.

Land surface contours first determine the degrees of gradient and curvature that are feasible on any route. But avoidance of valued man-made or semi-natural features become as significant in more precise alignment. Reduction in operating standards is a possible means of solving particular difficulties. However, on a railway this would be a last resort if it entails speed reduction. Such is the momentum of a high speed train and the distance needed for braking or acceleration that

Photo 8.17 The two carriageways of the M2 motorway and the straight line of the Channel Tunnel Rail Link cross the River Medway on adjacent bridges. The railway runs directly up to a tunnel through the North Downs while the motorway takes a curved course up to a summit cutting. Lesser highways cross at steeper gradients.

any speed reduction, such as to allow increased curvature, increases journey time over a much longer distance than that where the restriction is necessary. It might be as economic to reduce the speed standard for the whole line or a major section of it. On roads also, reductions in standards are seldom popular with highway authorities. But reductions in speed limits or the width of lanes are perfectly practical given ample notice to drivers. The ability of motor vehicles to slow down or accelerate rapidly gives much more flexibility than is available to trains in similar situations.

Lateral variations of high speed alignments can have considerable longitudinal repercussions. For example, if curves on a high speed railway must be at least 4000m in radius then to move a section of straight alignment sideways by 100m will involve about 3km length of adjustment, taking into account straight-to-curve transition sections at each end of the circular diversion. Priorities may have to be set as to what should be avoided. Where avoidance is impracticable then the options, for example for heritage buildings, may be:

- to protect, by a retaining wall, all or some of structures at the edge of the new construction, or to incorporate them within the new route, such as between carriageways;
- to move complete structures to a suitable setting elsewhere (see p8); or
- to make records of a structure and then dismantle or bury it.

Figure 8.1 indicates the issues in alignment of, particularly high speed, trains. Parameters for roads are generally less strict because of the manoeuvrability of road vehicles.

Drainage and formation

Road and railway bridges over rivers are major structures in their own right (see Chapter 9). Lesser underbridges and culverts have to be provided over smaller streams and for drainage of local storm run-off. Thus, cross-drainage structures are main features of roads and railways through steep country, as are retaining walls and side slopes. Across undulating country, traditional route construction roughly balances fill against cut. As much earth is laid above natural ground level as is excavated from below it. Where routes are below ground level, in cuttings or tunnels, then some run-off must inevitably drain along trackside channels or through tunnels until it can be released to main drainage outlets.

On level ground, railways and permanent roads are generally built on low (1–2m) embankments, made up of soil excavated from adjacent borrow pits which frequently become lateral drainage ditches. On main roads across flat land with a high water table, these borrow pits become ponds, welcome

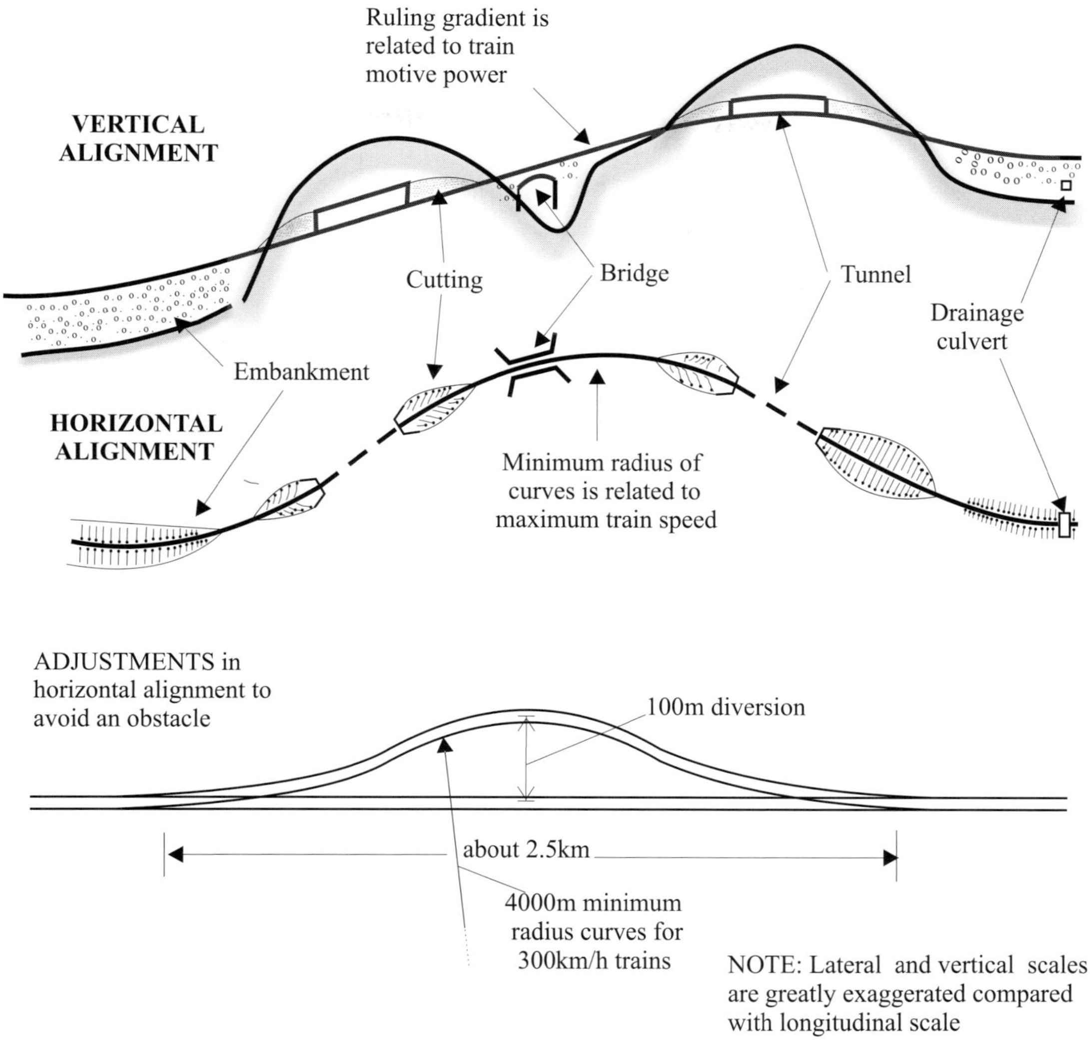

Figure 8.1 Alignment of railways

habitat for domestic ducks or wilder creatures. On marshy ground, transport embankments can alter wetland quality. Low-lying routes can choose different levels according to whether or not they need to climb out of the valley (Photo 8.18).

Highways and railways must rise above any seasonal inundation; but minor roads across flat land may be flooded for short periods after heavy rain. There are exceptional circumstances, however, in which even main roads might be used to relieve flooding. Sparks (2003)[2] refers to a tunnel, part of a Kuala Lumpur city relief road, which can be closed to traffic and used to carry stormwater after exceptional rain storms.

Transport or development corridors

Narrow valleys with special resources such as water power or coal were natural focuses for industry and settlement. These became development corridors through which roads and railways continued to be used or were upgraded for through traffic even after the original local resources were no longer relevant. In steep landscape beside lakes or rivers, railways, public promenades and dwellings vied for waterside space while towns expanded up the slopes or across less steep hinterland (Photos 8.19–8.21).

In more open, flatter territory, paved roads – intended for traffic between towns – also

Photo 8.18 Alternative railway routes – across level riverside ground or rising on a viaduct – near Banbury, Oxfordshire.

Photo 8.19 While the railway, main and local roads and a pedestrian promenade squeeze into Lake Geneva's shore development, a motorway cannot be discreetly hidden in the steep scenery above. However, construction of separate viaducts for each carriageway has made what some might accept as a strip of contrasting scenery rather than serious spoliation of steep wooded slopes.

Photo 8.20 A railway/road/waterway corridor by the River Labe (Elbe) in the Czech Republic ...

Photo 8.21 ... and the River Meuse in Belgium.

encouraged roadside 'ribbon development' as well as wider settlement around junctions or crossroads. In medieval times – or as late as the 18th century in stretches of country where paved ways had not been constructed – the passage of wagons and coaches on soft ground created ruts and sloughs. These could be avoided only by following new courses within a wider route corridor that might anyway best be cleared to a considerable width to make ambush more difficult for lurking highwaymen. Some of these wide rights-of-way proved to be valuable land assets in the 20th century, as did some abandoned railways.

In the second half of the 20th century, numerous motorways and high-speed railways were constructed and in many countries this construction continues. These routes take up land space, often of considerable commercial or amenity value; land units may be severed so that either special connections have to be constructed or land ownership boundaries and related activities have to be reorganized. There are opportunities to reduce the effect of greenfield land-take by creating routes adjacent to those that already exist. However, there are obstacles in the form of features that the original routes had carefully skirted past, or of built development that has taken place along these routes or at their junctions; also precisely adjacent routes may not always be possible, for example because full speed alignment parameters are more severe for railways than for roads.

So needs arise for skilful alignment of new routes within existing transport corridors and in planning for use of incidental space. That space comprises:

- islands between routes or carriageways; even if isolated, such space may still provide habitat for birds or small mammals; and
- land required as a buffer between a railway line and buildings but within which a less than tranquil but pleasant footpath or narrow park area might be fitted; or where a wildlife corridor could be practicable.

Waterways, railways and roads define the hard boundaries of corridors. Construction of connections *across* corridors may be the key to coexistence of transport, pedestrians and nearby residents.

Design for aesthetic enhancement

In *views from afar*, it is hoped that visible transport routes merge into the surrounding landscape. Tracks or paving can probably not be seen, but the earthworks of wide dual carriageway highways may be hard to hide. Visible traffic may be the obvious indicator, depending on the number, size and colour of vehicles and the extent to which sight of them is obscured by vegetation. Blatant blemishes may be notches on the skyline, if a route cuts directly through an escarpment, or scars on hillsides, where cut faces have not weathered or excavated debris has not been removed. Subtly discordant may be discontinuities in adjacent earth shapes or field patterns. In some circumstances, it may be possible to shape earthwork in a manner resembling undisturbed ground. More often, discontinuities can be concealed by trees or bushes. The choice of trees and siting and shape of additional woodland can help to achieve a suitable juxtaposition in semi-natural scenery.

Closer views, from buildings, nearby open space or vehicles on the route itself, are characterized by the appearance of overbridges, cutting faces or growth on verges. Railway companies and highway authorities should implement objective policies for planting and management of vegetation on green verges. Wild areas and formal flower patterns both have their place, as does well-designed detail of retaining walls or slope stabilizing structures. 'Furniture', such as road signs, crash barriers or railway overhead electrification equipment, can be provided from a range of alternatives. Lighting on highways, where it is necessary, can also be chosen so as to minimize its effect on welcome darkness in surrounding areas.

8.3 Land at transport nodes

Transport nodes are places where routes join, cross each other or terminate. Some are intermediate junctions where moving traffic changes direction. Others are areas where people (at railway stations), goods (at freight depots), vehicles (at bus terminals or car parks) or all three plus aircraft (at airports) have to be loaded or unloaded, stored or serviced and transfers made to other modes of movement. There is more choice as to the layout and sometimes as to the location than in linear route alignment but the total land areas covered at each node can be considerable. The impact on the built or semi-natural landscape depends largely on how operational layouts fit onto available land and how ancillary functions can then be fulfilled in any remaining space.

Junctions

Where traffic is slow and closely controllable, junctions may occur at abrupt angles. Slow moving trains can negotiate tight curves where lines join, often outside main city stations; or road traffic can turn through a right angle at signalled or clear priority intersections.

Fast bifurcations on two-track railways incorporate a flyover bridge in one direction, a spur on the other, so that there is no possibility of trains in one direction crossing the tracks used in the other. The amount of land needed for such connections does not add much to that of the linear routes concerned. Rather more land is involved if there are connections in more than one direction, for example, to and from Brussels, Paris or Calais at the TGV junction south of Lille.

Motorway grade-separated junctions can achieve similar one-way connections with tighter curves (Photo 8.22); but road users expect to be able to take any direction out of an intersection; so cloverleaf and other fascinating structural solutions arise (Photos 8.23a and b).

There is functional beauty in the curves and pattern of junction spurs and bridges. A car driver is aware of ease in manoeuvring smoothly and pleasantly from one route to another in light traffic and can appreciate the architecture of bridges seen from beneath. The land space enclosed by the various spurs provides opportunities for earth shaping, planting, watercourse or drainage features and possibly for productive or conservation

Photo 8.22 High speed road junction at approach to a bridge at Coimbra, Portugal (photo by Robin Carpenter).

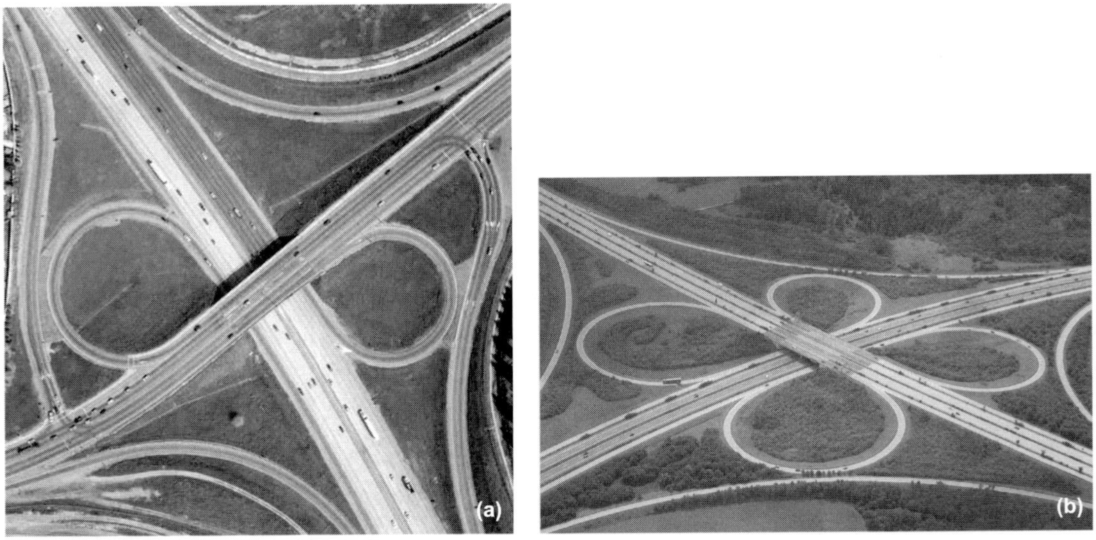

Photo 8.23a, b Alternative highway grade-separated intersection patterns (both photos from www.istockphoto.com).

purposes. This incidental space can also be useful during construction for temporary location of equipment or stockpiles for which there is not always room along linear projects. The same is true of any land adjacent to the alignment that may be allotted subsequently to railway maintenance or rolling stock depots or to motorway service areas or toll plazas.

Large railway stations

Nineteenth-century travel heritage is exemplified by the continental city central station or '*hauptbahnhof*' (Photo 8.24). In operational terms, these are important junctions or terminals sporting numerous long platforms. The layouts were planned for complicated locomotive movements and train assembly rather than the much simpler operation of modern reversible electric or diesel multiple-unit trains. Meanwhile, in human terms, stations remain places of gathering, whether on the spacious concourses or in the more intimate confinement of railway carriages.

Location of city stations was most often at the edge of a mid-19th century, densely built-up nucleus, such as at Prague, or adjacent to the medieval walls, such as at York. In large cities, different main line terminus stations might be located just outside a defined periphery line, such as what became London's underground Circle Line. Many main stations have become hubs of integrated city transport.

Extensive space was once needed for tracks, platforms and public concourses and also for goods depots and rolling stock storage yards, although the latter were often moved further out as the volume of traffic and land values increased. Rationalization of the configuration of trains and stations has made some platform space and even more rolling stock storage sidings redundant. Any space released can be used for green areas, bus termini, car parks or buildings, according to city priorities.

The outside appearance of stations can be appreciated from open foreground space. Frontages – of railway offices or hotels – sometimes achieve exotic splendour in facades and spires such as those displayed at London St Pancras, Mumbai or Kuala Lumpur. However, overall, often semi-cylindrical, roofs are best seen from inside the stations, as at Cologne or York, where the spaciousness they afford is best appreciated. From outside, most of them are obscured by buildings and, where they can be seen such, as from the harbour at Amsterdam, resemble 'train sheds', the functional name given to them in Britain by their Victorian designers.

Railway stations are by no means universally recognized as paragons of city architecture.

Photo 8.24 Prague Central Station retains basic 19th century layout with extensions from the original high-roofed train shed to additional platforms on the right. The formal station entrance buildings (centre) are a handsome end to a vista stretching down the long Wenceslas Square to the old city.

Indeed, some citizens are now hardly aware of their existence. Unfortunately their better features have often been replaced or obscured by less splendid structures in front or on top. New York's magnificent Pennsylvania Railroad station was demolished in 1960, before the architectural heritage value of such buildings was recognized. But, by 1978, Grand Central Station in the same city was declared immortal by a Landmarks Reservation Committee set up in 1975.[3] Characterless office blocks and shopping centres were built in 1969 above what became dismal dark platforms at Birmingham New Street. Hopefully current redevelopment will be inspired by the magnificent transformations that have been achieved at London's Liverpool Street and St Pancras stations.

Freight transfer yards

Railway freight sidings and marshalling yards are mostly history. 'Wagon-load' freight was a system whereby individual railway trucks were loaded up at a factory or port siding. They were sent via a number of sorting points and as components of different trains until they were shunted off at a particular destination. On each of ten main lines out of London there was once such a yard covering 20–30ha. Today's far fewer railway freight-sorting yards tend to occupy somewhat more space. The relevance of redundant yards is in how that space is converted – to modern transport use, to building development or to more exhilarating landscape. Freight handling facilities at ports have been introduced in Chapter 7. There are also 'inland container depots', typically covering about 10ha, where train or lorry-conveyed containers are transferred or stored for short periods. Otherwise, loading and unloading of vehicles, containers or bulk cargos takes place on the land space provided at mining, industrial or retail sites.

Planning of most port and inland freight or vehicle transfer yards concerns land space rather than visible landscape features. However, wagon

Photo 8.25 Cheriton rail terminal for road vehicles using Eurotunnel services between England and France.

shunting, container crane operations and roll-on roll-off ship transfers can be viewed with interest from any bridge across the yard or from adjacent high ground (Photo 8.25).

Airports

Whereas most railway yards cover less than 50ha, the land space required for international airports typically exceeds 1000ha. Aircraft maintenance hangars, passenger terminals and parking areas, although individually large or extensive, are equivalent to similar facilities for surface transport; but, since they fit mainly into otherwise open space, they more resemble an industrial complex placed in a rural or suburban setting than the stations or freight depots that have been squeezed into cities.

Where airfields are assigned for occasional or emergency use many of the usual ground facilities may not be necessary (Photo 8.26).

Considerable buffer space has to be kept clear up to a certain distance on the verges of aircraft movement areas; but beyond that, where there is less risks of emergencies, there can be space for vehicle parking or service areas, water or sewage treatment works. Nature conservation, agriculture or horticulture can be practised on the rural sides of the airfield, both within or outside the perimeter, always providing that structures and vegetation do not impair aircraft operations.

Reshaping the land is necessary to provide appropriate formation levels, first for runways and then for other pavements or structures. The orientation of runways depends upon prevailing wind directions and practical flight paths, the latter sometimes being adjusted to minimize noise disturbance. The length, width and spacing of runways is determined in accordance with anticipated aircraft characteristics and numbers, ambient meteorological conditions and traffic control arrangements. The operational configuration is then completed by

Photo 8.26 This airfield in Switzerland is one of many kept for occasional strategic or emergency use. The pasture and farm buildings are for livestock or hay.

layouts of taxiways, aprons and terminals located to suit the remaining space and landside connections. An important construction consideration is the means by which any watercourse is channelled beneath runways (and possibly contaminated by oil spillage which can soak into the soil beneath aircraft apron areas) or is rerouted around the airfield. Another factor is the amount of earthmoving that has to be undertaken for new or extended runways. There may be a need for borrow pits for extra fill or for disposal of surplus – on the site or outside it and possibly affecting the land drainage pattern.

Car parks

Any vehicle needs somewhere to stand when out of use. Buses, trams and trains are used intensively during the day and are assembled at terminals, sidings or sheds for a few hours at night. Private cars are much less efficiently used; indeed, at any given time, most are out of use. Their purchase price, new or second-hand, is within the scope of many people, as are running costs while petroleum remains plentiful.

In car parks, each vehicle needs about 25m^2 of space for access, manoeuvring and standing. Photo 8.27 shows a typical configuration. Parking presents little difficulty in spacious suburbs, where each house has sufficient front or road space for one or two stationery vehicles. But parking space is a problem in cities, at work places not served by public transport, and at places of attractive scenery. Restrictions on access can be applied. Private cars have to be left at a village lower down the valley from Zermatt, a Swiss resort well served by rail. Day visitors to St Ives in Cornwall can travel thence by a train from a pleasant car park about 5km away. Thousands commute the same way into cities from suburban railway stations or use park-and-ride bus services into medium sized towns.

In cities, large car parks must either be underground, often as basements integral with building foundations, or they must rise in tiers above. Multi-storey car parks require geotechnical skill below ground, structural ingenuity above.

Photo 8.27 Car park at Heysel, Brussels.

Although the appearance of multi-storey car parks may resemble that of a building block, there are fundamental differences, not least in that the structural elements are exposed to public view. So design should present clean lines, discreet joints and good drainage. Facades can be fashioned to blend or contrast with adjacent buildings (see Photo 12.30 in Chapter 12). Roofs may be the top parking layer as well as viewpoints.

Suburban car parks usually occupy ground level space. Paved car parks may well look more attractive with colourful vehicles in them than when they are empty. In either circumstance, the stark paving can be relieved by narrow, planted or grassed, islands or borders. Paving should not be so spacious or impermeable that the majority of rainwater run-off flows directly into drains; as much as possible should penetrate the ground, as it would the natural soil.

Rural car parks should be as unobtrusive as possible (Photos 8.28 and 8.29). Preece (1991)[4] recommended well-drained grass, gravel or stone block paving as well as informal kerbs and boundaries to support a principle that car parks need not look like car parks when all the cars have gone. Trees or 'ha-ha' ditches can contain cars within their assigned space. Close woodland can be both a screen and a pleasant surrounding to footpaths out of the parking areas (Colour Plate 1).

For long journeys, car park space is needed at petrol stations (beside ordinary highways) or service areas (on motorways). However, on scenic routes, there may be opportunities to afford short-term parking for contemplation of fine views – at discreetly located smaller service areas, French *aires de repos*, or ordinary road lay-bys. Stopping at such places should be one of the pleasures of travel. However, when in motion, car drivers are culturally undeserving and should concentrate on driving. It is the passengers in every mode of transport who deserve the maximum visual pleasure that their journeys can afford.

8.4 Balancing benefits and impacts of transport systems

The economic value of good transport systems, or the need for better ones, is commonly asserted. Less often debated are the wider implications of easy rapid access on land resources and on human and natural communities.

Economic issues in transport and travel

Transport planners seek to provide:

- efficient despatch of goods to regional or global markets, enabling movement of perishable products quickly or stable commodities more slowly – for example by sea; and
- rapid travel so that people can visit distant places – to gain livelihoods or for pleasure – or make more regular shorter journeys such as commuting to work.

Freight (transport of merchandise) provided the main impetus for ocean trade and then for the construction of inland waterways and railways. It enables goods to be transferred when the cost of production and transport can meet the prices affordable and the quality required in a regional or

Photo 8.28 A rural car park at Carisbrooke, Isle of Wight ...

Photo 8.29 ... is concealed by trees in distant views towards the castle.

global market, or according to a central plan where different items are built at a few main centres and the costs of distribution are absorbed in a national price, as in the former Soviet Union.

Travel is journeys to make contact with places and people. Whitelegg (1993)[5] maintained that the contacts are more important than either the means or the speed of travel. Greater speed in sufficiently convenient cheap travel enables us to make similar contacts but over a greater distance. Travelling further, for example by inter-city trains or inexpensive airlines, loses some of the time saved in travelling faster.

Convenience is particularly relevant in choosing private transport to travel from one's home to nearly anywhere and at any time. Car ownership is the first goal of many people aspiring to middle class status in developing countries. They long to avoid the discomfort and inflexibility of public transport. However, when they are successful, they tend to forsake their bicycles – one of the most useful travel inventions of recent centuries. Costs of transport may become more expensive as energy shortages develop and energy costs rise; and the flexibility and convenience of private powered vehicles for daily travel create problems of congestion on roads.

Drawbacks of easy transportation

World trade and transport routes to previously less accessible inhabited regions are widely regarded as of mutual benefit to all but the least productive communities. But perhaps an ultimate penalty of *cheap freight* is that it encourages distribution to places where similar commodities might more beneficially be produced locally; and roads created into remoter areas – for example for logging, mining or tourism – encourage exploitation of land that might better be conserved.

In civilized cities, *passenger travel* is mostly on public (suburban or metro) railways, light rail (trams) and buses. However, in rural areas and many medium sized towns, private vehicles dominate. In developing countries, car ownership becomes a sure indicator of inequality. In richer countries, cheap car travel has encouraged people to live even further from their workplaces, schools or shops in communities that could become unsustainable for the poorer groups, or for the majority should car-operating costs rise sharply.

In terms of better use of land, energy and human resources, modern communications as well as rational urban and suburban planning should enable more people to live in communities where they can walk or cycle to work on most days. For those who must continue to commute, we should continue to develop equitable competitively priced sustainable transport systems. There are plenty of models of such systems already. In the Netherlands, there is reputed to be a friendly ambience on trains; for journey sections beyond the trains, bicycles can be hired at main stations. Some station car parks are even free on rainy days.

Pleasures of travel

On 18th century 'grand tours' of Europe, hired carriages were expensive and sometimes impracticable. Mountain routes might have to be tackled more arduously on horseback or as a pedestrian. Adventure and cultural experience rather than recreation was the motive for journeys in all weathers on roads of very variable quality and through hazardous, sometimes dangerous, territory. From 1815, steamboats plied between the River Clyde's estuary ports and islands. Many of the passengers travelled for the views as well as the comfort and reliability of the new form of travel. Tourism became popular as railways were established. The earliest excursion trips were to experience views from trains. Such were 1830 journeys on the Liverpool and Manchester Railway to view the Sankey Viaduct, a wonder of engineering on the route (Simmons, 1995).[6]

The beauty of valleys like that of the Susquehanna River in Pennsylvania was not fully appreciated until the New York and Erie Railroad was completed in 1848, making the region accessible. Cropsey (1865)[7] painted the valley landscape together with the Starrucca Viaduct, a notable element of the railway. According to Wilton and Barringer (2002)[8] the viaduct became a popular viewpoint where 'trains stopped regularly to allow passengers to admire the scene'. To travel was becoming as significant an experience as to arrive. By the 1880s, tours were commonly advertised. For example, *Bradshaw's Railway Guide* for August 1887 offers combined coach, steamer and train trips based on Oban, 'interesting to tourists' and through 'scenery surpassing grand'.

Pleasant travel encompasses:

- interesting views from railways or roads, partly depending on the alignment of routes and the state of their immediate surroundings;

- vehicles with windows designed to enable these views; and
- amenable conditions at railway stations and motorway service areas.

Views of interest to travellers by any mode include:

- grand scenery – mountainsides, lakes or coasts – whetting the appetite to return and explore on foot;
- less grand but intriguing rural countryside and buildings; and
- city views and vistas, as much for cultural interest and rear views (seen from trains) as for the building facades that commonly dominate street-side architecture.

Bus journeys offer similar distant aspects to most of those from trains, although viewing is complicated by less stable motion and competing traffic. Aeroplane journeys are reasonably comfortable except when delayed at airports. Window seats offer superb views of geographical features – from Himalayan snows to Aegean islands or Saharan oases.

Underground railways offer no scenery outside the confines of the carriages; and the sheer density of passengers accounts for rush-hour discomfort. The best solution may be large carriages and spacious platforms after the style of the Moscow Metro.

The environment for travellers at all stations depends on that same spaciousness, unimpeded movements along the interchange passages, and light – if not from the sky, at least throughout whatever lofty cavern can be contrived.

Disturbance and intrusion

Public consultation on transport projects has focused considerable attention on disturbance of people living under flight paths, near railway lines or adjacent to highways. Public concern often hinges on the short-term disruption that arises during construction; but that is seldom a valid objection to the form of the final structures. There are feelings that the *sight* of trains or vehicles will diminish the distinctive quality of a place or create 'blight' on the value of property, but later the visual effects are no more felt than is the sight of aircraft over beauty spots or stately homes such as Windsor Castle. As for intrusion, people have always been able to see into private garden space from high walkways, the top of double-decker buses or suburban trains. However, the main issue, on which objectors to transport projects focus, is the *noise* of passing trains, road traffic or aircraft.

The propagation, transmission and measurement of transport noise is well understood; it has been explained in detail as to its generation and propagation (Nelson, 1978)[9] and as to its implications in transport infrastructure planning (for example, Carpenter, 1994 and 2001).[10,11] Noise occurrence and intensity are complex; so loudness is usually assessed as an equivalent continuous noise level at affected places. The nature of noise is also complex. Different sound frequencies arise, for example from different types of engine, ancillary motors, fans, vibrating or rattling parts of vehicles, or as impact of wheels on wet or dry roads of different pavement material or on new or worn rail track, jointed or continuously welded. The nature of noise propagation determines how sound is transmitted as does the absorption of the ground or structures across which it passes.

Where noise is a real cause of concern and quieter vehicles or running surfaces are not a sufficient solution, it is first necessary to determine where noise levels matter and whether barriers can be effective in reducing the impact. Then the most appropriate form of barrier has to be selected both as to its effectiveness in reducing disturbance and its consequences on views from or towards the route.

Meanwhile, there are occasions when transport planners adopt unnecessarily drastic solutions as a practicable way of overcoming the vociferous objections that are common features of public consultation. Such short-term expediency can have unfortunate long-term consequences. For example, cheap opaque wooden fences may not be as effective as lower or transparent, acoustically and aesthetically well-designed barriers. Nor may such costly general solutions as lowering the level of the running surface, creating more cut than fill, be really necessary. Extreme examples of negative design include the enclosure of railway bridges, giving passengers no inkling of what sort of territory they are crossing.

Mitigating the effects on biotic communities

Flora and fauna exist in local ecosystems – natural ones in wilderness or large nature reserves, semi-natural ones within land areas devoted to human agriculture, settlements and transport infrastructure. Optimum conditions for semi-natural habitat can be planned and managed at modest cost. Features attractive to wildlife can include:

- green motorway verges, railway cuttings or embankment slopes and other unused but fenced-off space, or nearby wetland or watercourses;
- wildlife corridors connecting some of these features; or specially designed wild animal crossings where such corridors need to pass over or under transport routes; and
- niches in buildings, underbridges or even in electricity transmission equipment or signal gantries.

Transport ideals

Transport systems join human settlements on routes dictated by geographical features. The systems are main determinants in how land is occupied, developed or conserved. They provide the economic benefits of carrying freight to markets and the pleasures of affordable travel.

To avoid excessive impact on resources, the construction of infrastructure enabling high volumes and speed of traffic has sometimes to be restrained:

- to limit further development in regions where conservation rather than exploitation is needed; or
- to limit the use of the more extravagant forms of transport and to encourage more efficient yet pleasurable communal travel.

Notes and references

1 NCE (*New Civil Engineer*) (2009) 'German bridge cost Elbe its World Heritage status', *New Civil Engineer*, 2 July.
2 Sparks, A. (2003) 'Malayan road tunnel to double as storm drain', *New Civil Engineer*, 13 November.
3 Alistair Cooke in one of his 'Letter from America' BBC broadcasts in 2003.
4 Preece, R. A. (1991) *Designs on the Landscape*, Belhaven, London, p185.
5 Whitelegg, J. (1993) *Transport for a Sustainable Future*, Belhaven, London, p76.
6 Simmons, J. (1995) *The Victorian Railway*, Thames & Hudson, London, p272.
7 Cropsey, J. F. (1865) *Starruca Viaduct*, painting illustrated in Wilton, A. and Barringer, T. (2002) *American Sublime*, Tate Publishing, London, p141.
8 Wilton, A. and Barringer, T. (2002) as Note 7, p140.
9 Nelson, P. (ed) (1978) *Transportation Noise Reference Book*, Butterworth, Sevenoaks
10 Carpenter, T. G. (1994) *The Environmental Impact of Railways*, Wiley, Chichester, p130.
11 Carpenter, T. G. (ed) (2001) *Environment, Construction and Sustainable Development*, vol 2, Wiley, Chichester, p609.

9

Bridges and Crossings

Bridges are features of transport routes and, as such, are elements in the general impact of transport infrastructure on land resources. High bridge approach embankments cut across the land. But the main impact of bridge structures is visual. From the days of Roman aqueducts, a host of bridges have become monuments to extraordinary construction achievement. Others have been recognized as splendid, elegant or controversial works of art. Many are neutrally functional, a few clumsy or, to some eyes, even offensive. The more a bridge displays the elegant logic or the daring ingenuity by which it stands up, the more it is likely to be seen favourably.

Bridge design concepts concern function, load transfer, available materials, foundations and methods of construction (9.1). Whether a bridge is ugly or beautiful then depends on the form that is adopted (9.2) and how various forms fit into different landscape situations (9.3).

9.1 Design concepts

Function

The primary purpose of bridges has always been to enable transport routes to cross gaps – occasionally chasms, sometimes wider valleys, usually watercourses. At first, modest streams were the obstacles, then major rivers, eventually estuaries or even sea straits.

For wheeled traffic, especially railways, it became necessary to limit the gradients on bridge approaches. This need could be partly met by earth embankments but multispan viaducts permitted freer drainage and saved ground space. For railways there has also been a need for bridges over or under roads at what, in the motorway age, are labelled grade-separated crossings.

Multifunction opportunities have often occurred, such as joint road and railway bridges or construction of bridges in conjunction with river barrages or flood protection works. However, to achieve a balanced structural form, each function must be integrated into total design. Tilly (2002)[1] points out that many historic bridges have been spoilt by late attachment of utilities such as pipework. Beauty has tended to result from designs that most simply combine all functions.

Load transfer

A bridge structure bears dead (permanent) and live (transient) loads. These stresses are transferred through the superstructure onto abutments or piers and their foundations. The weight of the bridge deck and traffic can then be:

- borne directly, at the end of spans of simply-supported, continuous or cantilevered beams (including girders or trusses), on abutments at the ends or on intermediate piers;
- borne on arches, which transmit purely compressive forces at an angle onto abutments or are partly balanced by thrust from adjacent arches at piers; or
- borne by tensile pull on cables suspended from towers or 'pylons' on suspension or cable-stayed bridges.

Figure 9.1 and Photos 9.1 to 9.6 show these basic bridge concepts.

This simple explanation concerns only gravitational loads – the dead load of the structure itself and the live load of traffic moving across the bridge. The compressive and tensile stresses can be

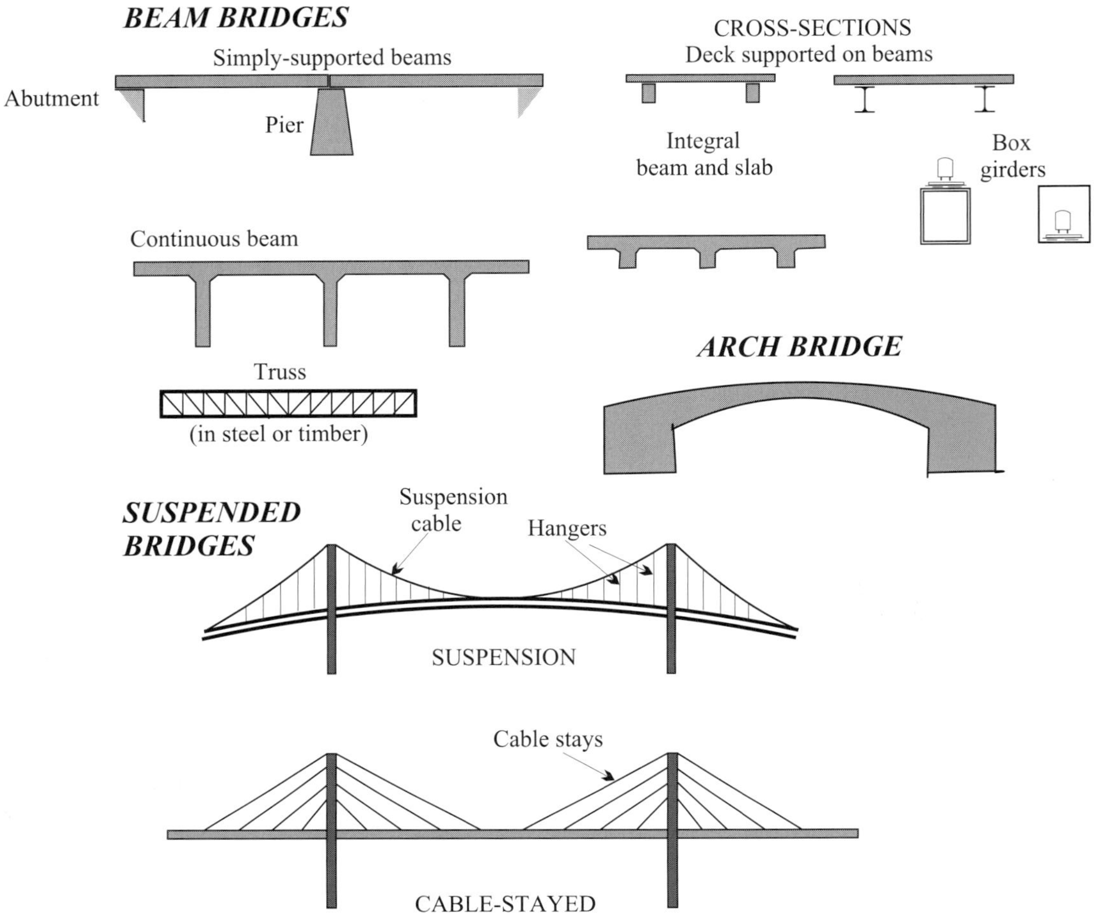

Figure 9.1 Bridge concepts

accommodated even by slim structural members if their materials are strong enough. Complications arise where structures are subject to transitory non-gravitational loads such as arise with high winds or excessive vibration. Adequate stiffness has to be provided to resist or damp twisting, shearing or resonating stresses, possibly at the expense of aesthetic slenderness. The trusses of the second Tacoma Narrows Bridge were wider and four times as deep as those of the first thinner structure which was torn apart by a moderate 67km per hour wind in 1940. Since then, considerable advances have been made in the strength of materials, the aerodynamic design of bridge decks and the understanding of structural response to vibration. These advances permit safer yet more slender bridge concepts.

Other temporary loads that must be allowed for at bridges result from floating ice or debris and accidental impact of vehicles or waterborne vessels against their piers. Most significant of all are the particular stresses that occur – by design or accident – during the erection of a bridge.

Construction materials

Structural elements of bridges have to:

- carry or transfer tensile and compressive stresses, needing greater strength for longer spans;
- be ductile and shear-resistant, rather than brittle, in the face of irregular distorting stresses; and

Photo 9.1 Continuous beam concrete footbridge over the M4 Motorway, England.

Photo 9.2 Steel truss railway bridge over the River Meuse, Belgium.

Photo 9.3 Forth Railway Bridge. Massive cantilevers and modest connecting trusses. A wondrous sight mainly because of its size.

Photo 9.4 Masonry arch bridge on former packhorse route, Minllyn, Wales.

Photo 9.5 Golden Gate suspension bridge, San Francisco.

- survive predictable collision, wear and weather or be easily repaired or replaced.

Timber has only modest strength in relation to its weight; but can withstand both compression and tension and is reasonably resilient. Larger or stronger wooden elements can be formed by assembly in trusses or trestles or by lamination. So old bridges have to be carefully preserved or reconstructed. Wood can be fashioned with artistry and connected with ingenuity, as for Swiss roofed bridges (Colour Plate 8); or it can be used in random pieces for the simplest temporary crossings (Photo 9.7).

Masonry (cut and shaped stone) utilizes arch action in bridge spans and mass in their abutments. *Mortared brickwork* can do the same. Both materials give scope for skill to complement the elegance of arches in connectional detail or facing patterns. *Cast and wrought iron* were used in trusses, beams and cylindrical piers for bridges but have been succeeded by *steel* – now the generic name for alloys of iron, carbon and other elements produced as ingots by a variety of heat treatments to meet a range of specifications for strength, hardness,

Photo 9.6 The cable-stayed Rio-Antirio bridge across the Gulf of Corinth, Greece (photo by Robin Carpenter).

Photo 9.7 This timber bridge is picturesque and certainly impermanent.

malleability or corrosion protection, and rolled or drawn into shapes or wires for particular applications. *Concrete* is used in mass form in bridge foundations or as infill in thick piers. Otherwise it is reinforced or prestressed by steel in more slender, load-bearing members – cast in situ or assembled as prefabricated (precast) units.

Modern construction materials include adaptation to new circumstances of all the traditional materials, particularly steel and concrete, plus:

- metal alloys of ever greater strength or specially suited for particular or variable loading conditions;
- built-in properties or sensors that cause structural members to take their own protective measures in response to changes in loads or ambient conditions;
- protective or decorative coatings; and
- plastics (such as fibre-reinforced polymers (FRP)) which may have structural and visual similarities with natural materials but their scientifically controlled production can provide properties not found in nature.

Foundations, piers and abutments

Vertical loads on piers or inclined forces on abutments have to be carried by adequate foundations. Hard mother rock is ideal. Otherwise, load transfer to artificially densified ground or to deeper, firmer layers through piles can prove satisfactory in fully investigated conditions. Foundation conditions affect the location and spacing of piers and thus the span, type and appearance of bridge superstructure; and the cost of hidden bridge substructure is typically of the same order as that of the visible part.

Particular difficulties affecting bridge foundations are:

- that they often have to be constructed in rivers or in saturated conditions; in alluvial river beds, deep scour around piers can be expected during floods throughout their lifetime; and
- on valley slopes, rock may be weathered to a considerable depth; or there may be potentially unstable landslip material.

The need to reduce these difficulties and the number of foundations in deep water has been a main reason for choosing long spans. Another, with high clearance, is freedom for navigation. The visual impact of abutments on urban surroundings can be related to that of the bridge itself as well as adjacent structures. For example, suitably massive abutments may complement arch action in the bridge superstructure or the style of nearby retaining walls or river front buildings. In rural surroundings more modest abutment appearance may be appropriate. On already wooded valley slopes, new trees may hide them. The most attractive abutments for long single spans are sheer rock faces.

Construction

Bridge building strategies require practicable solutions for:

- diverting rivers and keeping water out of working areas;
- constructing foundations in deep water, below the water table or in unstable conditions;
- assembling the superstructure and holding it in position before it is self-supporting, for example before an arch is complete; and
- eventually, from a few decades to a century later, replacing worn or inadequate elements or reconstructing the bridge.

Chosen solutions may affect design of the structure and influence its appearance.

Aesthetic design

Discussing the Pont du Gard aqueduct in France, Wheeler (1964)[2] sees the apotheosis (top) of the arch as the means by which Romans transcended basic utility and elevated engineering to the level of an art. In the Italian Renaissance, bridge building was indeed regarded as a high art form (Bennet, 1989).[3] Civil engineering and architecture were the concern of painter-sculptor polymaths. Beauty and function were united. Bridges came to be regarded as civic works of art in the same way as public buildings, churches and monuments.

Much of the classical tradition, perhaps best exemplified in elegant arch construction, was continued in many 18th century river crossings and 19th century railway viaducts. It was probably iron construction that gave early grounds for certain bridges to be seen as unsightly. Particularly during their experimental development, iron frameworks could be mechanically clumsy. Early prototypes were adapted to suit a variety of conditions without full understanding of the capability of the materials or the ways in which they could most effectively be manufactured and assembled.

In addition to the splendour of great bridges, there are countless lesser situations where unobtrusiveness may be regarded as a positive quality. This is comparatively easy at deck level where most of the structure is unseen, more difficult to arrange looking up from below. Yet there are opportunities that can be taken to harmonize the way in which overbridges spring from their abutments beside roads or from riparian strips beside waterways.

For great bridges, quality of appearance concerns their approaches and foreground as well as the elegance of their main spans. Along their whole length it is the skyline profile that shows up against surrounding natural features. The Highways Agency (1996)[4] emphasizes that 'the horizontal and vertical alignment of a bridge relative to its topographical context has more effect on the bridge's appearance than any other single factor'.

9.2 Forms

Basic parameters for a bridge concern the loads to be carried, the foundation conditions and necessary or acceptable span lengths and heights. The chosen form must then suit available construction materials and means for erecting each part.

Timber bridges

Today a few old timber bridges remain where refurbishment has been possible, some as heritage structures, some still carrying traffic (Photo 9.8). New timber construction is mainly for footbridges and often in situations where design and construction quality is exposed to examination at close quarters. The visual attraction of timber bridges is in workmanship, joints of piers, deck, or balustrades and sometimes roofs.

Photo 9.8 Timber piers support replacement precast concrete deck beams of Glaslyn railway bridge, North Wales.

Arch bridges in stone masonry or brickwork

Stone in rocky areas and brick clay in alluvium are obvious local building materials. Square-cut stone (ashlar) can be shaped at the quarry; bricks are baked in kilns beside the clay pits; and these materials from whatever source can be formed into robust structures if all applied stresses are in compression. Arches meet this requirement; and, according to the setting, the elegance of arches and features of their facades promote their visual modesty, function or splendour (Photos 9.9–9.12).

At close quarters, elegance in masonry arch structures concerns the clean artistry in dressed stone courses and their quoins (external angles) or the complex pattern of brickwork, especially in skewed arches for 19th century railway bridges. In the middle distance, a profounder beauty arises in sweeping curves, symmetry, proportions and clear functional interfaces in the piers and the superstructure.

Iron and steel bridges

Many cast iron piers, wrought iron beams or trusses in both materials still survive, particularly in railway bridges. Modern steel bridges comprise girders, trusses, arches, trestles and various ties, all connected at fixed, hinged or sliding joints. The types of framework adopted depend on the length of span required and the ground conditions. For example, on ground that is soft behind the abutments but sound in the river bed, a deeply curved segmental arch may be more suitable than a flat arch with greater horizontal thrust (Bennet, 1989)[5] Photos 9.13–9.17 illustrate a range of late 19th to mid-20th century iron and steel bridges.

Simple girders were found suitable across short spans, such as for railways over urban streets where maximum clearance is desirable between building lines. Plain but comparatively deep crossings of this type are obstructive in that they block vistas that might otherwise be more complete. Some bridge faces were ornamented by city coats of arms or fancy ironwork that remains appropriate on conserved heritage structures; but modern equivalents aim for simplicity, minimal depth and bright yet sympathetic colouring.

For medium-length multispan crossings over rivers such as the Seine or the Thames, numerous plain heavy girder bridges survive. When new, these

Photo 9.9 Classic parabolic masonry arch, later strengthened by lateral tie bars, shown by X-shaped end plates, Aylesford, Kent.

Photo 9.10 Ancient pointed arch bridge, northern Iran.

Photo 9.11 Circular arch, reflected in River Wear, Durham.

Photo 9.12 Eynsford viaduct, Kent, strengthened by tie bars with round end plates.

bridges seemed intriguingly modern and were prominently included in views of riverside life by painters such as Monet.[6] Today such bridges are tolerated but rarely extolled, sometimes benefiting from bright fresh paint.

For longer spans, the type of steel structure best suited to a location depends on that span, the required headroom, the nature of support, the cost and availability of steel components and the expertise and techniques needed for construction. Aesthetic considerations also become increasingly significant as the size of structures makes them more prominent.

The possible combinations of structural function and shape give opportunity for mechanical ingenuity and artistic arrangement; or there may be clumsy adaptation or ill-conceived caution. Burke (1998)[7] reviewed a century of bridge aesthetics literature and concluded that five bridges were particularly notorious for their ugliness. Four of these were wholly in iron or steel, the fifth was in its functional parts.

Today new medium span steel bridges probably look best if:

Photo 9.13 A stark iron truss railway bridge near Oxford.

Photo 9.14 Tower Bridge, London (1894) has to raise its deck for ships to pass. Its bizarre features are not out of keeping with the great variety of ancient and modern buildings nearby.

Photo 9.15 A single-span steel lattice bridge over the River Douro near the Portuguese border, west of Zamora, Spain (photo by Robin Carpenter).

Photo 9.16 Craigellachie Bridge, River Spey, Scotland. The effect of providing the necessary massive abutments would probably be brutal without their castellation.

Photo 9.17 Railway bridge over the River Rhine, Cologne. It has footbridges on both sides, almost constant passing trains and joins quiet suburbs on one side to the city centre, station and cathedral on the other.

- they demonstrate unashamedly how they function, emphasizing arches and decks but also revealing, rather than concealing, abutments, nodes and interfaces;
- they are economic in use of materials, minimizing mass and dead load and with as few structural members and as much transparency as strength and ingenious design allow;
- struts in trusses are parallel, vertical or diagonal at 45°, or in 60° equilateral triangles between the horizontal members and as slender as is practicable without being too numerous; and
- curvature (in arches, tied arch bowstrings or arched frames) balances rather than confuses the straight members of structures.

Concrete bridges

Steel trusses are assembled in frameworks in which individual members fulfil specific load-carrying roles. Concrete is poured into a variety of monolithic shapes in which reinforcing rods, pretensioning wires or ducts for post-tensioning cables have to be incorporated. Tilly (2002)[8] referred to the durability of concrete bridges. By 2002, there were examples more than 100 years old and even a post-tensioned, prestressed concrete bridge was in excellent condition after 55 years. However, Tilly emphasized the need to design superficial repairs so as to be comparable in composition and appearance with the original concrete.

At *close quarters*, concrete's appearance is judged by its surface quality. A prime quality may be enduring stark cleanliness. Unshaded faces may be white or in a bright colour or there may be some form of surface treatment, such as paint, sand blasting, modest ornamental patterns or even stone facing. The *view from beneath* may be one of interest – in the functional solutions at abutments and in haunches, and in arrangement of slabs and beams; or there may be dominant massive low clearance beams and an overpowering dimness. *From further away*, along the road or river that is crossed, the silhouette of a concrete bridge is not always strikingly different from that of a steel one.

Photo 9.18 High level crossing for an autoroute over the River Meuse, Belgium.

Concrete beams may not, at a distance, be easily distinguishable from steel girders; but, because concrete is monolithic, its junctions and corners can be less abrupt and more curvaceous.

In terms of their place in the landscape, typical concrete bridges can be categorized as:

- short spans on beams of the sort that are usually designed to appear as unobtrusive as is practicable when seen as overbridges;
- long multispan viaducts over land or shallow water, often as approaches – in preference to embankments – to longer span main crossings; and
- high long-span concrete bridges that are more dominant and therefore of greater aesthetic significance (Photos 9.18–9.21).

Photo 9.19 Long-span concrete beam railway bridge ...

Suspension bridges

The *key structural features* of a suspension bridge are:

- the towers or pylons, over which the suspension cables run;
- the abutments, where mother rock or massive concrete holds the tensile load of the bridge transmitted by the cables; and
- the deck – on a continuous beam or truss, or as the top of a box girder – to carry the traffic load and to be stiff, light and aerodynamic enough to withstand wind loads or vibration.

Photo 9.20 … and concrete arch road bridge, both in Porto, Portugal (both photos by Robin Carpenter).

Photo 9.21 Motorway viaduct above Montreux, Switzerland.

Photo 9.22 Forth Road Bridge, Scotland.

The means of preventing cable deterioration – to prolong their life – are not yet universally established.

The *key visual features* of suspension bridges are:

- the catenary curve of the suspension cable, substantial viewed close at hand, just thick enough to be seen as a line across the sky from the middle distance;
- the pylons at each end of the main span and visible from much greater distances or, in the case of the A90 road approach to the Forth Road Bridge, above the land horizon; and
- the thin flat or slightly upward curving road deck (Photo 9.22).

Pier foundations cannot be seen, nor can details often be discerned of the abutments. However, locations suitable for foundations and abutments are critical in establishing where a suspension bridge can be built and in what sort of setting its then largely predetermined form must fit.

Cable-stayed bridges

Consequent upon development of rigid box girders and reliable strong multi-wire cables, it became possible to support bridge decks directly from pylons rather than from a high cable spanning between the pylons. The vertical load is transferred via the diagonal stays directly to the pylons. The pylons need massive foundations in any case but, for the stayed solution, heavy abutments may not be needed as the horizontal components of the loads in the inclined stays are usually balanced and resisted in the deck itself. Cable-stayed bridges have become cheaper solutions than suspension bridges but so far are impracticable for spans exceeding about 1km in length. So most notable examples are multispan (Photo 9.23).

Both suspension and cable-stayed bridges continue to require tall pylons for the cables and long stiff decks for the traffic. The most obvious visible difference is the substitution, for the stayed bridge, of almost straight diagonal stays instead of the inverted arch suspension. In place of a perfectly parabolic cable and light vertical hangers on a suspension bridge, a stayed bridge has families of straight stout inclined cables. A wide range of geometrical patterns is available whereby stays secure decks to pylons – see also Highways Agency (1996),[9] Bennet (1989).[10] The cables may even be suspended from an arch rather than a pylon (Photo 9.24).

Photo 9.23 Millau viaduct, France (photo by Robin Carpenter).

Photo 9.24 Cable-stayed bridge over the Río Navia, Asturias, Spain (photo by Robin Carpenter).

9.3 Situations

This section concerns the location of crossings, the types of bridge that are appropriate and their interaction with various natural and man-made landscapes. To recognize the issues involved it is perhaps logical to proceed from numerous modest route crossings to bridges that have to be long or spectacular:

- The most common bridges, over or under roads and railways, are hardly noticed as you cross them. From below they may offer either an attractive frame to a continuing vista or dismal blockage of might once have been a pleasant prospect. In urban situations, railway bridges are features of greater built-up patterns.
- Bridges over rivers follow long traditions as structures of functional beauty in both rural and city surroundings. Respect for both rural landscape and urban aspiration have to temper measures for meeting new traffic demands. Footbridges over water offer opportunity for inspiring light designs.
- Longer crossings – such as across flood plains or estuaries – are structures that both dominate the wide landscape in which they are constructed and influence the lowland morphology.
- Awesome natural situations, if man has to venture into them, require awesome construction strategies.

Grade-separated route crossings

Grade-separated crossings are those where one route crosses over another on a bridge. Commonly new roads were built over railways in this way; or road bridges replaced level crossings. Grade-separated junctions are those where spurs or flyover viaducts connect the crossing highways or joining railways.

In flat country, railways and highways are usually built on a modest (1–2m high) embankment. Minor roads cross *over* them on bridges high enough to clear the trains and electrification equipment, incidentally providing a view of the surroundings. This is much more pleasant for minor road drivers, cyclists or pedestrians than passing through a long narrow opening *beneath* the wide main route.

In undulating country, railways adopted courses requiring roughly equal volumes of cutting and embankment. The choice of overbridge or underbridge crossing is then determined by the topography, preferably choosing locations where the difference between road and rail levels is substantial. General cut-and-fill balancing was also adopted in the construction of high-speed, high-capacity motorways. It is then preferable that a local road or pathway should pass over the wider highway by a long slender bridge minimizing visual intrusion. Topographically this is not always possible. Waterways and associated footpaths or valley bottom roads usually lie below the level of major highway routes. To avoid a dark and forbidding tunnel ambience being cast by wide multilane overbridges, designers can strive to:

- maximize the span and height of overbridge openings (Photo 9.25);
- separate the major road carriageways allowing light to penetrate between two narrower overbridges; and
- shape the soffit and facade features to express the functional elegance of the overbridge span.

Bridges over railways are seen only fleetingly by train passengers. It is more important that they can see clearly the views over the parapets of underbridges. On some Train à Grande Vitesse (TGV) lines, opaque noise barriers block these views.

Motorway interchanges may have two or more intersecting levels and possibly a waterway at the bottom. The viaducts are the visibly striking

Photo 9.25 Junction of rural highway (B4009) with motorway (M40) in Oxfordshire.

structural elements, seen briefly by travellers at the lower levels of the junction.

City railway bridges

Early railways had to cut through any already built-up areas. In city approaches, the tracks were elevated on bridges over the streets and on viaducts across the intervening cleared ground. Many of these routes became permanent main lines.

In the 19th century, according to Binney and Pearse (1985),[11] 'the town viaduct was considered to lower the tone of the neighbourhood'; but 'parliament preferred them, because they avoided wholesale street closures and level crossings'. So towns such as Stockport or Mansfield and parts of south London and Manchester were dominated by viaducts. Today, the tall brick arches of Stockport viaduct are perhaps as admirable and less arrogant than many of the buildings looking up to them. Durham viaduct is not an unwelcome feature seen from the top of the cathedral tower (Photo 9.26); and the longer lower viaducts of south London and elsewhere have afforded views to railway commuters as well as accommodation for a variety of small industries (Photo 9.27).

The visual impact of railway girder bridges across city streets was often obstruction of a vista. Many have survived and still carry trains. Some have been removed; but one, over Friargate in Derby, although closed, is now listed Grade II for preservation. According to Simmons (1995),[12] engineers such as Sir John Hawkshaw, writing in 1863, stressed that 'these bridges spring from the shop front on one side to the shop front on the other without intermediate supports and at a uniform height above the street' but that 'girders cannot be adorned with any advantage'.

In cities such as Chicago or Berlin, urban railways ran above the streets. Even on new metros, such as those of Miami or Hiroshima some lines are still elevated – usually on steel or concrete portals which leave more open space below than did brick arches. The same double-storey strategy has been adopted for city motorways, such as west London's Chiswick Flyover or

Photo 9.26 Viaduct carrying the East Coast Main Line railway past Durham.

Photo 9.27 These 'underneath-the-arches' workshops are on the approach to Windsor station and castle, Berkshire.

Westway. Consequences of such efficient use of scarce land space include effective separation of faster through traffic but also a need to maintain well-lit conditions rather than threatening dimness beneath; if the underneath is not actual road, there may be opportunity for other uses such as open-side walkways with underside shops, bus stops or car parks.

Bridges across rivers

Rivers are ancient determinants of land surface topography and current features of observed landscape. For land transport, they are obstacles, although their valleys provide route opportunities. Bridges overcome the obstacles to routes either across the valleys or along them.

Bridge sizes and strength must respond to the traffic across them; their clearance above the water must accommodate any vessels or floods; and their openings and faces must please riverside walkers or people in boats. Particularly in rural situations, the visual impact of a bridge depends on whether it dominates the scenery or can blend into it. The Highways Agency (1996)[13] asserts that 'the drama of a bridge and usually the general appearance are enhanced by great height – and height is emphasized when length is reduced. Therefore, it is better to have a high short crossing than a longer lower one.' One might infer that high bridges are dominant, suited to dramatic situations, while long low bridges are potentially intrusive. Apparent length of bridges can be reduced by planting near the abutments; and clusters of trees can break the sight of long low approach viaducts.

Thus the beauty or lesser attraction of bridges depends both on their harmonized design and the riverside setting at each end. Photos 9.28 and 9.29 show fine and dismal examples respectively.

Many towns grew up around ancient river crossings. Some are now substantial cities with

Photo 9.28 The current Westminster Bridge in London (Grade II*) was opened in 1862. At that time the new Houses of Parliament (Grade I) were still being built and construction was just starting on Bazalgette's Thames Embankment – incorporating an underground railway and an interceptor sewer, behind a river training wall and beneath a wide road and promenade.

Photo 9.29 A clumsy combination of road, pipeline and power line crossings of the River Warta on the outskirts of Poznan, Poland. The empty riverside land offers great scope for improvement.

numerous bridges across the same river. City river crossings are essential parts of urban layout and architecture. As such, they are illustrated further in Chapter 15. Here it remains to be noted that, in practice, there is considerable mêlange, even mêlée, of structural and architectural styles, in a city's buildings even more than its bridges.

One stretch of the River Tyne is crossed from Gateshead on one side to Newcastle on the other by steel bridges of four widely different types for road, rail and pedestrian traffic, constructed over a century and a half. These bridges constitute more a working museum than any integrated concept of aesthetic city planning. The view from each bridge of the others is probably as fine as any towards the city's buildings.

The most recent of the Tyne bridges is a footbridge of spectacular design enabling it to be raised for river traffic to pass. Such structures can be of light construction because of the modest loads imposed. Yet they need to be of long span because of river traffic as well as for aesthetic reasons. The combination of light construction and long spans provides opportunities for ingenious design that the Highways Agency (1996)[14] says 'should be slender, elegant, dramatic and unusual' and that gives what Dyckhoff (2003)[15] describes as a 'thrilling experience'.

Thus, in cities, it seems that the functional design of bridges pays little attention to the built surroundings, providing they do not obscure each other or are uncomfortably adjacent. It is at the connection with each river bank that aspects of the structure's fitness to its immediate vicinity become relevant – how the bridge approaches, footways and abutments unite with river embankment walls, riverside promenades and city boulevards.

Tidal and wider water crossings

River valleys widen into alluvial plains and estuaries. Their topography is determined by geological features such as rock outcrops, by the dynamic morphology of erosion and depositions of waterborne sediment, and by human construction. We attempt to influence river flow, to protect certain areas from flooding or to reclaim them from the water, or to take advantage of the firmer features to construct crossings. Because the land is mostly flat, landscapes comprise vegetation, structures, people and animals.

Needs and opportunities for long road or rail crossings over tidal and sea water arise:

- across tidal flats or rivers of varying width;
- for causeways to be built across the shallow parts of estuaries; and
- for fixed links across the narrowest parts of wide deep water, either as long span bridges (as across the Golden Gate or proposed over the Straits of Messina) or deep bored tunnel (from Honshu to Hokkaido).

The Medway River, 3km upstream of Rochester, England, is tidal but passes through a comparatively narrow (1km) gap in the North Downs hills with a normal river flow width of 300–400m. The gap was first crossed by a high viaduct for the M2 motorway in 1963 (Photo 9.30). Then, early this century, two parallel viaducts were erected – one for the Channel Tunnel Rail Link, one for additional lanes of the widened motorway. Great care in planning was taken, particularly for the high speed railway, to direct the new alignments so that they crossed the river at the same height and that the piers of the two new viaducts were precisely aligned with those of the first (Photo 9.31). As a result, the 152m main span of the railway crossing is the longest in the world carrying high speed trains.

Solutions may vary, both as to options adopted in a total strategy for a whole crossing and for different types of structure for each section of a crossing. The greater part of the 600m-wide Conwy river estuary in North Wales has been crossed since the 1820s by a causeway (Photo 9.32). The remaining gap between the causeway and the small walled town of Conwy has been crossed by a suspension bridge (Telford, 1826, now for pedestrians), a wrought iron box girder (Robert Stephenson, 1849, for the railway) and a steel arch (1958, for an improved road). To avoid traffic congestion in the narrow streets of Conway a completely new crossing concept was required by the 1990s. Since 1991, the A55 trunk road provides a bypass, crossing further downstream through a 1km long submerged tube tunnel. Such crossings are increasingly common beneath estuaries. They are easier to construct than tunnels excavated

Photo 9.30 First Medway M2 motorway bridge complete in 1963.

Photo 9.31 Three precisely positioned Medway bridges in 2009.

in soft, water-bearing formations. The practicability of submerged tube crossings depends partly on adequate waterside space being made temporarily available for construction and launching of the immersed tube elements; these have to be floated out to the site where they are sunk onto prepared underwater foundations in a dredged trench.

For long crossings over varying topography and foundation conditions requiring different spans or navigation clearance on different sections, a linear series of solutions may be adopted. In Japan, where most long bridge building has recently taken place, single routes have incorporated suspension, cable-stayed and steel arch bridges for main crossings as well as beam-and-slab viaducts and embankments on the approach sections.

Photo 9.32 Conwy crossing, North Wales. Out of sight in the left background is the entrance to the 1991 immersed tube tunnel bypass road. The causeway in the middle of this view leads to the three earlier single span crossings – from left to right, the 1958 steel arch, the suspension bridge of 1826 and the wrought iron box girders opened to rail traffic in 1849.

Awesome situations

Stark natural beauty is best conserved. Awesome scenery most likely to be crossed by a road or railway is that of steep sided valleys – rocky chasms, or wooded gorges with at least outcrops of protruding rock, and possibly white water flow along the rough bottom. Often, particularly in limestone country, gorges are dramatic incisions into the surface of plateaux or gently undulating territory that otherwise offers relatively few natural obstacles. In mainly wider but still steep valleys constituting routes through mountains, a railway may have to climb steep sinuous sections by a series of spiral tunnels within the rock, emerging briefly at each chasm crossing.

Over some narrow gorges, road or railway bridges may be able to cross the very top in an single span (Photo 9.33). In such situations it may be argued that the scenery is best conserved in its pristine sublimity. However, if such gorges abound in the district, crossing needs must be met somewhere. Then there is no hiding the bridge. Dramatic situations call for dramatic structures.

Steel or concrete arch bridges spring directly from abutments formed by steep chasm or valley walls. Preferably they cross in a single span or on slender piers whose verticality somehow balances that of existing trees and cliffs. The drama of the bridge is perceived in profile against the skyline and the cliffs; and views from the bridge can be of the intimidating grandeur of the gorge itself.

A different, more one-sided vertical landscape is that of long steep escarpments, for example above lakes or coasts. Railways or corniche roads have been built along these with tunnels or hairpin bends and occasional panoramic viewpoint parking places. But wide highways constructed on high speed alignments by cut and fill along the hillside would cause extensive landscape damage. Tall viaducts can carry separate carriageways over rocky, wooded and ravined landscape with minimum

Photo 9.33 The gorge on the left is awesome scenery. The slender span of the bridge does nothing to dispel its steepness. The roofed timber bridge on the right complements the period buildings in charming, rather than awesome, and completely contrasting foreground (Fribourg, Switzerland).

permanent ground level disturbance especially if the steepest bulwarks of the mountainsides are penetrated by discreet tunnels.

9.4 Bridges as scenery

Bridges are only features, albeit often outstanding ones, of transport routes. If a bridge is a misfit in the landscape, possibly there should be no route there at all. But most untidy bridges were probably intended to be temporary, being constructed in a hurry; or they have been adapted, widened or put to new uses with short-sighted inexperience. Many permanent bridges are handsome structures; yet, within the principles of route selection adopted in their location, their design is largely functional.

There is seldom any positive means by which a bridge blends with natural grandeur. However, clean, high, wide span crossings can be as sublime, in a completely contrasting way, as the chasms that they cross. Among man-made landscapes, such as city waterfronts, bridges may be as eclectic or as conforming as the architecture of the surrounding buildings. A middle category is typified by semi-natural river landscapes in which arched masonry, brick, steel or concrete bridges or viaducts follow a tradition of gradual aesthetic development which has become widely acceptable to admirers of natural beauty and engineering elegance alike.

Notes and references

1 Tilly, G. (2002) 'Historic trail', *New Civil Engineer*, 22 August, p36.
2 Wheeler, M. (1964) *Roman Art and Architecture*, Thames & Hudson, London, p149.
3 Bennet, D. (1989) *The Creation of Bridges*, Quintet Publishing, London, p21.
4 Highways Agency (1996) *The Appearance of Bridges and Other Highway Structures*, Her Majesty's Stationery Office (HMSO), London, p13.
5 Bennet (1989) as Note 3, p74.
6 Numerous impressionist or neo/post-impressionist painters of the Seine railway bridges included Manet, Monet, Seurat, Signac and van Gogh. Most of the bridges shown are plain, even ugly, but the

paintings depict Parisian riverside life rather than landscape.

7 Burke, M. P. (1998) 'Aesthetically notorious bridges', *Civil Engineering*, vol 126, pp39–47.

8 Tilly (2002) as Note 1, p38.

9 Highways Agency (1996) as Note 4, p61–64.

10 Bennet (1989) as Note 3, p92.

11 Binney, M. and Pearse, D. (eds) (1985) *Railway Architecture*, Orbis Publishing, London, p143.

12 Simmons, J. (1995) *The Victorian Railway*, Thames & Hudson, London, p166.

13 Highways Agency (1996) as Note 4, p13.

14 Highways Agency (1996) as Note 4, p71.

15 Dyckhoff, T. (2003) 'Take me to the bridge', *The Times*, 25 November.

10

Military and Industrial Construction

10.1 Functions, appearance and land use

Military construction is undertaken for defence or to enforce political will. Industrial structures are set up for continuous economic production. In both cases, organization is paramount, public access is restricted and environmental issues are seldom seen to be of primary importance. Nevertheless, industrial processes or military manoeuvres may incorporate understanding of land and nature conservation or even social justice. The shape of equipment or structures affects their appearance in the landscape; their location and the activities that take place determine the way in which land is used and how natural resources are affected. Because of the exclusive and extensive ownership of land, its disciplined use may be a safeguard against too rapid and less sustainable speculative development. Even if there is some pollution, spare land space may be most useful for further development for future populations when land values and needs may be different from what they are today.

Military structures and functions

Ancient military structures are earthworks, now irregular features of the land surface, or the more upright remains of castles or walls. Techniques for bulky wall and tower construction led to more elegant structural frameworks in civilian dwellings and public buildings. Meanwhile, even bereft of wooden floors and roofs, castles remain impressive monuments – best viewed from surrounding or interior grassland cleared of extraneous paraphernalia. Defensive walls may still define old city boundaries whilst many serve as fine high walkways. Along the coast of Southern England, Martello towers are forts built in case of French invasion 200 years ago. Today these attract interest on sea fronts or as navigation features.

Modern military functions include:

- offensive action (war) – from naval bases, airfields and missile launching sites;
- training, for example, on tank or artillery ranges;
- urgent civil engineering works, such as bridges, harbours, airfields, fuel depots or pipelines, necessary for military action or to repair damage caused by war or natural catastrophes; and
- routine or emergency peacetime logistic support to civilian authorities.

Practice battlefields may require remote or strictly detached sites; but permanent barracks or depots for civil support are often located within or close to urban or suburban communities.

Industrial production

Manufacturing or materials processing was practised mostly on a small scale until the late 18th century. Rapid industrial development dates from about 1760. Certainly in earlier periods, water power had been harnessed for milling, charcoal for casting iron was produced in woods, and trees were cut for building houses and ships. Even steam power had been in use for pumping water out of mines. But few single industrial plants employed more than 100 people.

Wider application of steam power in factories and locomotives was the catalyst for large-scale industrial development, firstly near coal mines and then further afield when coal could be carried by canal or railway. Production was typically of textiles, pottery, iron and machinery. Where there had been settlement and production near the sources of minerals and along river valleys, there was soon conversion of countryside or market town areas into conjoined smoke-ridden 'black country' of closely spaced foundries, factories and workers' accommodation. The social consequences have since aroused much adverse comment but were not entirely retrograde – witness certain pioneering industrial and housing communities created in Britain that were, in December 2001, given World Heritage status at New Lanark (Scotland), Saltaire (West Yorkshire) and Derwent Valley Mills (Derbyshire).

Many black industrial regions developed – in the north and Midlands of England, in South Wales, southwest Belgium, the Ruhr basin of Germany, Upper Silesia in southern Poland and Pittsburgh, USA. The blackness was air pollution, both as smoke from combustion or iron working and as dust blown from coal or ore stockpiles. At the same time, untreated effluent from chemical processes was poured into watercourses. In due course, the ground itself was contaminated by effluent and spilled toxic chemicals such as lead, arsenic, mercury, coal derivatives and hydrocarbons.

In the first half of the 20th century, industrial development was related more to clean electrical technology and safely contained chemical engineering. Progress, for example in steel manufacture and rolling, enabled mass manufacture of sophisticated consumer goods. Internal combustion petroleum engines engendered huge advances in transport and mobile machinery. Cheap freight transport resulted in location of factories far from the source of all but the most bulky raw materials. Production of electricity expanded rapidly with rising demand for industrial, transport, municipal and domestic use.

In the second half of the 20th century, traditional industries, such as steel manufacture or textiles, were rationalized. In long-industrialized countries, this often meant reduction in the face of wider global competition. Large-scale industry continued to expand in petrochemicals but was approaching a peak in manufacture of vehicles and domestic machines, except in still-developing countries. In light industry, specialized assembly of white goods, communications equipment and computers, called for clean, comfortable working conditions and a skilled labour force, as did the servicing of people's numerous mechanical aids.

Appearance of industrial and military structures

Processes determined the shape of blast furnaces or 'beehive' pottery kilns as much as the structural functions of walls, bastions and lookout towers had defined earlier forts. Operational requirements called for distinctive shapes, such as the vertical shafts of grain silos, nearly horizontal cement-grinding cylinders or pressurized fluid storage spheres. Modern furnaces, steel rollers, textile and other factory machinery are conveniently accommodated in rectangular buildings while, except for rocket weaponry, military workshops or garages tend to follow designs developed for civil applications.

Twentieth century materials enabled longer spans and larger-dimensioned, box-shaped factories, vehicle maintenance depots or aircraft hangars than had been practicable in earlier brick or masonry construction, even of cathedrals. The facades of 19th century pumping stations commonly incorporated Gothic or Renaissance features to resemble castles or palaces. But modern industrial construction tends to favour plain unembellished rectangularity, although curved or sloping wide span roofs offer stylistic opportunity. Binney (1996)[1] observed: 'Some of the best architecture in the new towns lies in their factories. In Washington (new town, Durham), the new factories are set in landscaped grounds along the main arteries echoing the Art Deco plants that once lined the Great West Road in London.'

It may be decreed that secret or ugly things should be concealed or disguised. Concealment may be achieved by hiding behind natural or constructed mounds or by partially or totally burying a structure. Disguise in war has been achieved by camouflage where an installation is painted, reshaped or covered to mislead the viewer, on the ground or in the air, into thinking it is unexceptional. Where concealment or disguise is impracticable or unnecessary, then the significance of a structure in the landscape will lie somewhere between prominence and assimilation.

Probably a great deal of the diversity of shapes and layout is random and came about as a matter of available space, commercial demand and continuing technical development. Sometimes there was a deliberate input of industrial landscape architecture. Certainly there is scope, in initial plans or subsequent extensions, for consideration of aesthetic concepts or artistic patterns. These may concern contrast or similitude, symmetry or asymmetry, colour or sobriety, lustre or matt, as seen in views from different directions.

In distant views, the skyline profile is the main opportunity to create a stimulating sight. But any isolated object 'must be simple, wild landscape must lap unbroken to its base and no other dominating man-made erection must appear within its zone of influence' (Crowe, 1958).[2] Weather conditions – and particularly sunshine reflection – affect the perceived colour and texture and thus the drabness or brightness of structures. Design should probably strive for the optimum effect in the brighter condition; but anything that can be done to relieve the monotony or menace perceived in a duller light remains welcome (Photos 10.1 and 10.2).

Land use

Whereas 19th century industrial towns arose around factories or mines, more recent industries have relied on cheap transport of raw materials and tend to be located near where settled but locally mobile communities seek employment. However, oil refineries and grain depositories may still need to be adjacent to deep-water ports where bulk loads are delivered; and military forces need empty land for their exercises.

Photo 10.1 This Scottish whisky distillery is brightly welcoming in sunny conditions. Its roofline is fascinating in any weather ...

Photo 10.2 ... but perhaps the subdued brickwork of this, now closed, chocolate factory by the River Avon was well-suited to a rural setting.

One reason for the relatively remote location of certain military or industrial installations is the danger of explosions or release of toxic emissions from weapons or chemical processes. Another reason can be secrecy. Much of the land appropriated may be wilderness, attractive to wildlife, human recreation or both. Wildlife may suffer if habitat is damaged but benefit if human disturbance is limited. People will object on either count. Strategies to overcome these objections must make best use of land already assigned to or spoilt by military or industrial use and, where greenfield sites have to be occupied, should safeguard significant natural features.

10.2 Military installations and land

Medieval castles or city walls were constructed as robust defence against the onslaught of enemies or insurgents whose strength was in their numbers as much as their weaponry. Modern military establishments tend to be more spacious in area, aware of the risk of aerial bombardment but protected from unauthorized ground level entry only by fencing and alarms. Military land space and buildings comprise:

- barrack accommodation, offices, workshops and social amenities only partly restricted to public access: these townships have similarities with civilian equivalents, but a single command authority has opportunities for landscape design or conservation different from, possibly better than, those in more commercially competitive or politically active situations;
- missile 'silos' and underground or 'hardened' weapon arsenals and ammunition magazines, to some degree protected from attack or concealed from aerial photography;
- airfields and depots for wheeled or tracked vehicles, similar to some civil freight airports or transfer terminals; and
- tactical training and weapon testing grounds.

The last category is the most extensive in terms of land space. In England, infantry manoeuvres are undertaken on wide west or north country moorland or within the more restricted sandy heaths of northeast Hampshire not far from the London conurbation. Some heathlands were originally allocated to the army as land unsuitable for agriculture. Extensive attractive scenery was assigned for tank exercises in Dorset to the extent that any villages were evacuated many years ago and conditions are more completely wild; the land is subject as much to natural growth as to the occasional impact of shells or erosion by heavy tracked vehicles. Entry into artillery ranges in South Wales or Northumberland was barred to civilians, at least during firing. However, livestock roamed freely, protected hopefully by notices instructing gunnery officers not to 'engage sheep'.

Kitching (2004)[3] described how the needs of tanks and artillery have been met at Otterburn training area in Northumberland National Park. Landscape and environmental precautions taken to permit modern military manoeuvres include sensitively-coloured buildings of traditional stone and timber, and road widening that is not visually evident because the new formation is grassed over. Land drains are built in local stone and reed beds planted to stem the run-off from impervious surfaces. Some bunds have been constructed to reduce the noise of guns, and stone block forest tracks are laid on textile so that they can be removed when their use eventually terminates.

In the US and the former Soviet Union much larger and more remote areas are available for testing longer-range weapons or even for detonating nuclear explosions. An area of about 10,000km^2 (1 million hectares) was designated in New Mexico and one even more extensive in Nevada. The former includes the White Sands National Monument within its overall perimeter, and one part of it (Yucca Mountain) has been selected for deep storage of spent nuclear fuel.

Of the ten military training areas in Britain, with a combined area approaching 1000km^2, the largest (Salisbury Plain) contains the largest extent of wild grassland in the British Isles, which might otherwise have been ploughed up. Four other areas wholly or partly comprise Sites of Special Scientific Interest (SSSIs), military land being the largest national category for nature conservation. Features of conservation interest typically include:

- habitat in water-filled shell craters or (for spawning frogs) in tank track ponds;
- natural features such as sea cliffs, heathland, bog and marsh; and
- archaeological sites.

Formerly, military priority was paramount in any training area. Today, even national security is unlikely to be given exclusive choice as to what takes place on the land. Governments may have difficulty justifying armed forces budgets unless those forces fulfil important secondary functions when not engaged in or training for their primary combat roles. The world's natural resources are being depleted or damaged at rates which are not sustainable. Waging war is the most wasteful, destructive and polluting of human activities. But government-controlled military or paramilitary forces remain a valuable resource for assisting peacetime civilian effort in development and conservation.

10.3 Large-scale industrial production

This section concerns both heavy industries and production of lighter goods at single centres on a large scale. Crowe (1958)[4] pointed out that a 'fine working landscape can be exhilarating, provocative and with considerable grandeur. It is not therefore a tragedy that certain parts of the landscape take this form; but it will be a tragedy if it is not treated as a landscape in its own right, or if its influence is allowed to seep into the whole of the countryside'.

Types of manufacturing

Heavy industry comprises processing of raw materials and manufacture of machinery. Thus steel, aluminium and alloys are produced as ingots and cast, rolled or drawn into structural shapes, plates or wires and assembled, together with lighter components, into products such as locomotives or hydraulic equipment. Heavy industries are centralized and large-scale and tend to continue to be near water sources, transport links and labour forces. Space has to be assigned for stockpiles of raw materials and waste; and pollution control requirements are increasingly stringent. Large, usually box-like, structures enclose foundries, processing or assembly plants.

Petrochemical processes produce a variety of chemicals, fertilizers and synthetic polymers using a proportion of the hydrocarbon throughput of an often adjacent oil refinery. The combined industry seeks:

- relatively remote sites, because labour requirements are modest but processes are potentially hazardous; and
- access for delivery of crude oil and dispatch of petroleum products, usually by sea or pipeline.

The processes involve a variety of structures and shapes (see p195). These provide an intriguingly complex skyline that dominates flat, otherwise empty landscapes. Sulphurous smells at an oil refinery can be strong. While this is comparatively harmless and unnoticed by seasoned refinery workers, it is desirable that the prevailing wind should blow in a direction carrying the fumes away from any nearby town.

Construction materials such as bricks, cement, concrete aggregates and prefabricated elements are comparatively cheap but bulky. They tend still to be manufactured near to the sources of raw material because of the quantities involved and for consistency. Brick or cement kilns or crushing plants sit among pits and ponds to suit the layout of the works and the sequence of operations (Photos 10.3 and 10.4).

Small goods manufactured in large numbers include textiles, food and drink products and electric or electronic devices for which

Photo 10.3 Chinnor cement works, Oxfordshire, viewed in 2001 from a nature reserve on the Chiltern escarpment. Chalk – for lime – was excavated from a pit in the escarpment, clay – for silica – from near the bottom of the slope. The cement works has since been demolished but …

Photo 10.4 … the nearby chalk pit remains, hidden from most views by the trees surrounding it.

sophisticated or expensive processes require economies of scale.

The *pattern of future industrial production* in any region depends on:

- human resources and a competitive regional economy;
- availability and price of natural resources, including land space and fuel; and
- pollution caused by industrial processes and combustion and the costs of preventing these.

Industrial land use and reuse

Land is one of the most valuable resources in densely populated areas of the world. New industrial plants are an occasional necessity in viable economies, but there is often strong resistance to use of space not previously allocated to industrial use. Textile mills in northern England in the 18th and early 19th centuries were located within tight spatial constraints. In steep country, they had to be constructed within the narrow confines of valley bottoms so as to be near to running water and to road or early railway routes. Because of the shortage of ground space and to accommodate the many machines, mills were constructed on several storeys. In more open terrain, factories were the focus of urban development that had to take place within an area extending only as far as the workers could walk.

From the mid-19th century, choice of factory location was less restricted. Water could be piped to most sites and steam boilers were fired by coal brought by rail. Nor, in the 20th century, did factories have to remain near to the labour force because of improved transport facilities. As to necessary ground space, mechanized processes in factories and warehouses are best suited to one-storey operations.

Today, there are two main alternatives in choosing a site for new industrial development. One is a greenfield site, often apparently less expensive in initial investment but only justified for strong and enduring technical or political reasons. Isolation may be necessary because of potentially hazardous processes; coastal or riverside locations may be needed for delivery of imported raw materials; or there may be a demand for employment in rural areas. Rural land is normally much cheaper than town land. In spacious flat areas, industrial use may not necessarily damage land resources more than intensive agriculture. Nevertheless, even if land is not officially protected, intrusion on rural countryside is likely to be resented. If development proceeds, its layout, process design and appearance must be planned with maximum sensitivity to the surrounding landscape and communities.

The second alternative is to use abandoned industrial space, full landfill sites or disused railway yards or docklands. Such brownfield sites have often been degraded or contaminated by earlier industrial operations. If it is determined that the ground is significantly polluted, remedies that can be adopted include:

- *removal* of contaminated soil for off-site treatment and disposal, requiring allocation of space elsewhere to do this;
- *in situ treatment* by *biological, chemical* or *physical methods*; this may take time and need to be coordinated with programmed clearance of hard obstacles, ground strengthening or foundation construction; and
- *containment* of contaminated liquids and soils beneath the site, for example, by impervious capping layers or by hydraulic barriers constructed around the affected zone; such solutions may be combined with cut-off walls needed during construction, or as part of the permanent foundations, or may form part of the containment system for dealing with effluent or leakage that may arise from the proposed new industrial processes; groundwater protection is usually a priority; so drainage characteristics at or around the development may be critical.

Most brownfield land is regarded as suitable for new industry or at least redevelopment; and most greenfield sites are cherished countryside or at least open space of possible greater value in the future. Whereas industrial or commercial entrepreneurs might be encouraged to clean up and occupy brownfield sites, fiscal penalties might be levied on any greenfield exploitation. These penalties could reflect the long-term damage to or diminution of land resources and discourage it by reducing the net benefits of development.

Incorporating industrial layouts in the landscape

In any land development, location and layout should be those least damaging to fragile local ecological and cultural resources as well as scenery. *Location* has to take into account: the necessary process, storage and ancillary space; access for raw materials, equipment, finished products and people; effects of emissions and effluent; and accommodation for employees in nearby settlements or company townships. Often some of these issues have already been addressed in regional government plans or in designation of specific zones for each type of development or conservation. But formerly, as Crowe pointed out in 1958,[5] there was 'a dangerous tendency for industry to send out scouts over the adjacent countryside in the form of housing or pylons, and then to consolidate the invaded territory on the plea that the landscape is already spoilt'.

Industrial layouts have to suit:

- efficient process sequences and space for stockpiles, railways, roads, conveyor belts and services;
- manoeuvring space for initial earth-moving and ground treatment and subsequent construction operations;
- needs for permanent open space and opportunities to provide this within the site topography – as paved, gravel or more pervious surfacing or as grassland, trees, water features and pockets or corridors of wild or wet land;
- the total industrial scene as it will be perceived in the wider surrounding landscape; and hence
- the visible pattern of the structures themselves.

The *process plants* and their layout design and method of construction must take into account possible future changes in technology or product demand; and space must be left for future expansion. *Construction space* needed beyond building lines can subsequently become open, paved or vegetated land. Planning for that land should be related to surface drainage, roads and their verges as well as space between buildings. There may also be a need for some sort of buffer zone – for safety against fire or explosions or to reduce impacts of noise or air pollution; but buffers need not necessarily be wholly within the formal perimeter of the enterprise. The *surrounding landscape* may therefore incorporate a buffer zone, or there may be a sharply defined boundary, such as a wall or fence, between the industrial plant and, for example, dense woodland; or there may be a less perceptible transition into a neighbouring built-up area.

The *visible impact of structures* arises from their close-up appearance, affecting the outdoor environment within the complex, as well as their profile as seen from afar. Key visible elements are structural shapes in a range from dispersed functional complexity to single box-like envelopment.

Functional shapes

Prolific examples of unashamed functional paraphernalia are the pipework, valves, chimneys, cylindrical or spherical tanks, fractionating towers and heat exchangers of oil refineries. Gushing steam and flickering flames of flared gas add to the strange components of industrial scenery. Or there may be a more disciplined line of brightness. In the mid-20th century, when petroleum plants were being widely set up, Crowe (1958)[6] observed, at Fawley refinery near the New Forest: 'Silver cylinders of oil tanks compare with the seascape of Southampton Water … a free composition given spaciousness by the expanse of sea and sky.'

Outdoor machinery, such as power station switchyards and earth or ore screening plants may be shapes bizarre to the uninitiated; but 'no well-designed machinery is ugly in itself' (Crowe, 1958).[7] Natural or contrived earth forms may provide foreground in these landscapes. Embankments can give thermal or noise insulation or protection against blast or spillage; or can provide visual barriers in front of stark structures, parking areas or activities that might offset finer scenic effects.

Box-shaped factories

A plain cubic or rectilinear block is the most uncompromising building shape, seen by some observers as structural purity, by others as architectural brutalism. However, such a shape is often operationally effective and structurally practicable, in which case perhaps the stark plainness in its appearance has to be its attraction. If a plain block is impracticable then variations or combinations will have to be introduced to satisfy both function and aesthetics.

Operational effectiveness is often achieved through layouts in which a pattern of equipment and operating space is laid out on a rectangular ground floor plan; storage and offices can then be fitted into spare space. The height of the block is determined by the vertical dimension needed for the tallest equipment or the clearance of any necessary overhead cranes or walkways. Under a flat roof, any unused upper space can also be allocated to more offices or laboratories by inserting a second storey where this space exists. If flat roofs are unnecessary or inappropriate, opportunities may exist for:

- use of exterior space at various levels for sun terraces or installations of solar panels or, conversely, to take advantage of shadows; and
- profile design in other forms such as sloping, semi-cylindrical or dome roofs, perhaps transparent to let light into an atrium.

Structural practicability of large industrial blocks concerns material and design of the load bearing members. Before the 19th century, loads were carried on thick walls of brick or stone masonry or, for relatively light loading, on timber frameworks. Floors and roofs were limited in span to whatever could be supported by timber joists, rafters or cantilevers or constructed as masonry or brick arches or domes. In some countries, it is still cheaper to construct factories from local materials than to import steel or other preformed members of structural frameworks.

Ever since the invention of wrought iron trusses, followed by steel and reinforced or prestressed concrete beams, large buildings have been constructed around frameworks capable of carrying both tensile and compressive stresses. Then long span beams permitted wide floor spaces between load-bearing walls or columns. But some of the walls, external or internal, can be non-load-bearing partitions with openings. External walls can incorporate windows or thermal insulation. Light internal partitions can be fitted or moved to suit changing requirements. Ingenious concepts in modern high strength or composite materials can be translated into designs through computer-assisted stress and strain calculations, enabling a range of more daring structural shapes to be constructed. However, the simplicity of design for rectilinear structures has led to the design of conventional framed block buildings more often than exciting application of state-of-the-art expertise in structural engineering and architecture.

The *appearance of box-shaped structures* at close quarters is determined by texture and any protrusions or openings, in the distance by its form, colour and brightness. Texture relates to surface roughness, for example, of brick or concrete, corrugations on iron sheets, or smoothness of plastic sheets. Deterioration can occur with time – as flaking brickwork or failure of its pointing, stained or spalled concrete, faulty or displaced cladding, or rusting of outside pipes or stairways.

10.4 Light industry and warehousing

Before the Industrial Revolution, cottage industries and small-scale factories were common in Europe. The introduction of steam engines enabled much larger scale operations, which rapidly became standard as industrialization proceeded. However, electric or oil-fuelled engines have now replaced most steam power throughout the world. In the 20th century, this enabled a gradual reversion to smaller scale production in certain ancillary industries. Some workshops were located in spare spaces found in cities. But new light industries, such as in electronics and many fields of micro-assembly, have often sought integrated location in industrial estates (Photo 10.5). Activities on these estates include:

- assembly or manufacturing of mainly small products;
- storage, packaging or repackaging and dispatch of products not necessarily made on the estate; and
- offices, either related to these light industrial or warehousing activities or completely unrelated administrative or planning activities for people who enjoy the environment of the new working areas.

Unlike heavy industry, which has to find a suitable location for a prime regional function, light industrial or 'business' estates are seen as focuses for varied productive employment that every town should offer. Planning authorities, encouraged by municipalities or entrepreneurs, have tended to allocate land for these estates at locations suiting regional development as a whole. Design of estate buildings is either speculative and general purpose by the developer of the whole site or specifically functional by a prospective tenant on a chosen plot. In either case the result typically comprises:

- box-shaped buildings which may be used for assembly or even manufacture but are predominantly warehouses; and
- office buildings, often in brick, with copious windows and prominent roofs.

There may be a common architectural theme, resulting either from forms conceived and proven popular by developers or enforced through planning permission policy. The estate tenants may wish to collaborate in particular facilities, such as local renewable energy, for example, from solar panels, and in selling some of that energy.

Layout of industrial or other commercial estates has to take into account:

- the number, space and orientation required for individual units;
- the space required between the buildings for roads and paths, drains and services, permeable surfaces to soak up storm runoff, green open

Photo 10.5 A light industrial and retail estate at Folkestone, Kent (photo by Robin Carpenter).

space and existing or created ecological features – all of which determine scenic quality;
- any communal buildings, such as shops or restaurants, unless these are already available near to the site; and
- parking space for lorries and cars.

The need for space to park lorries is obvious where products are being made or distributed. Car parks are both the attraction and the fault of these estates. Their attraction is to the great proportion of people who work there and who prefer to drive to work if they can avoid the crowded streets and shortage of parking space in typical urban situations. Their fault is both in this encouragement of unsustainable transport practice and in the great expanse of land needed, sometimes as much for car space as for buildings. The solution could be:

- to impose a tax on individual use of car park spaces;
- to provide adequate public transport to and within all large industrial estates; and
- to construct adequate walkways.

Meanwhile, some of the car park space should be regarded as in temporary use, perhaps eventually to be replaced by grassed amenity areas. Indeed, areas planned as amenity space have often been usurped as car parks in the past. All amenity, conservation and car park space should be planned together, with a planting and maintenance strategy, for a total visually pleasing effect within the estate and beyond its perimeter.

10.5 Electricity generation

Electric current is generated by rotation in a magnetic field. The rotational energy is provided:

- by the dynamic force of the wind or the potential energy of falling water; together with

photoelectric, chemical or thermal conversion of solar energy – these are generally known as renewable energy sources; or
- by the heat of combustion or fission of (non-renewable) fossil or uranium fuels, either directly or through the medium of steam – this is known as thermal energy.

Renewable energy, as electricity, currently comprises mainly hydroelectric and wind power. Landscape implications for these are examined in Chapters 6 and 11. In the future, solar power arrays may become profuse on roofs or on open plains. Other renewable sources include geothermal and tidal energy, also sharp temperature differences in the ground or at sea, but each is practicable only in particular climatic, geologic or marine situations.

Thermal energy can be derived by direct combustion, for example, in diesel engines or gas turbines. These applications are well-suited to isolated or small or medium-scale electricity production and can be closed down when there is no demand. But most thermal energy is converted to electricity by boiling water and using steam to drive turbogenerators.

Steam power

Turbine blades of electricity generators are driven by jets of steam released at high pressure. The steam passes onto blades of different size or shape as the pressure is dissipated in several stages, all the turbines driving a common axle which itself drives the electrical generator. Final exhaust vapour at near to atmospheric pressure can be released to the air or condensed to provide water that can then be recycled into the boiler.

The characteristics of different types and layouts of steam power stations depend on:

- the fuel or energy source that heats the boiler;
- the means of dissipating or using the surplus energy in any steam or hot water emitted at the end of the generating cycle; and
- measures that may be necessary to deal with the waste products of energy conversion.

Energy sources for steam power stations are:

- coal prepared and burnt in thermally efficient ways such as fluidized bed combustion;
- heavy (fuel) oil, from the bottom (viscous) end of the petroleum distillation process;
- sorted and baled or shredded domestic or industrial refuse, or agricultural biomass or biogas;
- heat from a heat exchanger connected to a nuclear reactor;
- hot exhaust gas from a gas turbine in a combined cycle power generation system; or
- steam from a higher pressure industrial process.

Energy dissipation of hot steam or water can take place:

- by channelling the hot water directly back to rivers, lakes or the sea;
- by releasing the heat into the atmosphere by open evaporation in ponds or from cooling towers in which the steam is condensed back into water for reuse in the boilers; or
- by using some of the exhaust steam or hot condensed water for urban heating (combined heat and power – CHP – systems)[8] or for other industrial processes.

Waste at power stations can be solid, liquid or gaseous emissions. A common solid waste is ash from coal combustion; as pulverized fuel ash (PFA), it may be used as a weak low-carbon substitute for cement, as an aggregate in lightweight concrete or, where it is especially abundant, as an earth fill material. Radioactive waste from nuclear reactors occurs in much less quantity but requires secure containment before transporting and recycling or long-term storage and monitoring.

Liquid wastes at a power station are usually confined to rainfall run-off. Profuse run-off from heavy rainfall can possibly be contaminated by stockpiles, for example, of coal or ash, and might require treatment before discharge to sewers or watercourses. Cooling water, returned to rivers or lakes, is not likely to be polluted but its higher temperature may affect river life. Large pike thrived in Trawsfynydd reservoir while the nuclear power station was operating. Some French inland nuclear plants have to shut down when river flow is exceptionally low.

Gaseous emissions from power stations include particulates (smoke and dust), sulphur dioxide (SO_2) and nitrogen oxides (NO_x) as well as the greenhouse gas carbon dioxide (CO_2). Local pollution can be reduced by tall chimneys. However, SO_2 and NO_x are carried a long way by the wind and are thus associated with long-distance contamination including acid deposition. These emissions may be significantly reduced by flue gas desulphurization (FGD) and selective catalytic reduction (SCR) processes. Particulates can be removed by pre-treatment of fuel or post-combustion scrubbing. These processes produce their own waste.

The location and appearance of steam power stations

The way in which a power plant fits into the landscape depends on its location, layout and the design of the structural elements. *Location* is particularly dependent upon a supply of cooling water and whether a remote location is required or one close to the area where the electricity or by-products are required.

Feed water may be supplied on open (fresh supply or 'once through') or closed (condensed steam) systems. In an open cycle, water is drawn from a large cool source, such as the sea or a large lake, and used as boiler feed; exhaust steam is condensed and returned as warm water to the source. Very large quantities of water are required and high capacity channels may have to be constructed to and from the power station. Therefore, for moderate-sized water sources, such as any English river, the demand for water would be excessive and the hot discharge would probably damage river life. In these circumstances, as at the inland power stations in England's Trent Valley, water from the turbogenerators is recycled and returned to the boilers after condensing in cooling towers. Nevertheless, the boilers have to be topped up from a local source to make up for any leakage in the cycle.

Remote locations are usually chosen for nuclear power stations. Wide river and coastal situations may also be preferred for ordinary steam power plants and their comparatively large land needs may have to be accommodated in rural areas and possibly near land of scenic beauty. However, especially in the colder climates of central, eastern and northern Europe, steam power stations are close to cities so that some of the waste heat can be utilized in CHP schemes. The total amount of space is typically 15ha for a city 1000–1500MW steam power station, 50–100ha in open territory.

Layout of steam power stations has to accommodate:

- boiler houses, turbogenerator halls and administrative buildings, typically occupying 4–5ha;
- space for additional generators to be installed as demand increases or for peak generation, for related industrial facilities or for special processes such as coal gasification;
- space for clean-up of waste emissions and disposal of sludge such as from FGD or SCR facilities and possibly later for carbon capture;
- fuel and ash storage areas with reclaim and conveyor equipment;
- cooling water towers, switchyards and transformers for conversion of electricity output to high voltage for efficient transmission; and
- ancillary buildings, parking, amenity and conservation space.

The latter has to take into account public and wildlife access to coastal, riverside and woodland paths in the locality.

Structural design commences at the foundations, which have to be substantial for the heaviest plant such as boilers or nuclear reactors. A single reactor unit may weigh 50,000 tons and needs very secure load-bearing foundations to avoid any possibility of fracture of the reactor structure or its connecting pipes and heat exchangers. Turbogenerators need foundations capable of absorbing their vibration.

Boiler and turbogenerator houses commonly assume a box-like form, largely determined by how the heaviest equipment is lifted and moved during installation or replacement. Tall chimneys of coal-fired boilers are the most prominent features of steam power stations; they may also be attractive in bright colours with sufficient contrast to make them visible to aircraft (Photo 10.6). Other elements provide further opportunity for variations on rectilinear design – domes at the top of nuclear

Photo 10.6 The chimney of this power station in Poznan, Poland, is pleasantly striking. The turbogenerator buildings gleam in the sunshine and the transmission lines are functionally acceptable. There are no cooling towers because waste heat is dissipated in district heating or the nearby river. The main blot in the landscape is the scruffy foreground; but that is no insoluble problem.

reactor housing or spherical liquid or gaseous fuel tanks. Most elegant of all can be the hyperbolic curves of natural draft cooling towers, the effect perhaps enhanced by clouds of condensing steam. Mechanical draft cooling towers may achieve a more efficient thermal cycle at a lower capital cost, but few built so far are visually attractive.

Chimneys, cooling towers and tall buildings, seen at a distance, should rise clearly from the ground. Ancillary buildings, switchyards and car parks can be kept to lower (even excavated) levels so as to keep them out of the main view; or they may be hidden behind hedges or woodland to the side of long vistas giving clear views of the more stately parts of the power station.

Non-steam thermal power stations

Diesel generators are needed for isolated communities that have no reliable energy source of their own or any transmission line connection. The diesel power house should blend with any other buildings and the absence of high voltage transmission lines should help to preserve the isolation. Gas turbines do not require large areas of space and can usually fit unobtrusively into the existing space allocated to steam power stations which they replace or augment.

Electricity production in regional economic development and landscape

Electricity is a main form of energy used by industry, rail transport and domestic consumers. It is also the principal means of harnessing renewable energy, derived in turn from engineering works that impinge on the landscape.

In rural areas, there is skill in selecting a site and developing a power station structure that will dominate its site without compromising any beauty in the greater landscape. Beautiful landscapes may be affected by structures in places

Photo 10.7 In the background, by the distant lake, is Trawsfynydd nuclear power station – now closed but still a national grid transmission line junction. In the foreground, and connected to that grid, is the power station at the lower end of the Ffestiniog pumped storage scheme (see also Photo 6.13).

that are remote (and may accommodate nuclear power stations) or mountainous (that suit pumped storage hydroelectric schemes). Photo 10.7 shows both these in combination.

At all sites, there is a need to conserve land space for future power generation, whether by constructing new nuclear stations at the same sites as those that are being decommissioned, by conserving the storage capacity of reservoirs in a hydroelectric scheme or by maintaining and possibly reducing the visual impact of the essential connecting transmission lines in line with any changes in supply pattern. Land-use planning should be integrated with that of light and heavy industries, energy production, municipal and military services, public transport depots, waste recycling and landscape amenity, some of them gathered within single 'parks'.

Notes and references

1 Binney, M. (1996) in a *Times* supplement on New Towns.

2 Crowe, S. (1958) *The Landscape of Power*, Architectural Press, London, p51.

3 Kitching, R. (2004) 'Undercover mission', *New Civil Engineer*, 2 December, pp20–21.

4 Crowe (1958) as Note 2, p56.

5 Crowe (1958) as Note 2, p27.

6 Crowe (1958) as Note 2, p52.

7 Crowe (1958) as Note 2, p43.

8 CHP is the use of a heat engine to generate both electricity and useful heat. In power stations, it works best in cold climate cities where there is a need for both electricity and district heating. But there is seldom an ideal balance between electricity and heat. Transmission of steam or hot water in pipes is much less efficient in energy terms than electricity along cables, unless the heat is used for a process very close to the power station.

11

Towers

High buildings may be constructed to save ground space in areas where population density is high but land is scarce. The structure of tall residential and office blocks is described at the end of this chapter, their role with other buildings in Chapter 12, and their displacement on the land as an issue in urban development in Chapter 15. Many other man-made structures rise high for a specific purpose other than saving space. Their main impact on the landscape is visual, often deliberately.

11.1 Structure and purpose

Tall structures dominate their surroundings. Inevitably they are principal elements of the scenery – perhaps the only element when isolated, part of a skyline when in groups. In flat country, towers are the only landmarks, at sea the only fixed features.

Natural rock towers are seldom abundant but usually spectacular. They are outcrops that have resisted erosion by ice, water or wind more stoutly than the surrounding formation. They can occur on otherwise flat peneplains (as in Monument Valley, Arizona), along mountain arêtes (the Chamonix Aiguilles) or in rocky coastal situations (as isolated offshore stacks, such as the Old Man of Hoy in the Orkney Islands). Their isolated grandeur should be respected.

Whereas natural towers are strong random remnants, man-made towers are constructed deliberately. Their structural integrity depends on adequate foundations to support their weight, and on sufficient strength in their superstructure to withstand lateral (wind) loads or seismic stress that might precipitate their collapse. Structural types of tower include:

- solid or almost solid, mainly monumental, structures;
- more open masonry structures, using arches to avoid tensile forces or timber beams to withstand them;
- framework towers or pylon-supported structures, in materials with tensile strength such as steel or reinforced concrete; and
- masts whose weight is borne on a single point and all horizontal stresses by cables guyed to outlying ground anchors.

Figure 11.1 shows various concepts for vertical structures.

Solid monuments can be formed in shapes from broad-based pyramids, in which stability can be ensured by sheer mass of material, to slender obelisks, columns or giant statues of more complex construction. Within substantial monuments, shafts or passageways can be constructed for access to the inside or to the top.

Masonry towers depend on sound foundations below, and on permanently compressive forces for stability in their superstructure. Foundation failure is comparatively rare. Expensive engineering[1] was applied to the base of the 56m-high, 19m-base diameter Tower of Pisa to correct part of its 9 per cent lean and to prevent it leaning further towards collapse while preserving the great attraction of its deformity.[2] Design of such a tower today could take advantage of more revealing initial investigation of ground conditions and then of techniques for strengthening that ground or extending the foundation to spread the load of the tower more widely.

The stability of the main trunk of brick or stone masonry towers depends on its strength in

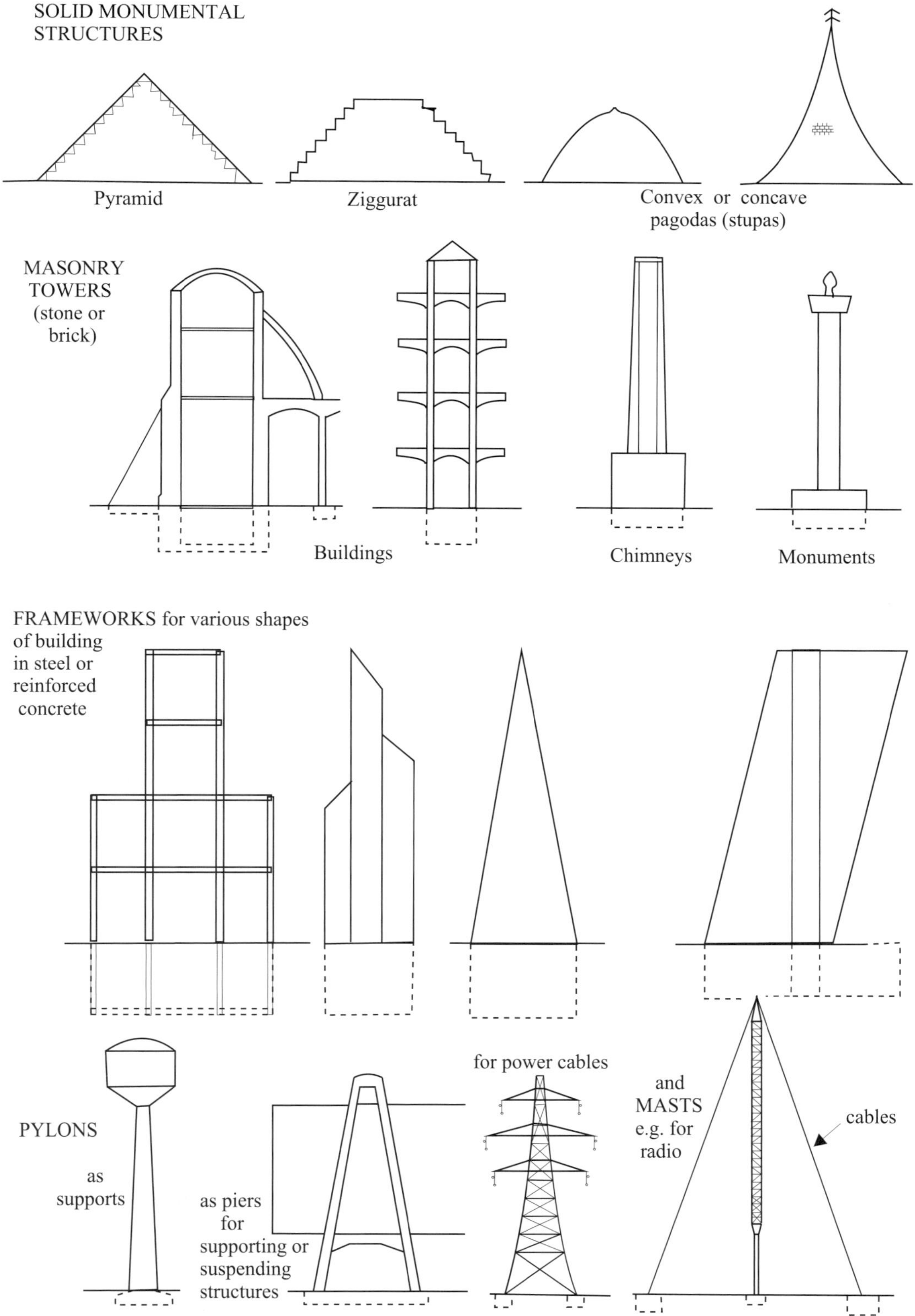

Figure 11.1 Concepts for vertical structures

supporting the load above each level and on its ability to resist overturning moments resulting from any side load. In slender towers, conical shape gives gradual weight reduction approaching the top while a circular plan gives horizontal arch strength, convenient construction and attractive shape. Both minimize wind load stress. In wider towers, resistance to overturning can be provided in the form of internal or external bracing.

Internal bracing can be provided in masonry towers by stout timber, including in floors or roof, or by masonry arches, struts or domes. External bracing can take forms such as flying buttresses supported at a wider lower level, or it may be the adjoining walls of castles or naves of cathedrals above which towers gain extra height and prominence. Buttresses add to the viewed shape. So, for the best visual effect, these supports should resemble extensions of the lower external structures and should not compromise the main tower's verticality.

Tall frameworks are constructed in steel, reinforced concrete or composites either as supports for high level platforms or suspension points or as the main load-bearing structure of high-rise buildings. Supporting frames may be steel lattices for electricity transmission lines or radio aerials, stout portals carrying bridges, or more pleasing – often curvaceous – concrete for overhanging water tanks or airport traffic control rooms.

High-rise buildings maximize accommodation on restricted ground space. They are towers if they are tall rather than long or wide. The taller the tower, the stronger must be the foundations and the more lateral loads have to be evaded or absorbed.

Masts are particularly light and slender functional structures. Their lightness may help their functionality to share the landscape rather than to blot it.

The *purposes* of towers can be visual inspiration; navigation, communications and other functions requiring height above ground level; electricity generation or transmission; or high-rise accommodation. Each is associated with particular characteristics that affect their appearance in the landscape.

11.2 Remembrance and inspiration

Tall monuments commemorate events and lives of the past or enthuse people for the future. Monuments can be considered in three categories:

1. pointers to heaven, adjuncts to religious or other prominent buildings, inspiring pious pursuits, worthy living or municipal pride;
2. obelisks, originally Egyptian commemorative symbols; and
3. statues, depicting people or personified ideals.

Towers and spires pointing to heaven

Church bell towers, mosque minarets or clock towers are not, in fact, divorced from utility; they resound with periodic calls to worship or chiming reminders of the time of day. But their age and symbolic status make appearance rather than function the prime test of their quality. *Campaniles* were bell towers conceived mainly as concomitant features of churches, but also as lookout posts or navigation beacons; even church construction was sponsored by temporal lords; so was that of public buildings. Before Florence's cathedral and campanile were completed, Siena already had a secular tower 102m high.

A few towers of remembrance can challenge campaniles as prominent features in distant landscape (Photo 11.1).

Minarets were conceived as lofty points from which muezzins could call. Modern loudspeakers can do so effectively, even at low level and on more modest mosques; but minarets remain admirable (Photos 11.2 and 11.3). They define the corners of Turkish domed mosques or open ones in warmer Pakistan. In Iran's Isfahan, they soar above glorious entrance facades. More commonly, minarets are single, tall, thin towers, akin to spires as distant stand-alone symbols.

Spires commonly cap church towers and occasionally more secular buildings in western Europe (Photos 11.4 and 11.5). Further east, onion domes cope well with heavy snow. Cologne has the highest masonry church tower and spire (156m). Salisbury cathedral has the tallest tower and spire in England (123m). The top of the latter can be

Photo 11.1 The 51m high funerary tower, Gonbad-e Qabus, in northeastern Iran was completed in 1006 in brickwork that would be admirable in any era. Shortly after the author's visit in 1957, a rival grain silo was erected nearby (de Schooten, 1963).[3]

Photo 11.2 Minaret in Shoreditch High Street, London.

Photo 11.3 Minaret in a small village in Anatolia, Turkey.

Photo 11.4 This church spire at Stamford, England, is the climax of a fine architectural ascension.

seen from several miles away where any evidence of the city itself is obscured by a foreground of high rural landscape.

Pagodas are the common Buddhist equivalent of Christian spires; indeed they are locally more numerous. The Burmese form of pagoda is a concave-sided stupa from a shallow cone at the

Photo 11.5 20th century tower and spire of Nuffield College, Oxford.

Photo 11.6 Sule Pagoda, Yangon.

bottom to a sharp spire at the top (Photo 11.6). The more splendid examples are covered in gold leaf and surmounted by a *thi*, a light metal mobile of metal or glass pieces tinkling in the wind. Almost all are solid, constructed of brick and, where not gold-faced, plastered and whitewashed. 'There are pagodas in every village, at most unexpected places along jungle tracks and especially on all conspicuous hills so that they form one of the most striking features of the country' (Dudley Stamp, 1955).[4]

Sri Lankan pagoda stupas are generally in a convex inverted bell form, rounded rather than pointed at the top. Chinese pagodas are completely different, being straight-sided balconied towers, probably more secular than sacred in context.

Obelisks

Tapering pillars, usually four-sided and of stone, spring up into the sky from ground level. They are memorials in public places or 'prospect' symbols in contrived landscapes.

Egyptian obelisks were shaped as single pieces at the quarry and hauled thence, sometimes over long distances, to wherever they were required. Some exceeded 500 tons in weight. The so-called 'Cleopatra's needles' were two 21m-high granite obelisks brought from Heliopolis to Alexandria and, eventually in 1878, one went to London and the other to New York.

The Washington Monument is an obelisk completed in 1884 and, at 169m, it is the highest masonry structure in the world. Iron frameworks within it support only the lifts and stairs. More numerous smaller American pylons are the high 'M' signs that indicate to strangers the location of a McDonald's eatery. These are now ubiquitous, in Europe being as welcome as more modest 'M' logos showing where to find Metro trains. In either case these structures often fit local scenery better than electricity or radio communication paraphernalia; if they are thought ugly, they can easily be removed.

Statues

Larger-than-life likenesses of people or ideals in human or animal form are common city centre memorials; they are significant in wider landscape only if big or isolated.

The Colossus of Rhodes (c.280BC) bestrode the harbour entrance until it was destroyed by an earthquake. The Statue of Liberty by New York Harbour (1886) was designed by F. A. Bartholdi (1834–1904) who specialized in gigantic and ostentatious statues. A gift from the French republic to the American people, perhaps it commemorates the centenary and success of the French and American Revolutions. More recently, a colossal statue of Jenghiz Khan has been completed on the spacious plains of Mongolia.

Whether the statue of Christ on top of the Corcovado, 712m above Rio de Janeiro, is an inspiration to people or an affront to natural splendour is a matter for conjecture. At the top of a steep funicular from the city, it might be argued that it is part of the city scenery rather than of the main background of mountains rising to more than 1000m. It may therefore be splendidly dominant rather than obtrusive in its setting, which is not all fine buildings and wooded slopes anyway, part is despoiled by slum shanty settlements; and the higher mountain skyline is now disfigured by tall radio towers. In other lands, roadside crucifixes or shrines are part of cultural foreground; but stark wooden crosses (Corbar Crags, Buxton) or more ornate iron symbols on central European peaks despoil the summits of peaks much more than small cairns of local stones. The Angel of the North, constructed at Gateshead in 1997, is a 20m-high human form with long wings in the form of steel rectangular sails, controversially claiming aesthetics in its size and structural engineering.

Some of the largest statues are not detached towers; for example they may be cut into the side of living rock as were four US presidents at Mount Rushmore, the recently spoiled Buddhas at Bamiyan, Afghanistan, and the four enthroned colossi at Abu Simbel, Egypt, which were raised above the Aswan Lake level in the 1960s.

11.3 Tower viewpoints

Discounting broad-based structures such as the Nile pyramids or Babylonian ziggurats, the first true towers were for defensive observation platforms in Mesopotamia, Assyria, Persia and the Roman Empire. Similar observation towers, with guardrooms at the bottom, were located to control access to strategic mediaeval bridges. Military observations have since been taken from balloons, in the 20th century from aeroplanes and now from various orbiting satellites. Popular appreciation of scenery from the air is now widely possibly from aircraft or less hurriedly and at closer quarters from baskets suspended below hot air balloons or from giant wheels (Photo 11.7). Meanwhile people can continue to enjoy views from high ground or tall structures.

In *hill and valley topography*, views from high ground looking downwards upon natural scenery are usually afforded without the benefit of constructed devices. However, towers have been built on high ground in particular circumstances, for example, to see over the top of dense woodland.

In *flat country* there are views only across open ground, and in forests or built-up areas, potential vistas may be blocked by trees or buildings. Observation towers are then necessary for panoramic viewing, although only exceptionally can the cost of construction be justified to that end alone. So, *in cities*, where man-made structures dominate anyway, viewing platforms are most frequently found on towers built partly for other purposes. Thus:

- the Leaning Tower of Pisa, completed around 1370 as a campanile, later became a public viewpoint (Photo 11.8);
- the Eiffel Tower, at 300m, was the highest iron structure in the world when constructed in 1889 to demonstrate that material's capability and to celebrate a Paris exhibition (Photo 11.9);
- some medieval Italian towers were high home-fortresses or watchtowers as well as the summits of prestigious buildings. (Photo 11.10);
- the Scott memorial in Edinburgh is a monument to Sir Walter Scott; and
- the Empire State Building and most other skyscrapers are office blocks.

Photo 11.7 The London Eye provides wide views of the city as it slowly revolves.

Photo 11.8 The Leaning Tower of Pisa (photo by Robin Carpenter).

Photo 11.9 The Eiffel Tower, erected in 1889.

Photo 11.10 View of Siena Cathedral from the 102m-high Torre del Mangia (photo by Robin Carpenter).

Other towering monuments with viewing terraces or galleries are the Arc de Triomphe in Paris (50m high, finished in 1836) and the Gateway Arch at St Louis (192m, 1966). Many very high towers have been constructed for radio or television (see below). Some have incorporated public observation galleries – at 350m on the 553m-high CN communications tower at Toronto (1975) or, more recently, in the 612m-high 'TV & Sightseeing Tower' in Guangzhou, China.

11.4 Navigation and communications

Navigation structures are typified by lighthouses and airport control towers. Other communications facilities on taller towers, specifically designed for the transmission or receipt of radio waves, are either robust freestanding structures or lighter stayed masts. Finally, in a variety of configurations, there are ground facilities for extra-terrestrial investigation by means of telescopes and radio transmission.

Lighthouses

Where land scenery is rugged, so may be the adjacent seascape of isolated rocks or hidden reefs, hazards to shipping which call for warning measures. Lighthouse foundation construction often required heroic efforts in wet and stormy conditions. The tall superstructure was commonly of robust stonework or concrete in an elegant tapering or concave shape. Lighthouses at sea are elements of seascape, together with oil platforms, wind turbines and ships.

In some situations, it has been practicable to site distantly visible lighthouses more accessibly on the coastal land, for example:

- at harbour entrances on breakwaters or quay walls;
- on the shore line, as below the cliffs at Beachy Head;
- near the top of cliffs, as near Newhaven, where lower structures have adequate height but may be threatened by cliff erosion; or
- further back, within and overlooking a cliff-top town such as Southwold in Suffolk.

The last example is in already built-up landscape, the others are more isolated landmarks. But, in the comparatively remote location of some land-based lighthouses, there may be opportunities for part-use of (operating or redundant) lighthouses or coastguard stations as bases for wildlife observation, especially of sea birds – for example at South Stack near Holyhead.

Air traffic control towers

The very substantial amounts of money associated with construction and operation of large airports imply that associated buildings, particularly spacious terminal buildings and tall control towers, can afford to combine advanced construction technology and functional excellence with outstanding aesthetic design. Air traffic control towers do not need to be exceptionally tall, as long as the controllers are high enough to have a clear view of the sky and the runways and taxiways. At the same time, they are usually prominent enough to be seen as landmarks by travellers on land routes leading to the airport and of an appearance complementing the other buildings, seen both from the air and the ground. Commonly these towers are of concrete, slip-formed stems supporting mushroom or ellipsoidal cabins. Design, construction and subsequent maintenance of the structures should ensure against discoloration or blotching of their clean, bright appearance.

Telecommunications towers

Powerful long-distance radio transmission aerials often exceed 100m in height. Particularly in the environs of cities, where the landscape is already largely urban, advantage has sometimes been taken of these height requirements to build a tower with viewing facilities and public accommodation. Such towers have typically been even taller but much more slender than skyscraper buildings; and have been round in horizontal section, most of the stem being constructed by slip-form concreting but to a considerably greater height than most airport control towers. Those at Seattle, Toronto or Guangzhou have become dominant city symbols.

Many lesser radio communication towers, often in steel latticework, are interesting rather than elegant. When they become redundant, they can be dismantled and removed, perhaps to the sorrow of farmers who have received good rent for the modest land areas concerned (Judge, 1999).[5] More commonly, mast solutions have been adopted in powerful radio transmission. Slender rod or lattice structures can be tall and striking without unsightly visual preponderance. The relatively modest vertical load is borne at a single point central foundation. Guy wires secured at encircling ground anchors stay lateral forces and any tendency to buckle. The most difficult part of providing these masts is in their erection or dismantling.

Extraterrestrial communication

Traditional astronomical buildings are squat towers with domed roofs through which telescopes can peep. Remote locations are preferred to avoid disturbance by artificial light. On Mount Teide in Tenerife the buildings are strikingly bright white amid a wild volcanic background. Thus they are mildly pleasant landmarks in the already accessible foreground of fine natural scenery. Modern radio-telescopes are of a size and shape that cannot fit into towers. Individual components are widely spread on the ground; but each components may still be prominent in comparatively empty landscape. If their shape is still considered discordant then, when their usefulness is over, they can be dismantled.

11.5 High tanks, chimneys and heat exchangers

The purpose of elevated tanks is to store liquid (usually water) at a height from which it can flow by gravity through distribution pipes and out of taps. Chimneys dissipate gaseous emissions that would otherwise cause dense pollution around factories or power stations. Visual forms of heat exchangers or dissipators include petroleum fractionating and water cooling towers.

Water towers

Elevated tanks are not necessary in steep country where a surface reservoir on a slope or hilltop can provide all the necessary head to command

Photo 11.11 Water tower at Berinsfield, Oxfordshire, a village of 2700 people that, with its own supporting infrastructure, was built as a completely new settlement in 1960. The tower's appearance is somewhat marred by telephone aerials.

a town or village below; nor are they required in major water supply systems in flatter territory if it is deemed more economic to pump water, and thereby maintain pressure, directly from major reservoirs. Water towers are common on comparatively thinly populated flat areas or plateaux; or they provide a secure short-term gravity supply of water to isolated settlements (Photo 11.11). Early or mid-20th century towers commonly support masonry, concrete or steel tanks for farms, factories or railway stations; often these structures illustrate architectural motif; sometimes accommodation floors and windows are inserted below tank level.

Modern concrete water towers come in a variety of geometric shapes on slender slip-formed shafts, structurally similar to some airport control towers. They are often welcome features seen beyond field foreground, over trees or in contrast with church spires.

The pressure in water supply systems is a hidden element of power which can occasionally be released as a decorative fountain. Geneva's Jet d'Eau was designed as a safety valve to release excess pressure in the city's water system. Today it is a spectacular municipal symbol soaring up to 140m above Lac Leman.

Chimneys

Tall brick chimneys, built from the industrial revolution until the early 20th century, were intended to create a draught for furnaces as well as to reduce smoke in the vicinity of mills and factories. Later 20th century factory, refinery and power station chimney stacks, usually in concrete or steel, were constructed to a height (even beyond 300m) judged sufficient:

- to dissipate sulphurous and other pollutants at least to beyond the immediate region; and
- to pierce any low-lying blanket of temperature inversion which might occur in certain weather conditions.

Brick chimneys secured their strength through circular (compression) brick courses of a thickness diminishing towards the top. Some taller concrete chimneys taper similarly.

Brick chimneys proliferated in industrial towns in the 19th and early 20th centuries. They were the core of industrial landscapes such as those of South Lancashire, recorded in L. S. Lowry's paintings and demolished or restored by Fred Dibnah. Later steel or concrete chimneys are generally brighter and undecorated. Large capacity single stacks tend to be isolated and locally dominant. However, if multiple stacks are appropriate, their configuration can be aesthetically interpreted. For example, there may be a rising set of chimneys emitting at different top levels where wider dispersion is desirable.

Heat dissipation or transfer

Power station cooling towers in Eastern Europe are often plain trapezoidal towers, sometimes constructed in wood. 'Forced draft' cooling structures elsewhere may be similarly austere. The more graceful 'waisted' hyperbolic shape of western natural draft cooling towers has been described as anticlastic – concave in one direction, convex in that at right angles. Such shapes can incorporate straight lines and can be constructed of straight pieces or with cables stretched tight over the surface, thus serving a constructional need as well as those of aesthetics and efficient convection. Plumes of white steam amplify the scenic effect in certain atmospheric conditions.

Heat exchangers and fluid distillation cylinders in chemical, petroleum or power plants are often vertical. Their geometry, however, is complicated by connecting pipes and valves, a near or middle distance sight of functional fascination. In the distance they are weird features of intriguing industrial landscape. In the future, extremely high chimneys may be used in some methods of harnessing solar energy.

11.6 Wind machines

Harnessing wind power

The force of the wind can propel a rotating shaft. The propeller may be the sails of a traditional windmill or the aerofoil blade of a modern wind turbine. The shaft can drive through gears into mechanisms for grinding corn or raising water; windmills were in use by the 14th century, draining marshes and lakes in the Low Countries; or the shaft can rotate an electricity generator. Electrical output may be used directly, for example to drive a nearby electric pump which can operate whenever the wind is blowing; or it can be transformed to a suitable voltage for transmission to some larger power system.

Some old windmills were constructed so that the whole structure could rotate around its base to make best use of the wind's direction. Modern horizontal axis wind turbines can also do this. The sails of old windmills were as large as could be practicably constructed. The mill tower then had to be high enough to house the rotating axle at the centre of the sails. Obsolete, now sail-less, towers are a picturesque feature of central Norfolk and, often restored complete with sails, in the Netherlands (Photo 11.12).

Wind electricity generators are no recent invention. There were 2500 in use, with a total generating capacity of 30MW, in Denmark by the end of the 19th century.[6] In the 1930s, many thousand supplied US farms not yet connected to grid supplies. But it was the late 20th century's global awareness of impending energy shortages and of the environmental and climatic hazards of fossil fuel burning that led to greater attention to wind power and other sources of renewable energy. In the present century rapid expansion of wind 'farms' is taking place (Photos 11.13 and 11.14). Ultimately, the justification for large-scale harnessing of wind energy depends on the cost of transmitting that very intermittent energy and the capacities of regional power networks to make use of it as it occurs.

Photo 11.12 Dutch windmill.

In countries such as Britain, which have relatively shallow coastal waters, wind energy is increasingly being harnessed offshore. However, vast regions of windswept land are remote from the sea. There, land-based wind power will probably remain a principal opportunity for generating renewable electricity. The extent to which windy land space should be made available becomes a

Photo 11.13 Wind farm from the front, in sunshine ... (photo from www.istockphoto.com)

Photo 11.14 ... and from behind and under an overcast sky (photo by Robin Carpenter).

controversy, partly between renewable energy and environmental (conservation or landscape) interests.

Modern wind energy generators

Most wind turbogenerators are driven by propellers rotating on a horizontal axle perched on a tower. The energy of the wind is proportional to the cube of its velocity; and generally the faster winds are higher up. The height of the rotors, and hence the supporting towers, should therefore be as high as is economic and practicable. The blades, which typically rotate at 200km per hour at the tips, should also be well clear of any people, animals or vehicles that might run up to the tower base.

On land, the size of individual turbines and their disposition in wind farm groups depends on the local wind occurrence pattern, the topography and accessibility of the sites and the available designs of generators and tower structures. At the beginning of this century, a typical 250kW generator was propelled by 10–20m-long blades on a 25m tower, more recently 2–2.5MW with 30m-long blades on a 65–75m tower. Note that generating capacity is quoted in peak megawatts, in other words, only when the wind is blowing hard.

The spacing of generators may be as close as does not cause wind shadows to substantially reduce the output of each turbine. In 1995, Otway[7] mentioned early research in British conditions indicating that 'the spacing of wind turbines in clusters, where the wind is variable in direction, should be about ten diameters of the wind turbine rotors.' More recently, MacKay (2009)[8] cites general expert opinion that 'windmills can't be spaced closer than five times their diameter'. Whatever the ratio for optimum spacing, there is no gain in varying rotor size at the same height. So the main reason for building large wind turbines is for economies of scale. In early wind farms with relatively small machines, for example at Altamont Pass in California, 2ha of space was allowed for each turbine tower. A recent scheme near Glasgow allows nearly 40ha each for much larger turbines (MacKay).[9]

Although wind farms are proliferating rapidly, the optimum type, design and configuration of turbines may not yet be finally determined. There may yet be changes in the shape of blades, their axle directions or the ways of controlling them in differing conditions, and possibly use of several smaller turbines rotating at very high speeds on a single framework. For any design, important economic and logistic considerations include the size of tower and turbine elements that have to be transported, assembled and erected at different sites.

Towers for on-shore turbines are usually founded on cast in situ reinforced concrete footings, on a central boss of which the tower or mast is fitted. Hopefully the footings are of such a size and arrangement as to suit replacement or improved machines later; and the footings should be reducible so as to render land suitable for other uses if any wind farms eventually become redundant.

Environmental objections commonly concern the land space taken by wind farms, hazards to wildlife, noise and visual intrusion. Actual *land take* is ground space for tower foundations, any associated buildings and vehicle access. But the total land space affected and allocated is all that within each wind farm's boundary. Because of the wide spacing of towers, there is usually land left between the towers for agricultural or other purposes. On remote moorland – such as in central Welsh uplands – the land continues to serve as rough grazing. However, across arid stony Californian hillsides, the cutting of access tracks to each relatively closely spaced tower causes surface disturbance across a wider area of what is admittedly more sterile ground. On the other hand, on the North European Plain – where no site may be particularly windier than another – arable agriculture is practiced right up to the wind tower bases.

Hilltop locations are frequently of archaeological importance. If foundations for towers and road bases are constructed carefully their excavation may provide evidence that can be used subsequently to investigate wider land areas. Elsewhere wind towers may be sited – sometimes in lines – where there are already suitable foundations, for example, along high ridges, lowland embankments or offshore rock spits or breakwaters.

The main *wildlife hazard* is the danger to birds, which can be killed or confused by rotating blades; this is similar to the more lethal and more widespread danger to birds of road traffic. Wind machines can be designed to reduce the perching and nesting opportunities for birds. On the windswept peninsular of Kintyre in Scotland, eagles were encouraged to move away from the area of a proposed wind farm to a new equally remote grouse moor habitat, created for them at a safe distance (English, 1999).[10] Meanwhile, further observation continues to establish the long-term effects on birds of wind generator propellers.

Noise of wind generators comprises mechanical (gearing) sound, aerodynamic effects and the swish as each blade passes the tower. Some of these sounds may be modified by design. At a few hundred metres from both, the continuous noise of a wind farm is less than that of a highway in the daytime; or, in windy conditions, the noise of wind in the trees will drown out that from the wind farm. There are occasional reports of undoubted distress caused at night but, as far as effects on land resources are concerned, there is little evidence that wind farms have any influence on house prices.

Visual intrusion is a more contentious issue in establishment of wind farms. Regarded by some as unwarranted intrusion on long-esteemed scenery, it is also argued that the middle distance appearance of wind generators is of that of elegant windmills spinning in the breeze (Photo 11.15). Furthermore, being bright, they shine attractively in the sun amid surroundings that are usually open and unspectacular, even on high moorland. Musgrove (2003)[11] explains that British turbines in the early 1990s were white because that was the preferred colour in Denmark, where the turbines had been pioneered. 'Most subsequent wind farms have used light grey turbines, so as to minimize the contrast with a variety of backgrounds, but against a dark sky background they appear white.'

On flat, uninteresting country in continental Europe, coloured warning stripes (for aircraft) on turbine blades brighten scenes as much as they do on power station chimneys (see Photo 10.6 in Chapter 10).

In England, national parks, areas of outstanding natural beauty and heritage coastlines are zones specifically protected against most forms of development; and it has been generally accepted

Photo 11.15 Can function still be beauty?

Source: *New Civil Engineer* (by kind permission)

that wind farms should be among the developments banned within their boundaries. Similar restrictions are suggested for such recognized zones in most countries. Meanwhile, outside the limits of protected landscape but still in relatively remote semi-natural situations:

- the scenery is, almost by definition, less spectacular than steeper mountain slopes or wilder woodland;
- comparatively few people make recreational use of these areas, except to seek solitude (which is invariably a loser where remote development takes place); and
- the mechanical arrays may be visible from distant scenically protected mountain-tops but need be no more intrusive than other visible features beyond national parks and are often of interest to walkers who are there to see the view.

There is great uncertainty as to where future energy supplies are to come from. Wind power is being proposed as a possible substantial contributor. There remain many doubts as to its comparative economic viability but the present proliferation of schemes may serve to publicize the problem and the unlikelihood of other 'something-will-turn-up' renewable energy. In case wind power proves inappropriate in the future, present planners and constructors should ensure that the structures are dismountable and that each site might be restored to some semblance of its former state.

11.7 Electricity transmission lines

Efficient electricity transmission

Most electricity is generated and consumed at relatively low voltages; but, for long distance transmission, the higher the voltage the less the transmission losses. Effective electricity transmission along cables requires:

- insulation so that electric current cannot escape to the ground or interfere with or induce current in nearby electrical or electronic equipment;
- dissipation of heat generated by resistance to current along the cables; and
- reduction of the resistance losses to acceptable proportions.

These requirements are most easily met by cables suspended in the air. Insulators are inserted in the equipment suspending the cables from pylons, heat is dissipated in the air, and losses are reduced by raising voltages as the distance of transmission and volume of transmitted energy are increased. Therefore, high voltage cables (typically 110–765kV depending on the distance between substations) are predominantly carried overhead between pylons and are prominently visible.

If high voltage cables are buried in the ground, they require substantial insulation as well as cooling equipment – in special buildings at intervals along the line or through heat-dispensing material in continuous troughs. Construction, operation and maintenance is considerably more expensive. Typically, buried high voltage cables cost at least ten times as much to install and operate as overhead ones.

Low and medium voltage cables are for short and medium distance power distribution. Heat generated and cooling and insulation requirements are less than those for higher voltages but, nevertheless, the cost of buried cables is still 2–4 times that of overhead lines (EEI, 2010).[12] Power distribution lines are usually overhead in rural areas, reduced insulation requirements requiring masts or poles sufficient only to clear any trees, buildings or vehicles along the route. In well-planned urban areas electricity distribution lines are usually concealed in ducts below pavements. However, where there is uncontrolled development, they are commonly carried in a maze of overhead wires between buildings and across streets, accentuating the generally random and often unsightly layout of buildings and services.

Meanwhile, the main cited impact of electricity transmission is in the appearance of the high voltage overhead lines and their effect on the surrounding landscape.

Appearance of high voltage overhead power lines

Power grid transmission lines are steel-reinforced aluminium cables suspended between pylons. Typically six (three-phase double circuit) lines are carried, three wires being suspended from arms on each side of the pylons. Grierson (1994)[13] illustrated a variety of power pylon shapes from the traditional tall British model shown in Photo 11.16 to the guyed V-shaped pylons and lower height 'gantry' towers as are common in continental Europe and North America. Hayes (2005)[14] explains the various features of transmission lines in more detail.

The loads carried by the pylons are the weight of the cables and of the pylon itself, usually of relatively light steel lattice construction. Wind forces are largely dissipated by this open form but under certain circumstances there can be exceptional ice loads on the cables and pylons. These led to widespread collapse of pylons and power lines in Canada in 1998, but the risks are generally not such as to justify bulkier design.

At close quarters, electric power pylons are inevitably dominant – although, because of their see-through lattice construction, not necessarily visually obstructive. Their appearance results from design choice, typically on a national or wider regional basis, for the configuration and shape

Photo 11.16 High voltage power transmission line, Oxfordshire, England.

of the pylon structure, its cross-bracing and the cantilevered arms carrying the cables.

The height of pylons depends on the spacing between them and the sag in the suspended cables. Typically a pylon may be 50m high on open ground. Pylons on grid routes spanning wide rivers may have to be as high as 200m. The latter are seen almost as towers standing on their own, whereas ordinary power transmission routes are recognized as lines of pylons across the countryside. The base area of each pylon typically occupies a 12m by 12m square while the span of the cables between pylons might be 400m.

The sight of lines of electricity pylons has been widely decried since their invention. For example, Sylvia Crowe (1958)[15] claimed that the interposition of grid (pylons) into rural scenery 'robs us of that feeling of contact and identification which is one of the virtues of the countryside; the grid adds one more to the barriers which our machines are erecting between the human body and its natural environment'. She also noted that lines of pylons across open land 'effectively unite the urban areas on each side and thus extinguish the Green Belt' (Crowe, 1958).[16] However, her experience and positive ideas were recognized and, in 1961, she was appointed to advise on routes for British high voltage power lines.

In practice, the routes and precise location of pylons have to take into account particular factors in each stretch of country. These might include the situation at viewpoints and the skylines affected, any competing or contrasting vertical features such as trees or other masts, and the lines of walls or roads or broader swathes of woodland. Thus a line of pylons keeps a few hundred metres clear of the roughly parallel upper Thames path.

Across valleys and in higher spaces with a background of outstanding scenery, a fine view may be spoilt by a line of pylons. Such situations have occasionally justified the considerable expense of buried cables (Photo 11.17).

In attractive upland landscapes, there may be no alternative but to cross fine scenery. This may be the case where power stations are remote or large consumer centres are in coastal situations. The routes should be through gaps, restricting skyline impacts, and along valleys of ample width or least remarkable scenery.

Possibilities for concealed cables lie in construction of short tunnels beneath particularly sensitive cols or through longer tunnels should these already exist and be available, such as Woodhead (former railway) tunnel beneath the Pennine Hills (Photo 11.18).

11.8 Tower buildings

The location and appearance of office or residential buildings is reviewed in the next chapter. This section concerns the function, structure and shape of all high buildings.

Function

A basic reason for multi-storey buildings is to increase accommodation on scarce land space. This was the case on Manhattan Island, New York, in the early 20th century and later for residential blocks in crowded cities such as Hong Kong. However, tall buildings are sometimes constructed for reasons other than land shortage – for example, to look

Photo 11.17 A national grid line across hills of North Wales connects into a buried cable for the 6km crossing of the Glaslyn Valley and views into Snowdonia.

Photo 11.18 Power grid line is transferred to an insulated cable through the former Woodhead railway tunnel under the Pennine hills.

prestigious or to offer particularly fine prospects to well-heeled occupants. Investors may anticipate some special national, corporate or personal glory or speculate on some greater floor space value reflecting the building's exceptional shape or height. In the city of London, it has been said that 56–60 storey buildings are economic in their use of height; but higher buildings are 'statements'. Elsewhere, in towns where no high-rise buildings have been built before, an apartment block of even a dozen stories might command exceptional prices because of the views (for example, the tower block in Montreux shown on Photo 2.6 in Chapter 2). Perhaps the most impressive yet economic structures are those designed to provide adequate indoor space within challenging or inspiring exterior shapes and facades.

Load-bearing structures

The first high structures, such as ziggurats, were solid mounds with steep-stepped sides, providing accommodation only in buildings on the top platform. Roofs and floors within buildings became possible using timber beams and then masonry arches. Wider timber spans could be constructed with devices such as hammer beams. Higher roofs were built on masonry columns or supported by flying buttresses and spreading of loads onto lower storeys.

The introduction of steel and then reinforced concrete enabled early and many modern skyscrapers to be constructed on rectilinear frameworks, the loads of floors and walls being transferred equally by the columns. More recently the loads have often been concentrated. For example, in 'hull-and-core' construction, the core is a vertical load-bearing box, often also containing the main lift, stair and service shafts, and the hull is the outer structure, including the facades, floors, walls and any additional load-bearing columns. Alternatively, the main loads can be carried on an external framework carrying the inner floors and partitions.

As towers rise to greater heights, so they are increasingly susceptible to lateral sway due to wind or seismic loads. Tanks or columns of water or self-activating dampers at high level can reduce the amplitude of sideways oscillations. Circular or indented shapes rather than hard corners reduce the impact of wind. So do 'exoskeletons' or 'diagrids', the framework meshes around London's 30 St Mary Axe ('Swiss Re), Beijing's Bird's Nest Olympic Stadium or Abu Dhabi's leaning Capital Gate.

Foundations must be deep and wide enough to carry the heavy direct and lateral loads of the high superstructure. Additional subsurface accommodation space may then be created within the foundations.

Shapes for tower buildings

The plan of each vertical section of a tower can be rectangular – as in London's Docklands blocks – to suit the optimum provision of room space, or it can be polygonal, round or elliptical. The vertical shape may be a straight-sided block or cylinder, a pyramid or a straight or hyperbolic cone – each for the full height of the tower or at its summit climax; or the whole mass may be spherical or ellipsoidal, as a balloon or an airship, with an overhanging outer hull or more abrupt cantilevered out-thrusts. Considerations influencing architecture and structural engineering may be individual or competitively judged inspiration, relevance to local building themes or standards, or suitability to layout of surrounding ground space.

Tower profiles may be stepped, with roofs at different levels at each reduction in floor plan area; or completely separate towers may be joined and braced by interconnecting bridges. Inlets can be let into facades or complete floors may be open to the air between the structural core and columns. At ground level, these open floors may be paved vehicle or walking space or allocated to public transport systems; higher up, they may provide additional breathing space to complement balconies and roof gardens.

In Moscow between 1947 and 1953, Stalin's regime commissioned seven huge buildings (including hotels, apartment blocks, a ministry and a university) all with high central towers in similar Russian Baroque/Gothic style on US-type skyscraper frameworks but with different designs for the projecting base wings. Photo 11.19 shows

Photo 11.19 The Palace of Culture and Science in Warsaw – originally a gift from the people of the Soviet Union to the people of Poland.

a similar tower building in Warsaw, Poland. These structures remain strikingly impressive – whatever their political connotations – mainly because they are so much higher than any adjacent buildings. A new one (the Triumph Palace) was built in Moscow at the beginning of the 21st century when it was Europe's tallest building at 264m. Two taller buildings have been constructed since, both in Moscow.

As for the very varied, very high and often extraordinary towers recently built or planned worldwide, different opinions see these as art forms, as iconic fetishism or as expressions of the architect's machismo. Cheek (2004)[17] is 'all for drama in every art' but believes dramatic architecture should be 'tempered with a little humility'. She suggests that some modern buildings are, like tabloid news items, tediously shocking. The matter may be one of personal perception, for example, as to the shape of the Swiss Re tower building in London (shown in the left-centre background in Photo 12.1). Popularly called 'The Gherkin', Cheek evidently sees it as an arrogant male symbol, some of us find it a not-unattractive fir-cone embodying the revolutionary curvature of two contemporary but more spherical London buildings. Perhaps a more rational objection by Cheek is that The Gherkin is grossly insensitive to a site that she asserts is one of 'cohesive grandeur'. However, others would claim that the architecture of that part of London has been wildly eclectic and far from visually cohesive ever since Sir Christopher Wren was thwarted in his plan for a completely new and spacious city.

Modern structural capability enables a huge variety of aerial sculpture. Deliberate planning of a high-level city silhouette may constitute a huge municipal work of art. More usually, the skyline is a fascinating but ill-disciplined melange.

Notes and references

1 Oliver, A. (1999) 'Leaning Tower of Pisa', *New Civil Engineer*, 14 January.
2 Owen, R. (2008) 'Tower saviours inclined to predict 300 years' stability', *The Times*, 29 May, p15.
3 de Schooten, U. (1963) 'The Turkoman Steppe', *The Geographical Magazine*, August.
4 Dudley Stamp, L. (1955) 'Pagodas', in *Chambers's Encyclopedia*, vol 10, p307b.
5 Judge, E. (1999) 'Mobile mast removal to bite into farmers' incomes', *The Times*, 8 October, p53.
6 (2002) 'Wind power' in *M'Graw Hill Encyclopedia of Science and Technology*, vol 2.
7 Otway, F. O. J. (1995) discussion on 'Development and construction of ... wind farms', *Civil Engineering*, August, p133.
8 MacKay, D. J. C. (2009) *Sustainable Energy – Without the Hot Air*, UIT, Cambridge, p265.
9 MacKay (2009) as Note 8, p33. Actually 140 turbines (total 522 MW) on 55km^2.
10 English, S. (1999) '£2m feathers rare eagles' nest', *The Times*, 24 May.
11 Musgrove, P. (2003) letter to *The Times*, 29 July.
12 Edison Electric Institute (2010) 'Underground vs overhead distribution lines', www.eei.org/ourissues/electricitydistribution, accessed 24 February 2010.
13 Grierson, R. (1994) 'Towers for tomorrow', *Landscape Design*, June.
14 Hayes, B. (2005) *Infrastructure: A Field Guide to the Industrial Landscape*, W. W. Norton & Company, New York, pp230–246.
15 Crowe, S. (1958) *The Landscape of Power*, Architectural Press, London, p18.
16 Crowe (1958) as Note 15, p27.
17 Cheek, M. (2004) 'Building – it's a boys' game', *The Times*, 17 January.

12

Buildings and Settlements

Buildings provide shelter from weather and extremes of temperature, places for people to live, work or perform community activities. They also accommodate fireplaces or stoves, furniture, bedding, books, consumables and machinery. Construction of buildings is commonly related to regionally abundant materials and traditional techniques. The consequences can be recognized as vernacular styles of architecture in ordinary houses but may be combined with more exotic concepts at grander municipal levels.

Civil engineering is applied in the foundations and structural frameworks of large buildings, but it is more universally responsible for the layout, design and construction of the supporting infrastructure – such as roads, water supply and drainage. The layout of these supporting systems is intimately related to land features and is reviewed, for urban development, in Chapter 15.

Settlements often imply established or temporary villages. Robustly built villages may burgeon, coalesce or be reshaped into towns. Temporary settlements may be for political refugees; more often they are extensive shanty towns which arise around cities where rural people seek economic refuge.

This chapter concerns:

- the populations and activities that have to be accommodated (Section 12.1);
- housing of different types and their need for land space (12.2);
- similar considerations for public buildings (12.3); and
- strategies for rapid or urgent settlement (12.4).

12.1 Accommodating people

The purpose of buildings and community land

Human activity in buildings comprises:

- home life, such as sleeping, eating and bringing up children;
- working in factories, workshops or offices in buildings that are comfortable for people engaged primarily in administration, planning, design or commerce, including in support of activity that takes place mainly outdoors; and
- recreation and cultural pursuits.

Homes are family shelters, usually provided in the most economic way of using available resources. The quality of homes depends also on the space available and on whatever built-in comforts and pleasant surroundings householders can afford or can contrive. *Buildings for working in* have to provide whatever conditions are necessary to attract workers and ensure effective operation and production. They are also social centres in that workers form acquaintances that sometimes extend beyond the work pattern. *Recreational and cultural amenities* attract a wider community – under roofs or in open space. People come to them often or occasionally, make close acquaintances or none. Parks and possibly museums or libraries may be free for access to all. Playing fields are available for organized sport. The success of restaurants and entertainment centres depends on their charges, quality and accessibility.

Except in unauthorized 'informal' settlement, the way in which available land is built on is determined by the local planning authority or the architects of whoever is able to exploit the land commercially. But the recognized quality of built-up areas also reflects the extent to which different sections of the domestic, working, playing and governing communities influence the pattern of building. This influence can be first in the choice of where it is pleasant to live or work, and then in democratic planning processes concerning further development.

Population density and land space

Most people live in close proximity to their neighbours and many are uncomfortably overcrowded. However, while half the world's population is now widely reported to be urban, this does not mean that all that half lives in big cities. National data is based on a variety of criteria for defining 'urban'.[1] In several countries, settlements of more than 1000 people are classified as urban; but communities with populations exceeding that threshold might, in reality, be little more than large villages.

As a rough indication for urban and suburban communities, 4000 people live in 1000 houses on $1km^2$ of land. This area includes non-residential buildings and paved or open public space and does not imply that actual residential space is as generous as 40 people per hectare. Equally roughly, 'built-up' land might be that on which buildings occupy more than 10 per cent of the gross land area.

In many farming areas, almost everybody depends on a livelihood in agriculture. But in the open spaces of the American Midwest and many parts of Europe, agriculture supports less than one person per hectare; whereas in some fertile but crowded areas of South Asia, several people attempt to eke out a living on each hectare. In such straits, the capability of the land to support its inhabitants is more critical than the space needed to build dwellings. However, there may be good reasons, such as security or the threat of floods, why settlements should be compact.

In modest-sized towns, suburbs, city dormitory country or in spacious interstices of conurbations, few people actually practise agriculture. They live in one or two-storey houses to whatever standard of building and extent of land they can afford. The population density in 2000 of the former English county of Middlesex, with many houses for London commuters but also pockets of land not built on, was about $250m^2$ per person, in other words, the same 40 people per hectare as speculated above as a global indicator for gross urban areas. Housing in Middlesex is mainly two-storey detached or semi-detached buildings with front and back gardens, considerable public amenity space and, furthest from London, adjoining some remaining agricultural land. The area is perhaps typical of some elements of open conurbations which, with considerable modifications, is suggested in Chapter 15 as the direction for 21st century land development in well-inhabited countries. Without any agricultural elements, MacKay (2009)[2] mentions $160m^2$ per person as the density of a typical English suburb (62.5 people per hectare).

In pre-industrial and early industrial communities, towns comprised mixtures of public buildings, houses and workshops. In planned industrial and post-industrial cities, however, factory, office and residential buildings tend to be grouped in their own zones. Recently, population has decreased in many European inner cities because of obsolescence of certain industries. The relatively poor work-related housing was replaced by better units elsewhere; and city building concentrated on offices which people occupy only in daytime. Thus the 315ha central area of London known as the City had a resident population of over 100,000 in 1950 but housed only a few thousand in 1994. But the daytime population was by then 320,000 (or 1000 per hectare) – see Photo 12.1.

Sherlock (1991)[3] notes that Paris has, for a long time, had a population density double that of London while, close to the centre, the ratio approaches three. Both London and Paris have central park areas, but Paris's many large apartment blocks give it a resident population density of 207 per hectare – see Photos 12.2 and 12.3.

In sharp contrast, some eastern cities are much more intensively occupied. In 1976, Stephens[4] mentions housing estates in Hong Kong occupied by an extreme 5000 people per hectare. More extensively and perhaps more significantly, Tong and Wong (1997)[5] describe a strip of land along the coast of Hong Kong Island that accommodates about a million people on $22.5km^2$ – 444 people

Photo 12.1 Buildings in the City of London are occupied only during the day ...

Photo 12.2 ... whereas most of Paris's Left Bank blocks are residential but ...

Photo 12.3 ... with some green open space.

per hectare but including space for 700,000 jobs and all the usual city facilities. Apparently the area enjoys good accessibility, few roads but commercially viable public transport.

In a less satisfactory mix of permanent and shanty housing, Greater Mumbai's population equals that of Australia. Such extreme population densities are in multi-storey housing blocks. But great overcrowding can also arise in low buildings and shanty towns where everybody lives at one level. In refugee situations in the former Yugoslavia, Zetter (1995)[6] mentions cases where space occupancy was less than 3.5m^2 per capita – far below accepted local or United Nations High Commission for Refugees (UNHCR) standards but found necessary in the short-term to cope with the number of displaced people.

The areas of greatest population density are generally those where people live in least comfort and security – because their quarters are so cramped or because the whole community lacks outside breathing space. Steps must be taken to alleviate the conditions of the most unfortunate – by finding more space, by sharing more equitably and effectively any land that is already available, or by allowing the surplus population to move to less crowded regions or countries.

12.2 Housing

Residential buildings are the most numerous structures in built-up areas. Any housing has to temper the climate and provide the degree of comfort, security and permanence that people can afford. Ground space is then occupied in different ways by rural homes, institutional accommodation, urban dwellings and suburban estates.

Climate

Houses have to provide protection from adverse ambient conditions, especially cold temperatures and rainfall, but also high winds, dust storms and excessive heat. Provision of shelter, shade and insulation are inherent in many traditional building forms. People may then burn fuel or tap commercial energy for extra warmth in winter and try to minimize their physical effort at the hottest time of day in summer. Energy conservation can be achieved by building design (optimizing thermal mass and heat transfer), by grouping of buildings and, where appropriate, by systems such as combined heat and power. But the more wealth people command, the more they can afford to use up energy resources to create warmth in winter and artificial cooling in summer.

Wealth and comfort

Socio-economic strata are universally evident in housing. In relatively warm climates – where basic shelter can be minimal – there is a huge gulf between conditions in shanty towns and those in opulent villa estates. In colder climates, human inequality is less superficially obvious, especially where the majority of people are housed in large apartment blocks of similar exterior appearance. The social difference is more apparent inside – in the number and size of rooms, the number of families having to share each unit and the affordability of energy. Other significant factors are the standard of nearby amenities, such as schools or shops, and the equanimity or strife among the inhabitants. The reputation of places as desirable to live in depends also on the visual attractiveness of the houses and the ambience – particularly spaciousness and greenery – of their private (garden) or public (street) surroundings.

Construction options include residential modules of identical outside appearance but different capacity and price, or different strategies as to open or restricted access between living units. Choice of options may play a part in solutions to social problems such as:

- affordable accommodation for teachers, nurses or policemen in areas where otherwise exceptionally high earnings and housing prices predominate; or
- coexistence of well-employed with under-employed, and deserving with less deserving people on the same estate. Sherlock (1991)[7] mentions 'the uniquely Parisian custom of the poor living on the inferior floors above the apartments of the rich, but sharing the same entrance'.

Privacy or communality

In western residential systems, a majority of households live in single-family houses, typically wooden-framed in the US or brick in the UK. Substantial numbers live in apartments. Some people are lonely in their individual dwellings, especially the old, while some, especially the young, are over-exposed in the communal space to the pressures and influence of their least inhibited contemporaries. In each community, a balance has to be struck between individual privacy and secure property on the one hand, and freedom of movement and convivial liaison on the other. The consequences are reflected in residential landscapes that may be robust and austerely defensive or sensitive and welcoming.

Among town houses or apartment blocks, such amenity space as exists is either public or, occasionally, fenced off for exclusive residential use. Modest suburban houses may have secluded backyards but small forecourts or front gardens that are open to public view. Larger private front gardens may also be sufficiently visible to passers-by as to form principal components of the community landscape. There are also co-housing communities; for example, Denmark has 200, each with 10–40 individual houses in which privacy is valued but residents share a dining hall, gardens and recreational space (Gardner and Assadourian, 2004).[8] Meanwhile, Sherlock (1991)[9] regretted the demise, due largely to road traffic, of open residential streets where people can meet and children can play safely. He also explained how such conditions could be revived as had already been achieved in some Dutch towns.

Permanence

Climate, available building materials and skills, funds and security of tenure all affect the permanence of houses. There are also different approaches to long-term investment. In China, most people plan for the future and are concerned that their buildings should be well maintained and will look right. Western views are similar, if only because of the high sale value of houses. But 48 per cent of Indian families live without permanent houses (Savin, 2004).[10] In sub-Saharan Africa also, shelter tends to be more temporary (Photo 12.4). Indeed, when the head of a household dies, his family may disperse and his house may fall into disrepair. But the space will soon be occupied and the materials recycled in new construction.

Some pastoral families, for example, in the emptier parts of Central Asia, live nomadically in portable tents or yurts and migrate seasonally with their livestock (Photos 12.5 and 12.6). For far more people, living in regions such as the Ganga-Brahmaputra delta, it is inevitable that their homes are occasionally inundated or destroyed by floods. Where it is impossible to locate all settlements on safe ground, the houses may best be partially protected or dismantled as soon as flood warnings are issued, to ease their reconstruction when the water has receded. Where seasonal water levels are high but surplus water flow is not so strong as to be erosive, then buildings raised on stilts can be more permanent.

Available ground space

The population densities mentioned in Section 12.1 are for gross extents of land space. They apply to complete urban areas and may include industrial

Photo 12.4 Traditional African home, Tanzania.

Photo 12.5 Black woollen or goat hair tents at summer quarters for nomadic people in Iran ...

Photo 12.6 ... who move with their livestock to lower sites with permanent dwellings in winter.

or commercial as well as residential buildings. For housing, where space is tight or expensive, there are plot densities that are recognized as normal or ideal to balance space against demand. For example, for development of detached houses, the Campaign to Protect Rural England (CPRE, 2003)[11] notes that new estates in England typically provide 16 dwellings per hectare. CPRE proposes closer arrangements achieving 30 dwellings (with space for 143 beds and 32 cars).

Sendich (2006)[12] defines US low, medium and high density housing at 4, 16 and 48 units per acre respectively – equivalent to 10, 40 and 115 units per hectare or, at four people per unit, 40, 160 and 460 people per hectare. In London, 'net residential density' is a planning term that usually assumes the number of people in each dwelling will be equal to the number of 'bed spaces' provided and the area occupied by dwellings is the curtilage (area) of the site plus 6m or half the width of any community roads, whichever is less (Sherlock, 1991).[13] On this basis – excluding any allowance for community buildings or common open space – the net residential density of outer, intermediate and inner London was 247, 336 and 494 people per hectare respectively. In the mid-20th century, London County Council concluded that the intermediate figure of 336 people per hectare might be the optimum density for cities. It was assumed in the 1950s and 1960s that such a density required a major proportion of people to live in apartment buildings. High-rise structures were constructed, many later judged unsuccessfully. So, in the 1980s, many city planning authorities limited planned density to 247 people per hectare. Sherlock (1991)[14] maintained that not only was this figure too low for viable urban communities, but that densities higher than even 336 people per hectare, could be achieved with converted or new buildings of only two to four storeys with individual street entrances.

Meanwhile, English terrace house plots are typically $60m^2$ and the average new house plot in the UK is $76m^2$ (Butler, 2006)[15] compared to $100m^2$ in continental Europe. Taking account of adjacent road space and at four people per dwelling, these correspond to residential densities of about 400, 320 and 250 people per hectare.

High-rise apartment blocks have, for a long time, been inevitable to accommodate much of the world's burgeoning urban population. Half the people in Singapore live in high-rise buildings as do the majority of the inhabitants of Madrid. In Hong Kong, half the 1976 population of 4 million lived above the tenth floor and a single building accommodated 24,000 people per hectare, 2.4 people per square metre of ground space (Stephens, 1976).[16]

Singapore is a flat island but many built-up districts of Hong Kong are on hillsides, as are the environs of Rio de Janeiro. The amount of space necessary for construction may be different from that on level ground and actual layouts may depend on how the land may be shaped to suit construction of houses, roads, drainage and amenity space, or stabilized to support high or heavy structures.

Where expansion into countryside has to take place, the impact on that countryside's landscape may be considered less adverse if the ground taken had been devoted to arable farming rather than to locally scarce wilder land. The way in which the best natural features are conserved is probably more significant than the actual area of space taken.

Rural homes

Rural landscape is fields and woods interspersed with individual homesteads or villages – integrated settlements too small to be classified as towns. In many developing countries, the rural population is intimately linked to the rural economy and particularly to agriculture. In well-populated, developed/industrialized countries, villages also accommodate people who work in towns, although planners usually make an effort to conserve traditional village structural forms and to limit undue built-up expansion on to green land.

Individual homesteads of proven forms remain suitable where there is ample space and satisfactory water supply and waste disposal. Actual houses may be:

- temporary, such as those of nomadic people or refugees (Photo 12.7);

Photo 12.7 Central Asian yurt.

- fragile and basic – traditional forms such as sun-dried bricks, wattle-and-daub walls and thatched roofs (Photo 12.8); or
- permanent – built of brick or local stone, timber frames or floors and whatever roofing material is locally economic (Photo 12.9).

Photo 12.8 Timber-framed mud wall thatched roof house under construction, Iran.

Photo 12.9 Twin end-of-row dwellings, King's Lynn.

Photo 12.10 Permanent timber-framed building listed for conservation, Lavenham.

Steel or reinforced concrete beams or columns may occasionally be incorporated in multi-storey houses.

Among the fragile forms, including those with earth in their construction, are vast numbers of buildings – even in towns – that can be effectively permanent if properly maintained (Photo 12.10). Construction in compressed straw bales or rammed 'cob' (straw-reinforced mud) can be used to build comfortable houses that stay warm in the winter and cool in the summer without much extra energy input (Heinberg, 2003).[17] However, most earth structures or light timber frames have little resistance to lateral forces and readily collapse in strong earthquakes. If such houses have heavy roofs or upper storeys, people inside may be killed.

Groups of rural buildings may commonly comprise:

- farmsteads: separate or combined buildings for people, animals, machinery and storage of agricultural produce – inherent parts of agricultural landscape (Photos 12.11–12.14); or
- wealthy landowners' mansions with outbuildings, dependant housing and perhaps scenic parkland.

Rural settlements

Villages of a few hundred dwellings arise around a nucleus. In England, that nucleus could be a manor house, a church, a large farm or an early river crossing or railway station. In Africa, it may be the communal space around a water point or a prominent baobab tree. Subsequent development tends to lie along roadways. The layout and spaciousness of villages depends on the will of the inhabitants or the landlord, as well as the availability of building land.

The capability of villagers to shape their situation depends on the conditions of land tenure as well as topographical limitations. On extensive irrigation schemes, the canals and possibly the fields may belong to the government and, even if the landlord is absent, the village council may have a powerful voice in determining how development takes place – within the fixed layouts of the scheme (Photo 12.15).

Photo 12.11 Dutch farmhouse with barn and animal accommodation at one end ...

Photo 12.12 ... and a similar but larger Bavarian farmhouse, with solar panels on the barn roof (photo by Robin Carpenter).

Photo 12.13 Cornish farm buildings.

Photo 12.14 Estonian farm buildings.

Photo 12.15 Aerial view of a Punjabi village within an extensive irrigation scheme. Only the mosque's strict east-west orientation in the centre of the village is out of line with the rectangular pattern of fields, roads and water channels. There is very little 'incidental' space. The water at top-right of the picture and that diagonally opposite (bottom-left) are ponds for buffalo. The water at top-left is probably seasonal flooding.

On steeper ground, there may be space for a concentrated village at a level above the most fertile or irrigated land (Photo 12.16) or hilltop settlements suit incised plateaux if water can be supplied (Photos 12.17 and 12.18).

In less densely inhabited, open, gentle topography as, for example, in much of Africa, agricultural plots may be more haphazard, houses less permanent and scope for adaptation greater. Land tenure may be traditional, well-understood but unrecorded and therefore liable to predation by city land grabbers. In developed countries, where all land ownership is registered, it is easier to assign settlement land to whoever can afford the market price – within official land use restrictions.

Townships or colonies are large villages planned as residential parts of larger areas allocated for particular types of organized economic production, such as collective farms or mines. Layouts are usually contrived as straight lines and rectangles for optimum use of space and directness of communication.

Institutional living

Villages and townships are made up mainly of family units with a few communal buildings. Institutional settlements – rural or urban – have to combine more communal facilities with less private living quarters. *Refugee camps* are an important example of settlements where people have to be accommodated in large numbers, often suddenly. Communal feeding may be a fair and efficient (if expensive) way of distributing free food, sanitation and health care can only be effective on an organized basis and extra education and

Photo 12.16 A village in the Elburz Mountains, Iran, sits between the irrigated valley-bottom fields and the steeper hillside above.

Photo 12.17 Hilltop village Spain.

discipline may alleviate crisis stress. While the best political outcome is that refugees return to their homeland, this often proves impracticable. In fact, the likelihood of permanence and transformation into self-sufficient townships should be considered even in initial emergency planning.

Prison camps have to be more substantial, more secure and much more disciplined. Strictly organized working camps in comparatively remote but open surroundings may be a better solution than closed prisons for the conversion of both hardened criminals and unfortunate offenders. But means have to be devised for reconciling prison risks and activities with the economy and welfare of the indigenous population. One day it may be possible to construct camps from which convicts can contribute to regional economic benefit, rather than building closed prisons wherever local people make least fuss.

Military camps are, at their worst, bases from which surrounding regions are politically suppressed. More commonly, they are highly organized, competently staffed depots for provision of national defence. Hopefully, they are also suitable for emergency accommodation of displaced civilians. Military communities are necessarily orderly but can be humane and socially cohesive if their roles are predominantly peaceful, more like industrial townships than barracks.

Student accommodation is an element of university campuses, a most agreeable form of institutional

Photo 12.18 Hilltop village in Pakistan.

Photo 12.19 Brasenose College, Oxford.

life in fine historic (Oxford, Photo 12.19) or purpose-built (Bath) landscapes, for the duration of stay of those who can attain or afford it. Successful campuses can no doubt inspire improvement of less fortunately endowed institutional living; and collegians that lack dedicated residential accommodation can contribute to the communities in which they find lodging.

Photo 12.20 Early 19th century one-storey dwellings along an access alley too narrow for motor vehicles, Cullen, Scotland.

Urban dwellings

Town or terrace houses are lines of contiguous home units along streets or forming complete rectangular blocks. In the US these are also known as row houses where they are mainly built as low-income public housing for rent. Photos 12.20–12.22 illustrate development of small town houses in Britain. In European cities, town houses are common near to city or town centres, ranging from humble 19th century workers' dwellings, now mostly demolished, to high-priced luxurious homes built from the 18th century up to modern times.

Detached or *semi-detached* (duplex) *houses* – and some in rows – have a small front garden space, now sometimes paved over for car parking. Some estates have separate (including underground) parking areas, or cars are parked where permitted on the streets or pavements. Any space behind houses is usually part of the occupier's private domain; in closed blocks of town houses, only birds, domestic pets or small vermin reach whatever habitat backyards or gardens provide.

Large, usually multi-storey, *residential apartment blocks* are where the majority of many city populations live, for example, in modern China or continental Europe (Photos 12.23–12.25).

In Britain, life in some high-rise building estates has become unpleasant, but other experiences indicate this may be social rather than architectural failure. Indeed, well-managed, secure and maintained blocks of flats are often successful and sought after, even in Britain. In London and other towns there was, towards the end of the 20th century, widespread demolition of residential tower blocks that had been constructed mainly in the 1960s. This was undertaken as a matter of local government policy where the blocks were considered to have failed socially or through structural problems. Meanwhile, many cities in more crowded situations elsewhere could not possibly afford the extravagance of finding new land to replace unfashionable housing. Wherever ground space is insufficient for individual houses, then people need to live in tall blocks. They must be able to see out and will not be content to

Photo 12.21 Mid-20th century town houses with street parking in Stockport, England.

Photo 12.22 Late 20th century town houses with very small front gardens, Gallions, London.

Photo 12.23 Tall rows of apartment buildings, Lausanne ...

Photo 12.24 ... and Prague.

live or work underground or deep within broad pyramids.

One disadvantages of living in blocks is the need to share access lifts or stairways – with other occupants and interlopers. Another is confinement to a number of small rooms with no immediate outlet to garden or courtyards. To ameliorate

Photo 12.25 Tower blocks in Moscow.

the latter, balconies can provide small open air spaces with often interesting views. Communal recreational open space should be provided, preferably green and protected against incursion by commercial activity or car parking (both of which need space of their own – but not necessarily at ground level).

Suburban estates

Many towns in continental Europe have sharply defined limits. Everything within the municipal boundary is urban, everything beyond is rural. Other cities merge less abruptly into the countryside, through outlying residential suburbs. Some sectors of continental cities, within original or extended municipal boundaries, have the character of suburbs – rows of houses with small shopping areas and what might once have been village centres. There is further diversity in peripheral city situations, from spacious housing estates – that were initially built after railways made commuting by train possible, then later built near new out-of-town employment centres accessible by road – to crowded 'informal' housing that springs up, legally or illegally, around Third World cities to accommodate homeless incomers seeking work.

A typical North American suburban layout is shown in Photo 12.26. In England, 'leafy suburbs' emerged as the trees, shrubs and grass around houses and along streets reached a managed climax equilibrium, contributing significantly to the ecology and scenery of built-up areas. Biddulph

Photo 12.26 A housing estate near Calgary, Canada – detached houses on small plots (photo from www.istockphoto.com).

(2002)[18] described how the paving and trees of suburban streets can be adapted to accommodate more cars than they were designed for, yet without being dominated by vehicles or through traffic.

However, suburban estates need to expand if they are to accommodate all the overspill from decaying or overcrowded cities and people now dependent on car travel, often to work and sometimes to schools or shops. Sherlock (1991)[19] blames suburbanization (providing more space per person) for the decay of urban areas where communities are no longer sufficient either to support facilities reached within walking distance or to justify adequate public transport.

Informal city fringe housing varies from unplanned, unsanitary and often geologically unsafe shanty slums (Photo 12.27) to houses built by individuals or self-help groups on serviced plots provided by the government (for example, at Lusaka in Zambia). Self-help construction is a vital part of many economies but must be organized as to where and how it takes place. At best, it provides permanent housing for people who could not otherwise afford it. At its worst, the result is stripping bare of the total land surface, erosion of hillsides and disaster for both the inhabitants and the landscape.

Photo 12.27 Shanty town on steep slope, Haiti (photo from www.istockphoto.com).

12.3 Public buildings and office blocks

Common characteristics of non-residential buildings are their comparatively large size, frequently in block form, and their function for people to perform specific tasks, often during prescribed periods and requiring facilities for people to enter or leave buildings in large numbers.

Shapes and proportions of buildings depend on how their structural frameworks carry the loads and provide the necessary internal space. That space comprises rooms and access routes between them. Rooms vary from small offices through classrooms and hospital wards to the wide and lofty space in railway stations or factories already mentioned in Chapters 8 and 10. Access corridors, verandas, stairs and escalators are planned to suit the different ways that different people move about – teachers or students in schools, nurses and patients in hospitals, sales staff and shoppers in markets.

Features in the external appearance of structures have been introduced in Chapter 2 as shape, colour, texture and embellishment. Shape in particular depends on structural framework design leaving considerable architectural choice in features such as roofs, openings and facades (Photos 12.28–12.30).

Photo 12.28 Renovated 19th century roof and facade of St Pancras Station, now London's main continental rail terminal.

Outside ground space in non-residential landscapes may have a specific function. It can provide access routes into offices or railway stations, open storage areas, factories or playing fields for schools. Functional open air space is also contained within some structural perimeters, for example, at unroofed sports arenas or mosques in warm climates.

Office buildings

Many office blocks are higher than other buildings, incorporating lifts for access to the upper storeys. They need adequate entrances and outside access space. However, since most office work is sedentary, ideal walking distances for morning and evening commuting are not too short. Figure 12.1 indicates some aspects of office building shapes and their adjacent space and access.

At times, new office buildings have been built speculatively, encouraged by property price booms. Many buildings that now arouse public derision are offices or shopping malls that became unoccupied or were no longer properly maintained. Meanwhile some companies have chosen to occupy existing buildings, such as former residential mansions or even industrial premises, which are best conserved by converting them to new use.

Schools and colleges

Floor space indoors is primarily allocated to classrooms, but there is usually a need for at least one large-capacity hall, auditorium or gymnasium. Such halls call for long-span roof architecture in the tradition of 19th century theatres or chapels, mid-20th century cinemas or modern conference centres. Students also need adjacent outdoor space for playgrounds or sports grounds.

Ideally, children should walk or cycle to school. Where distances are too far or street conditions unsuitable, there should be school or public transport. Only as a last resort should children be delivered and collected by car. Teachers, however, should be at least as entitled to use their own cars as office workers, more so if school activities require flexible working hours. But this applies only in the common but unsatisfactory situation where transport economics make car travel cheap

Photo 12.29 Curved roof and bright facade of Musée d'Orsay, an art gallery and former railway station in Paris.

Photo 12.30 Modern buildings, including a multi-storey car park, avoid brutalist impact by geometrical variety and bright colour. Bracknell, England.

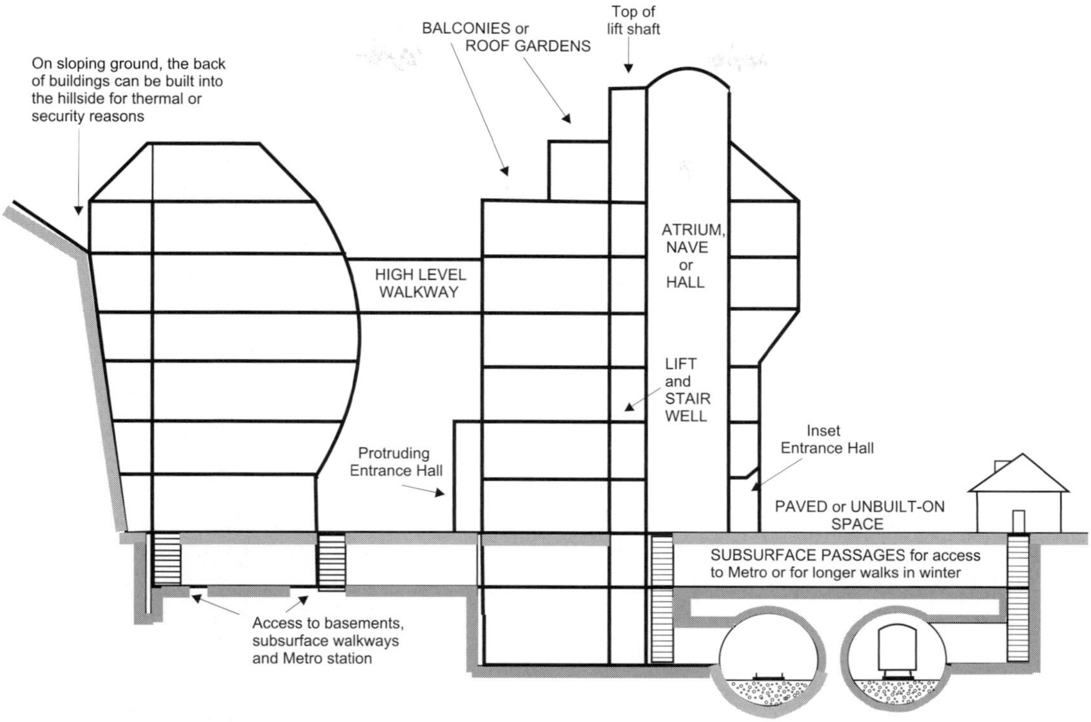

Figure 12.1 Building shapes, space and access

and public transport inconvenient. Until that situation can be corrected, some school staff will need to park on ground space that could one day be grassed.

Hospitals

People who are ill or injured are much less mobile than schoolchildren. But nurses and doctors require even more mobility than teachers, work under similar stress and have the same needs to get to and from work. Meanwhile, there are even greater needs in hospitals than in schools for specialized equipment to fit into wards, operating rooms, laboratories, corridors or service ducts.

Many hospitals or schools lie behind a stately facade of old buildings whose interiors have been reconstructed; or extensions have been built to accommodate changing needs of patients, staff, equipment, stores or vehicles. On suburban or rural sites there may be ample opportunity to conserve and create pleasant intervening green space. If there is space to build single storey accommodation this makes for easier access to invalids; and the bedridden can be wheeled out into the sun.

Shopping centres

Photos 12.31–12.34 illustrate traditional and modern shopping centres.

Shops must be accessible to the population that need them, mainly on foot with arrangements for retailers to deliver heavy items to homes. But supermarkets now dominate food retail sales, particularly on the edge of towns. Similarly located are furniture, white goods, office supply, garden and DIY stores. Their parking space and road access currently make additional demand on land resources.

The quality of landscape that results depends on the architecture of the store buildings, particularly roofs and entrances, and the way in which car parks are laid out within surrounding buildings or greenery. However, the long-term future of shopping by car at out-of-town shopping centres is uncertain.

Photo 12.31 Darjeeling bazaar, India.

Photo 12.32 Village shops in a valley in the Elburz Mountains, Iran, in 1957 (no road access to the valley at that time).

Photo 12.33 Inverary main street, Scotland.

Photo 12.34 Shopping precinct, Bracknell, England.

Photo 12.35 Optimum use of space for comprehensive athletic facilities within Paris's Left Bank tight layout of riverside, streets and high buildings.

Sports arenas

The ground space needed for sports stadiums comprises:

- the tracks, jumping pits or grass on which the events take place, set out to precise dimensions (Photo 12.35);
- 'stands' for spectators, pavilions and other accommodation buildings; and
- car parks and local transport terminals.

Arena roofs can be prominent or spectacular features in surrounding built landscapes. Wembley Stadium's iconic twin towers were replaced by an equally striking roof-carrying steel lattice arch.

Making the best of built landscapes

Prominent buildings are observed in the landscape as broad frontages viewed across a city square or boulevard, as distant rooftop, chimney and tower skylines seen from higher up, or at the end of vistas along streets or avenues of trees in city parks. Discordant features on the skyline or in the foreground may intrude upon views, but shared motifs can be comforting and deliberate contrasts can be stimulating. Random irregular skylines are so common as to be unremarkable, if not dull, but untidy foreground spoils otherwise admirable architecture.

Vistas may be obstructed by unfortunate alignment of road routes or building layouts. Views may be enhanced by taking advantage of clear or sloping ground, vistas by directing streets or openings towards pleasing distant features. Layouts of buildings and streets to achieve welcome sights and pleasant ambience are matters for urban planning, reviewed further in Chapter 15.

12.4 Planned settlement or urgent pragmatism?

Populations, development and settlement

The number of people in the world has been growing rapidly for two centuries and continues to do so. Even in countries where the indigenous population has been stabilized by economic and social development, that economic success attracts immigration from less successful more overcrowded regions.

Developed communities are those where advancement has been so effective as to raise the standard of living of all the citizens. In developed countries, the vast majority of people live in recognizably permanent forms of housing. Somebody – private individuals, corporations or the state – owns all land, but new building is usually restricted to what is deemed to accord with official land planning policy. That policy should take account of the land space available, the predicted changes in population and the corresponding forecast demand for such houses as people will be able to afford to buy or rent. As importantly, land-use planning should strive to ensure that settlements are located in places with access to employment and social facilities, and that the density of settlements is sufficient to generate or justify public services.

Parts of many towns are already 'developed' within countries such as China or India. These nations are still labelled as 'developing' because a great many of the rural areas are undeveloped and some of the people remain poverty-stricken. In 'less developed' countries, including several in Africa, increasing pressure of the expanding population has not been balanced by economic and political advance. A vicious circle of poverty, high birth rates and overpopulation results.

Among the great proportion of the world's population who live in the less developed communities, mobility – to find a livelihood – is as great a priority as secure shelter. People are compelled to move from their homeland – forcibly as refugees or economically because they will otherwise starve or live in greater penury as the number of mouths to feed increases.

As the surplus population moves, particularly to cities, they do not readily find vacant accommodation or they cannot afford it. Many of them therefore occupy such land space as they can squeeze onto and construct their own housing on it without any security of tenure. To help gain control on unauthorized or 'informal' settlement, governments can take pragmatic action, for example, to sanction and assist people who can build their own houses in designated areas, mostly near cities, and to accommodate refugees, mostly in rural situations.

Rapid low-cost urban settlement

The majority of people who arrive on the peripheries of developing country cities come from poor and increasingly overpopulated rural areas where they have failed to subsist in overmanned agriculture. They include desperate but physically capable and intelligent 'landless' or 'outcast' people. However, in cities, incomers may meet and establish bonds with other determined people from similar backgrounds. Thus, they may better face the difficulties in finding employment and then share in the spontaneously constructed, overcrowded and sometimes squalid accommodation which is the only shelter available.

Governments or municipal authorities cannot but recognize these symptoms of demographic change and their social and economic consequences. They then have a choice of policy between building formal settlements to accommodate the incoming people or leaving settlement to the initiative of the immigrants. Alternatively, they can take compromise action, for example, by providing services at sites on which the incomers can build their own houses.

Formally planned settlements may be the ideal solution for making the optimum use of land resources for appropriate long-term housing, the term being at least as long as the economic life of the buildings. Governments buy, own or allocate land and arrange for building and engineering contractors to construct the houses and services, or they permit private developers to build on a site in accordance with local planning policies and building regulations. This is general practice in developed countries where there is a ready market for housing of different sizes but no need for, or toleration of, informal, temporary or very low-cost housing at lower than normally prescribed standards.

In countries where there is a considerable need for more basic accommodation, government housing may make a significant contribution to solving the problem and influencing the way that less formal settlement takes place. The allocation of land and provision of municipal infrastructure can be made strictly in accordance with long-term plans; the building regulations might be modified to suit particular needs – to achieve performance thresholds rather than inflexible standards; and attempts can be made to constrain informal settlement to accord with the prescribed land plan and approved building techniques.

However, provision of government housing to needy people may only be practicable if the property prices or rents are subsidized, and, while such relief of genuine hardship is admirable, undeserving already well-heeled opportunists can exploit the system. They may buy up or rent properties at the subsidized prices and then sell or sublet them at great profit to themselves but waste of public investment. In any case, the number of impecunious people seeking accommodation in Third World cities is usually much greater than their governments could contemplate accommodating, nor may conventional construction companies have the capacity to fulfil sudden demand. The discrepancy can, at least partly, be met by the self-help capability of the incomers.

Informal settlement is undertaken by immigrant or homeless people entirely on their own initiative. They settle on whatever land space they can find, usually illegally and through sheer necessity. In practice, local authorities have to accept this necessity and overlook the legality where enforcement seems impracticable. Official policy may focus on:

- preventing any incursion onto existing built-up property or land of particular strategic importance; or
- retaining the right, ultimately, to evict the squatters and demolish their structures so as to redevelop the land.

In many cities, such as in India, up to half the inhabitants live in bustees or similar forms of spontaneous housing – structures that can be readily dismantled and re-erected. Incoming people may already be familiar with simple building techniques, but those that they have learnt in rural situations may have to be adjusted to suit the ground space, wall or roof materials and fuels available in towns.

Informal settlement does not necessarily imply shanty towns. Substantial houses may be built on vacant land space. For example, outside some Turkish cities, people build brick or block houses with metal roofs – it is said that if the building is completed in one night it has better legal status than does one seen to rise gradually on forbidden ground.

Without safe water supply and adequate sanitation, any crowded settlement can become unpleasant and dangerously prone to cholera and other diseases. Clean water is often lacking and risks of disease are high in many rural areas, although simple solutions such as installing sealed wells or adequate latrines are available. But in urban settlements, there is seldom any adequate alternative to municipal water supply and sewerage. Accordingly, no form of settlement can be satisfactory until these are available.

The *compromise solution* for rapid low cost urban housing development is for governments to provide the land, basic services and any other direction or assistance to enable incomers to build their own houses. Provision comprises:

- preparing the ground at any chosen location: stabilizing slopes, clearing or conserving trees and vegetation, aligning and constructing roads (for example, in gravel or soil-cement) and stormwater drains;
- constructing extensions and enlargement to the municipal water supply and sewerage systems;
- assigning plots for building, rents for tenancies and charges for services; and
- continuing liaison with the settlers in determining what additional facilities or improvements are needed and how these might be jointly financed.

Assistance to self-help housing may also include large-scale procurement of building materials. Different materials may be needed than were used in rural construction; some, which were free in the country, have to be purchased in the cities. With intensive construction for large influxes of people, shortages of particular materials may arise.

As for labour, immigrant peoples may be well capable of adapting and extending their own skills and of organizing themselves as small building companies or subcontractors. In fact, as a main source of employment for incoming people, building on site-and-service development land could be extended to conventional urban development for which labour may be short but skills can be imparted.

Employment itself is the main objective of people coming to cities. Self-employment, using skills previously acquired, can be an attractive option if there is a market for their products. Even in shanty towns, a great deal of productive activity is evident that cannot as easily be undertaken when people are rehoused in apartment blocks. On site-and-services schemes, incomers can plan and build structures on their plots to accommodate simple machines, clothing manufacture, paint and repair shops or depots for recycling refuse as an alternative to seeking livelihoods in city factories or municipal services.

Refugee situations

Refugees are people who leave their homes because of war, persecution or disaster to seek refuge somewhere else. Forcible displacement by other groups of people is the most common cause, but internecine strife may be driven or aggravated by too many people trying to live off too meagre land resources. Refugees must find accommodation wherever they arrive or are sent.

As far as planned construction is concerned, there need to be appropriate locations and layout of camps, provision of water supply, sanitation and drainage, construction of administrative and community buildings as well as assistance to refugees building their own shelter. The *location* of camps depends on:

- where refugees are already settling spontaneously;
- where refugee camp activity can be integrated with – or will least upset – the lives of the indigenous population;
- geographical features – watercourses and aquifers, vegetation and fertile land, narrow or broad sites for settlement; and
- the size of camps (preferably for not more than 20,000–30,000 people) and how many camps there will be – usually the more and the smaller the better, even if this is initially more costly.

Planned *layout* of camps has to take into account:

- the natural topography, particularly in steep country, making the optimum use of valleys by locating shelter on sterile or sloping ground and conservation of features such as watercourses (for surface drainage and water sources), fertile land (for any appropriate agriculture) and woodland (including as managed plantations for sustainable supply of wood fuel and structural timber);
- the sort of layouts which the refugees understand, for example 'decentralized and clustered disposition of plots and shelter, with clearly designated and usable open space' recommended by Zetter (1995);[20] but also
- the more geometrical layouts – including grids, concentric and linear patterns – for which there are precedents in economically successful permanent communities to which any settlement should aspire in a more peaceful future.

Essential services, from the beginning, are reliable water supply and sanitation. International non-governmental organizations (NGOs) are well aware of the needs and are well equipped to install adequate solutions wherever funds and access are available. Basic drainage works should be ready before the onset of heavy rain. Permanent roads and drains, essential power supply, more substantial sustainable water supply and distribution, and waste disposal or recycling systems will need to be constructed later to the same standard as is planned for the social development of (non-refugee) settlements generally in the region. Planning for regional development and refugee relief should be integrated, whatever the time scale.

Most *buildings erected by refugees* incorporate materials that will have to be supplied (such as canvas, metal roof sheets, structural timber and prefabricated elements) or that they are allowed to obtain locally (grass, earth, light timber). Houses may be extended to accommodate workshops or such animal husbandry as they are permitted to undertake.

Buildings erected by government and aid agencies or NGOs as semi-permanent structures may include:

- centres for registering incoming refugees and for distributing food, clothing and bedding;
- grain silos and other food warehouses;
- schools;
- medical centres;
- accommodation for relief workers (the majority may come from cities, some from abroad); and
- camp maintenance buildings and equipment servicing depots.

Construction of these buildings (by appointed contractors normally operating elsewhere) and operation of the camp facilities can be sources of employment for refugees; and this employment can be supplemented by other productive activity as the community develops. Camps should have an inbuilt capacity to become viable permanent settlements. For example, Zetter (1995)[21] suggests reception centres at emergency refugee camps can be converted to schools or clinics, food distribution depots into markets.

Upgrading

Buildings or complete settlements need to be improved if they do not reach appropriate standards in respect of accommodation (adequacy of shelter, living space, comfort and convenience, quality of construction) or municipal services and social infrastructure (public facilities enhancing people's lives in or outside their homes).

For temporary, informal or spontaneous settlement in developing countries, regional governments may have to take certain initiatives: firstly, in correcting any serious deficiencies in basic services such as safe water supply; next, by imposing order, accepting or rejecting ad hoc development that has already taken place; then by allocating land or insisting on adherence to building standards; and finally, by arranging construction of any housing or buildings that cannot otherwise be properly built.

Formal upgrading plans, drawn up in consultation with the inhabitants by local government or their development contractors, have to take into account:

- security of tenure of owned property (land and buildings) and, especially in rural areas, rights to common land or water resources;
- affordability of property for purchase or rent and what government assistance or charges are envisaged, initially or perpetually; and
- opportunities for employment within or beyond each community.

Construction activities implicit in upgrading then include:

- consolidation (more or extended buildings on spacious plots) or dispersal (reducing density on crowded sites);
- surfacing of roads, installation of drains and pipelines or increasing the capacity of existing networks;
- improvement of buildings – insulation, ventilation, functionality – with the most appropriate materials and techniques; and
- demolition and reconstruction, including replacement with taller buildings where land space is very scarce.

In all cases, further adaptation and improvements to meet different situations in the more distant future must be considerations.

Land's limited capacity to sustain people

There are probably too many people on Earth. This probability is evident both generally and regionally. It is general in terms of the 'ecological footprint' – human per capita resource use that, it is argued, exceeds the rate of natural regeneration. This is easiest to prove for non-renewables like fossil fuels; and is strongly apparent in the destruction of natural habitat. It is regionally evident wherever people are undernourished or fighting each other to survive, such as on degraded marginal land in regions like the Sahel. Perhaps the best we can hope for is that the world's population may stabilize at about 10 billion by 2100. Meanwhile we must urgently reduce the per capita footprint, especially in the rich wasteful communities.

Rural economies, certainly in the poorer countries, need to be rationalized and diversified so as to take on more of the industrial (especially agro-industrial), commercial and services activity currently performed by townspeople. The latter may then find it more difficult to monopolize easy (clerical) work or to afford luxuries, including the more expensive types of food.

The role of construction in righting the rural/urban imbalance includes building settlements – houses, services and amenities – appropriate to the land space available in cities, suburbs and expanding country towns, as well as providing the infrastructure – access routes, water supplies, irrigation schemes, flood plain engineering – necessary for efficient production on the land.

Civil engineers and other constructors can work with town and country planners:

- to protect unspoilt land features for nature conservation, recreation or possible new needs in the future (Chapter 13);
- in developing agricultural and mixed (small town) economies, mainly on flat or gently undulating productive land (Chapter 14); and
- in urban development, usually on limited land space, for the most amenable accommodation of people, many of whom will probably be unable to afford current wasteful levels of energy and other resource use (Chapter 15).

These, with some further speculation about demographic, economic and social issues in the more distant future (Chapter 16), are the subjects of Part III.

Notes and references

1 A review of definitions of 'urban area' (evidently often synonymous with 'built-up area') is found in http://en.wikipedia.org./wiki/Urban_area, accessed 13 January 2010. Parameters used include total population of clusters (1000 upwards), population density (200–4000 per square kilometre), maximum distance between buildings (for example, 50m) or by purely administrative designation. European countries look for urban-type land use indicated by satellite photos. Some less developed countries adopt land use and density thresholds or require that 'a large majority of the population, typically 75 per cent, is not engaged in agriculture and/or fishing'.

2 Mackay, D. J. C. (2009) *Sustainable Energy – Without the Hot Air*, UIT, Cambridge, p152.

3 Sherlock, H. (1991) *Cities Are Good for Us*, Paladin, London, pp216–217.

4 Stephens, J. H. (1976) *The Guinness Book of Structures*, Guinness Superlatives, Enfield, London, p101.

5 Tong, C. O. and Wong, S. C. (1997) 'The advantages of a high density, mixed land use, linear urban development', *Transportation*, August.

6 Zetter, R. (1995) 'Shelter provision and settlement policies for refugees', in *Studies on Emergencies and Disaster Relief No 2*, Nordiska Afrikainstitutet, Uppsala, Sweden, p39.

7 Sherlock (1991) as Note 3, p217.

8 Gardner, G. and Assadourian, E. (2004) 'Rethinking the Good Life', in *State of the World 2004*, Earthscan/Worldwatch, London, p170.

9 Sherlock (1991) as Note 3, p218.

10 Savin, J. L. (2004) 'Making better energy choices', *State of the World 2004*, Earthscan/Worldwatch, London, p27.

11 CPRE (Campaign to Protect Rural England) (2003) *Shout it from the Rooftops*, CPRE.

12 Sendich, E. and the American Planning Association (2006) *Planning and Urban Design Standards*, Wiley, Hoboken NJ.

13 Sherlock (1991) as Note 3, p216.

14 Sherlock (1991) as Note 3, p221.

15 Butler, S. (2006) in an article on shrinking plot sizes in *The Times*, 25 November.

16 Stephens (1976) as Note 4.

17 Heinberg, R. (2003) *The Party's Over: Oil, War and the Fate of Industrial Society*, New Society Publications, Gabriola Is, Canada, p209.

18 Biddulph M. (2002) 'At home in our streets', *Landscape Design*, May.

19 Sherlock (1991) as Note 3, pp102–103.

20 Zetter (1995) as Note 6, p56.

21 Zetter (1995) as Note 6, p56.

Part III

Planning Construction in Various Landscapes

13

Construction in Scenic Country

Chapter 1 identified the elements of scenery. Geology, vegetation and water tend to be seen at their wildest in mountainous regions, where the steep topography enables fine views, some with picturesque man-made features in the lower foreground.

So the first section of this chapter defines the scenery, resources and sensitivity of mountain lands. Section 13.2 then introduces recreation – people enjoying themselves in wild scenery – and tourism, which also encompasses admirers of that scenery observing it from trains, roads or places where they stay. The effects of various forms of construction on different scenic situations are analysed in Section 13.3, and the last section (13.4) summarizes principles for planning construction while balancing the needs of tourism, economic development and access.

13.1 Mountains and other fine scenery

Mountains as natural scenery

Medieval townsfolk regarded mountains as sterile, horrific and even ugly. But by 1856, Ruskin[1] observed that 'mountains are the beginning and the end of all natural scenery'. In this simple observation, 'beginning' could be the cataclysmic turmoil which folded sedimentary strata, threw up igneous intrusions and metamorphosed the layers in between, causing the rock faces and bastions of bold hard landscape. The 'end', in geographical space, could be a rugged coastline where even modest hills have been cut into abrupt shapes by the sea. The 'middle', in the continuing passage of time, could be variations, as climates change, in the covering of snowfields or deserts and effects of erosion by ice or water providing pinnacles, serrated ridges, valleys or gorges.

Having added coasts and gorges to the general heading of mountain scenery, questions remain as to whether there are any other forms of fine scenery and whether all mountain scenery is fine. Exceptional beauty can be found on landforms less bulky than mountains or slopes less steep than gorges, particularly where rock, water and trees make happy combinations. But most areas of recognized rural attraction have been influenced by man and are thus only semi-natural. For example, it could be said that stone walls define the scenery of the Yorkshire Dales. In England, a generally well inhabited and not particularly mountainous country, there are unspoilt pockets of esteemed landscape in nearly every county. Much of Wales and Scotland is steeper and more remotely scenic.

Considered by some as ugly, by others as raw nature, are examples of natural debris such as naked Alpine moraine or Icelandic lava. However, especially where these are foreground to splendour, they remain important elements of a particular type of landscape. As for unremarkable mountain or upland scenery, one could cite:

- bare plateaux, classified as mountain terrain only because of their altitude and vegetation – exemplified by the Bolivian Altiplano or much of Tibet;
- similarly sterile but steeper hills such as support wind turbines in southern California, somewhat austere compared with the wooded or rocky scenery found 100km away;
- better-covered but mainly rounded featureless moorland, such as the heather covered hills

of Central Wales, also sites for wind turbines but popular mainly for sheep and occasional solitary ramblers relishing exercise and breeze; or

- thickly forested – even richly biodiverse – tropical mountains, which cannot be described as scenic because the scenery is obscured by dense greenery.

Nevertheless, there are whole countries, such as Switzerland or Nepal, where the preponderance of fine mountain scenery is such as to give a reputation to the entire region. Such territories can be considered as conservation areas in which even the mundane parts should be managed to best maximize the national image.

Mountain resources

Besides striking scenery, mountain resources comprise:

- water-gathering of often heavy precipitation and the sources of rivers; opportunities to build storage dams or to harness hydroelectric power;
- minerals – deposits such as copper, lead, precious metals and hard decorative rock associated with igneous intrusions; sedimentary strata such as coal measures and limestone; metamorphic slates, alluvial fertile silts or coarser gravels;
- forest and grassland, and the timber or grazing which they provide; and
- situations suitable for settlement in mountain valleys.

The first three are resources primarily exploited for the benefit of the larger populations downstream. With respect to settlements, 'mountains tend to be inaccessible places and challenging environments in which to earn a living; they often have provided sanctuary to refugees, indigenous people and ethnic minorities. Mountain people typically live on the economic margins as nomads, miners or wage workers, or in households headed by women whilst men pursue seasonal work elsewhere' (Denniston, 1995).[2] Within the extensive mountain terrain and foothills of Central Asia, these tough conditions persist, aggravated by population increase and improved only at a few established comfortable hill stations or by some recent long distance roads. In somewhat less remote mountain areas, such as the European Alps, rural activity has been transformed to provide seasonal playgrounds for mostly more affluent lowlanders – at once a boost to the local economy, but a threat to the more fragile of the natural and scenic resources. In reviewing Alpine land systems, Briggs et al (1997)[3] note: 'Dwindling glaciers, active and fossil talus slopes and frost-weathered pinnacles support a sensitive ecosystem above the forest belt. Traditional farms and tourist hotels are uneasy neighbours and every component has an uncertain future in a warming and economically insecure world.'

Mountain sensitivity

Compared with human lives, snow-clad mountains, rocky pinnacles and the walls of precipices are robust and permanent. However, ice age cycles last only a few thousand years, significant climate changes – affecting, for example, the size of glaciers – are noticeable within a century and the continuous processes of rock freezing, thawing and splitting maintain a continuous attack on all but the strongest homogeneous rock masses.

Much more rapid and noticeable are the effects on softer hillsides of run-off, erosion and landslides. These occur mainly as sudden events, although they are often the culmination of longer periods of gradually rising groundwater or build-up of sediment. Such events can lead to slope failures or, occasionally, more violent occurrences such as the collapse of dams formed naturally by landslip material or glacial ice. After incidents of erosion and surface denudation, natural recovery takes place as drainage routes are re-established, earth masses stabilize and re-emerging vegetation binds the topsoil. But in the most geologically and climatically unstable zones, erosive and hydraulic processes may be so frequent and continuous as to make permanent man-made structures impracticable. More often, the suitability of mountain landforms for agriculture or construction depends on facilitating drainage and stabilizing slopes.

On the positive side, terraced hillsides are an achievement of farmers' efforts over the centuries. Somewhat less fragile are mountain highways, buildings founded on hillsides and dams controlling river flow and storing water for ever-increasing downstream populations.

Negatively, pressure of rural populations can lead to short-term erosive practices in forest clearance and non-terraced or poorly managed steep cultivation. Hillsides have been scarred by roads and built development, paved surfaces have accelerated rainfall run-off by reducing inflow into the ground and reservoirs have silted up as a result of aggravated erosion in their catchment areas. Even if each human activity can be undertaken without significant damage to natural landforms, vegetation and scenery, there is a finite carrying capacity for the land. More and more people are attracted to scenic areas with a demand for built infrastructure that can become excessive.

13.2 Recreation and tourism

Recreation, taken as walking in the countryside, took early advantage of the hills and moorland surrounding Lancashire mill towns. Beyond that county, such exercise spread to weekend fell-walking in the beautiful Lake District. Later, Glaswegians climbed in the southern and western Highlands, Midlanders explored Wales and the English West Country. Of those who could afford longer holidays and more distant travel, Europeans went to the Alps and the Americans to what became their national parks.

Tourism was born in such exploits as the Grand Tour, in which young gentlemen travelled to countries such as Italy, primarily to see classical buildings, cities and art and to meet people of other nationalities. En route, they passed – sometimes on foot or horseback – over high exposed passes or through valleys of impressive natural beauty. Later generations did the same by train or motor vehicle. Today there are groups or individuals who:

- tour whole regions in coaches, trains or cars, stopping at permanent view-points or places of interest by the main roads, and spending the night in cities or major resorts;
- stay for several days in particular resorts from which they can walk, ride or just sit about viewing the scenery when it is fine, shopping when it is not;
- take long walks or climb mountains, staying sometimes in cheaper accommodation, campsites or bothies close to the routes; or
- pursue Alpine winter sports (predominantly skiing) which have developed on a large scale in certain resorts.

There are also very large numbers of people, also considered as tourists, who visit the world's cities and seaside resorts to be gregarious or to enjoy the climate but not primarily for the scenery. They are relevant to city or coastal built development rather than to scenic areas. But for those to whom scenery is paramount:

- one must balance the fragility of landscape against the demands of people who wish to enjoy it;
- access must be provided to meet these demands or limited to reduce them; and
- refreshment and overnight accommodation must be provided to meet the tourists' needs in a way which is convenient to them but does not damage the scenery or its foreground.

Photo 13.1 shows how a long walk can gain access to high passes and ridges. Photos 13.2 and 13.3 show different types of compatible structures at popular and occasional viewpoints respectively.

Balancing resources against demand

The number of people who can enjoy the attractions may be maximized by classifying them as to the way in which they participate and then providing the type of access and accommodation which can best cope with each category and with them all, within the carrying capacity of the total landscape.

Figure 13.1(a) shows the types of tourist activity already mentioned. The numbers of each group that might be sustainably borne can be assessed in terms of their impact. For example, for people who spend their time on resort streets, formal promenades and local walks, relevant parameters may be:

- the number of people counted in fine weather at a peak holiday week or weekend – perhaps

Photo 13.1 Valley path, Champéry, Switzerland. The hard folded strata on the left are eroded only slowly to talus. The valley bottom remains stable where movements of people are restricted mainly to the path. The landslip (centre right) may be natural erosion or it may have been triggered by wear of human boots.

Photo 13.2 Visitor centre overlooking a bend in the Grand Canyon. Prominent structures can blend into their setting by means of local material and regionally historic design.

Photo 13.3 A church on Brentor, Devon, a local stone addition to a fine view-point for occasional visitors.

a figure rather higher than the comfortable optimum but tolerable for such short popular periods;

- the minimum number of visitors that must be attracted in the low season or dull weather to ensure the economic survival of the resort; and
- the impact of certain numbers of people on the public places, paths, footbridges, walls and fences and the resources needed to maintain these.

Outside the main centres, the sensitivity to human interference of the landscape elements – rock and soil, watercourses, vegetation and animals – becomes paramount.

Controlled access as a means for managing demand

Figure 13.1(b) denotes the means of access that can be used by people to reach scenic places. Many main roads are part of national networks. However, some of the more scenic trans-mountain routes are not well suited to multi-lane motorways. Fast traffic unrelated to that locality would better take alternative or tunnelled alignments, as do trunk rail routes. Part of the enjoyment of, perhaps slower, travel on scenic main routes is that of viewing the scenery from trains or roadside viewpoints, whether or not the destination is a scenic resort.

Valley roads that lead to specific tourist sites or resorts have a capacity necessarily limited by their width and alignment. Enlarging them may cause unacceptable damage to the scenery through which they pass and congestion at the destination. Instead restrictions on traffic can be imposed beyond points where car parks are provided, travel further up the valley can be limited to public rail or road park-and-ride services. Cars are better left idle where there is ample space than equally idle, for most of the time, where they clutter up the resort itself.

At these resorts and beyond, people travel about mainly on foot. The exception is ski resorts where cable lifts take people uphill, an intrusion on the scenery but one without which the local tourist economy might collapse. For summer enjoyment of sometimes different scenery from that of the winter pistes, there can be a range of footpaths, for example:

- paved or gravelled promenades within the resort itself;
- narrower, steeper, rougher paths, cliff-hanging galleries or precarious looking bridges giving thrilling experiences; and
- wider, obtrusive tracks used occasionally by service vehicles or, perforce in larger numbers, by longer distance walkers or climbers en route to higher more pristine zones.

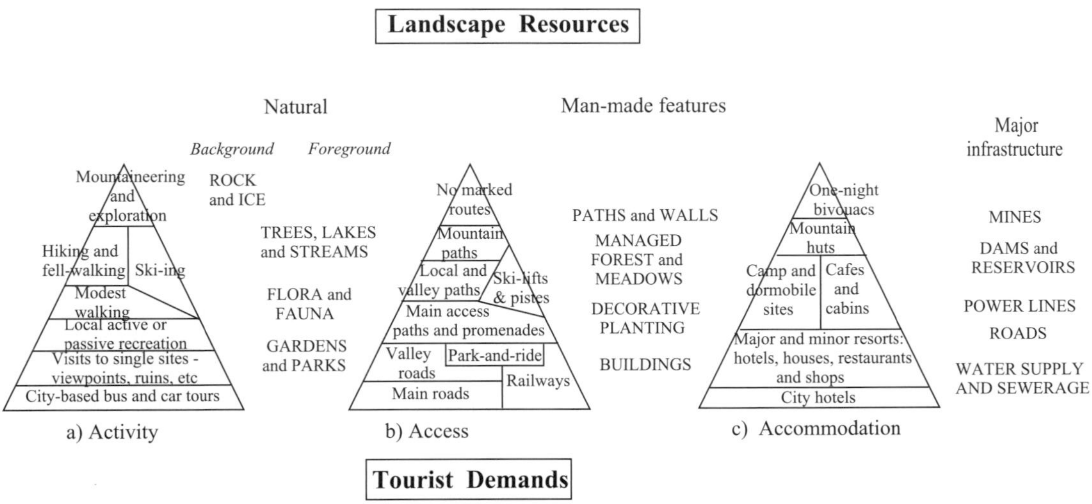

Figure 13.1 Balancing landscape resources against tourist demand

Tourist accommodation and facilities

Figure 13.1(c) indicates categories of tourist accommodation that can be provided. These range from large city hotels catering for coach parties, through resort guest houses, restaurants, self-catering accommodation and campsites, to climbers' huts or refuges at the top of hill paths.

Accommodation capacity – the number of hotel rooms, chalet beds, café seats, caravan spaces or tent pitches – is limited by the amount of space available at each resort and the viable extent of support services – water, sewerage and power – that can be provided. Resort towns and villages must not only accommodate visitors but must have the infrastructure to support the people who cater for these leisure-seekers and who undertake the other economic activities of a year-round community.

13.3 Effects of construction

Mountain scenery is not necessarily representative of all attractive situations. But the visual effects of structures may be similar, if at a different scale, amid more gently undulating hills and valleys, narrower but still steep coastal stretches or resort settlements in any of these settings.

Main effects of construction occur as land shaping (such as for agriculture), watercourse control (e.g. dams and irrigation channels), roads or paths, and buildings. These effects can be examined on such forms of landscape as:

- high rock peaks or snowfields in distant views, and gorges closer at hand, which constitute spectacular scenery and are very susceptible to any form of artificial modification, including in the foreground;
- hillsides – slopes, perhaps with rock outcrops or woodland, creating middle-distance scenery or suitable for agriculture, parkland, forestry, settlements and transport routes;
- valley bottoms with riverside features and perhaps meadow foreground scenery; and
- built settlements in which admirers of local scenery spend a lot of time and may hope to complement their more energetic pleasure with appreciation of regional architecture.

Peaks, snowfields and gorges

Man-made structures are most unwelcome where they intrude on awe-inspiring scenic features. Such features may be so spectacular or so unique that anything more permanent than a few climbers or a passing aircraft constitutes intrusion.

Rock pinnacles and precipices are sombrely forbidding and intensely wild. Snowfields are exceptionally bright and pristine, especially seen against a blue sky and with dark glasses momentarily removed. Such scenery cries out against anything that might tame its wildness or despoil its purity. The more prominent is a tourist restaurant on a mountain skyline arête, the more it spoils a sublime situation. The rock excavation, concrete foundation blocks and steel stanchions of a *téléférique* are a jarring blot on naked wild nature. Yet such ventures as the Jungfraujoch restaurant and viewing platform or the cableway up the Aiguille du Midi are commercial successes and, particularly in clear weather, afford great pleasure to people who would otherwise never see such sights first-hand. In mitigation of the intrusion, it might be said that neither despoil actual summits, that the top section of the railway to the Jungfraujoch is in tunnel and that there is plenty of unspoiled scenery left in the Alps. Nevertheless, these works are triumphs of hedonism over heritage. A hotel or restaurant viewing terrace should be in the valley or on a more ordinary hill, a more rewarding cable journey can be made on a lift up the Eiffel Tower or skiers wishing to take off from the finest peaks should carry their equipment to the top.

However it is difficult to extend this argument against mechanical access to high places of less than spectacular scenic or special ecological quality. There are railways to the top of Mount Washington and Pikes Peak in the US and many examples elsewhere. Most of these lines are to hilltops to which roads have also been constructed on routes no less practicable than those of lower level highways. Main access by rail is more interesting in its mechanisms and views en route and makes unnecessary the greater environmental disturbance and congestion with which road transport is commonly associated. But, where there is easy road access to open hilltops, it may be a fitting site also for a wind farm and a few discreet refreshment booths or shelters for people to view both distant scenery and rotating propellers.

Tourist railways to mountain-tops are by no means the most scenic train routes. Indeed, there are many spectacular journeys that can be made on main lines, for example, through gorges. The walls of ravines are best seen looking upwards. From trains, these views can be seen without noticing any damage to natural rock faces caused by construction of the railway itself. From any footpath along the gorge, the rail track can be seen and the trains may be the intrusion, as is traffic on any road. So an aim of scenic conservation should be to preserve unspoilt as much as possible of perhaps even narrower but equally spectacular tributary gorges.

Hillsides

Rural hillsides can support:

1 sparse low vegetation, some suited to low density rough grazing;
2 woodland – natural and often steep up to the established tree line, or planted for forestry up to a statutory or economic limit. For example, the British Forestry Commission's normal maximum for afforestation was 488m above sea level generally, only 244m for lower hills as in Herefordshire (Hackett, 1971);[4]
3 comparatively lush summer grazing on upland 'alps' (mountainside pasture);
4 ground crop agriculture (such as wheat or rice) or bush or tree cultivation (such as tea or coffee plantations, vineyards and orchards); and
5 settlements and roads.

Bare slopes (1) can stretch up to rounded hilltops or may be the foreground – neutral in summer, gleamingly white in winter – to sharp black pinnacles. In winter, some of the open slopes may be devoted to high level skiing. Woodland (2) is the natural cover to many hillsides or, again, foreground to exciting scenery beyond. Green forest uniformity can be varied by open glades or meadows (3), by rock outcrops or, in steep geologically young valleys, by erosion and landslips; or the forest mantle may be disturbed by man-made linear features – tracks, pipelines, power lines, cableways – or by ski pistes cut through the trees from the upper snowfields right down to the resorts.

Agricultural land (4) may have to be terraced on steep terrain, on milder slopes it may be interspersed with small scale grazing (3) and outlying houses (5) or can form attractive patterns of subtly contrasting crops. Villages and small towns (5) are usually found at the foot of slopes or on any expanse of less steep ground higher up. Individual houses may be dispersed more widely on accessible parts of the hillside. As buildings proliferate and land is cleared or paved, settlements may become increasingly at risk from avalanches, storm run-off and mud flows if the pattern of drainage and stability of the hillside is disturbed.

Forms of construction on slopes include:

- terracing of land for agriculture or buildings by means of excavation, fill, construction of stone retaining walls and provision of land drainage to ensure continuing stability of the ground;
- paved surfaces among buildings, and culverts to drain run-off from rainfall or thawing snow directly – rather than allowing it to be soaked up in the earth, so pavements and drainage culverts can exacerbate problems of sudden storm flow downstream;
- roads and tracks which cut across the hillside, also requiring culverts and retaining walls; and
- irrigation channels along the hillsides.

The lower slopes of hillsides may also be affected by construction of reservoirs. The depth of groundwater in permeable hillside material is raised to at least the level of reservoir water. Dam abutments or grout curtains seal off subsurface escape routes for accumulating groundwater, threatening the stability of slopes and, in some conditions, causing a risk of landslides into the reservoir.

If hillside construction is contemplated on slopes that may be inherently unstable or if the mechanics of drainage is uncertain, then estimates of long-term risk and appropriate levels of precaution should be made. Major highways and mainline railways are considered of such strategic or economic importance that their bridges, cuttings, embankments, drainage structures and retaining walls should be effectively permanent, as should be any protective galleries against rock falls or avalanches. However, such performance criteria might be unduly demanding and uneconomic applied to lesser mountain roads or irrigation water channels for which periodic failure and repair or replacement of particularly risk-prone sections is routine practice.

Valley bottoms

Natural features at the bottom of valleys are watercourses, their riparian features and any adjoining flat areas or meadows that may occasionally be flooded. Types of construction that might affect narrow valleys and gorges include:

- weirs diverting water, for example, through hydroelectric schemes bypassing steep river sections, reducing or cutting off flow in stream beds and over waterfalls;
- storage reservoirs which permanently inundate the complete valley bottom; and
- bridges, very visible features adding to river scenery – elegant or obtrusive according to their setting, dimensions and shape.

In wide valley bottom scenery there can be construction of buildings, roads and all the necessities of settlement. These involve drainage of the land, building on or paving of its surface and of protection for built-up areas against the ravages of floods. That protection includes riverside flood embankments, which may double as the roads and promenades that are central features of waterfront development. However, bank construction tends to canalize the watercourses, intensifying flow in their beds and destroying the informality of their banks. Riparian habitat could be lost unless bank protection profiles and alignments are sensitively chosen.

Existing settlements

Many buildings at the core of European resort towns and villages are old. They were robustly built to styles that were functional for the materials and craftsmanship then available. Homes or hovels that were less well-built have long disappeared. Survival of the fittest has resulted in churches, public halls, storage barns and dwellings which are fine examples of vernacular architecture, typifying what people respect and suggesting features or settings that might be considered fitting also for new development.

In the mountain valleys of Central Asia, the roughness and remoteness of the terrain and the sparse rewards of rural production mean that buildings have a functional austerity. Their design must conserve winter warmth and summer shade for people and animals with, for example, storage space on flat roofs where flat ground is at a premium. Yet changes may be rapid – for example, in use of materials such as metal roofing sheets or of machinery – when access roads or power lines arrive, often a century after these services reached more accessible populated regions.

Decisions on styles and settings for construction may lie within mature local jurisdiction. However, where there is no such influence, entrepreneurs seeing profitable opportunities in providing accommodation or facilities may seek to transfer concepts – massive blocks, towers or brutally functional bridges or culverts – to environments in which they are insensitive.

As relevant, in engineering terms, are the services demanded by modern lifestyles and densities of seasonal or permanent occupation. Paved roads, power, water supply and sewage systems, which were not necessary or affordable for the original rural population, became vital if visitors and new residents were not to overload and pollute land space and watercourses.

The planned ultimate capacity of settlements must not exceed that of the resources that are to be harnessed in any expansion. Then, for aesthetic concord, the layout must make adequate provision for space between buildings from which they can be admired and on which people can mingle in the open, and footways from which the surrounding scenery and its natural or contrived foreground can be even more consciously enjoyed.

13.4 Planning construction in scenic situations

Principles

If construction were planned primarily for aesthetic appearance, the following might be basic tenets:

1. Build nothing within areas of outstanding unspoilt scenery.
2. Fit into the character of any foreground in views towards that scenery.
3. When construction is judged necessary – in areas of fine but not the most outstanding scenery – ensure that the structures perform their function effectively and sustainably.
4. Make the structures sensitive to their semi-natural or built surroundings.

Regarding (1), it is usually obvious what are outstanding features or areas of scenery, especially if these have been officially defined. Hence, this principle can often be uncompromising. Contention arises in the fringe between exceptional and ordinarily fine scenery in any region. Claiming exceptional status may be a dangerous justification for rejecting the location for a tower or the route for a transmission line if any of these structures – already determined to be essential – is then located at a place where the impact is even worse.

The foreground (2) is both visible – a prospect stimulated by the background, for example of mountains – and it is geographical – a buffer zone for wild growth or organized activity appropriate to the transition. Imaginative planning, for landscape as much as for economic or social activity, needs to be based on inventories of needs, constraints, opportunities and priorities in each field of conservation or human aspiration. Location, layout and design of any foreground construction can then be planned within the principles discussed in Part II and the safeguard tenets (3) and (4).

In conjunction with these general principles, there are particular characteristics of each type of constructed development. Significant types in scenic country may comprise:

- settlements and infrastructure devoted primarily to recreation and tourism;
- productive facilities that may attract tourist interest but remain economically crucial to rural livelihoods;
- dams and reservoirs, essential civil infrastructure as well as significant features in river valley landscape;
- paths and transport routes through scenic regions; and
- construction in more accessible, already inhabited, only semi-natural but still fine scenery.

Tourism development

Scenically sensitive development is planned in both space and time. In space, some substantial areas must be conserved as natural beauty with such limited access routes as are appropriate. Other space of a certain character may have to be allocated to a particular recreation (such as skiing or golf) or some essential service facility (such as a reservoir). Then there must be room for settlements and economic activity.

In time, growth of trees and weathering of fresh rock faces takes a few years and land-use allocation has to take account of seasonal conditions. On coasts, there are beaches where people enjoy sunshine, sand and the sea in fine weather or there are cliff paths or promenades for hardy appreciation of wilder conditions. In mountain resorts, there are stark differences between summer and winter landscapes. Commenting on the comparatively modest development of skiing facilities in the Cairngorm Mountains of Scotland 40 years ago, Hackett (1971)[5] asserts: 'The damage to the mountain landscape by such facilities as car parks, roads, ski-lifts, etc. is very apparent when the snow has cleared away in the latter part of the summer.' With the covering of snow, any erosion is hidden and 'when the greater number of visitors are present, the landscape appears to be undamaged'. The contrast remains as serious in those Alpine resorts that rely on summer as much as winter visitors. Ski pistes have to be separated from rougher rocky or wooded slopes and designed and maintained so as to preserve grassy rather than eroded earth or paved sterility in summer prospects. Further, particular popular enthusiasms and perhaps climate may change in the future. So there should be scope for eventual relocation or removal of pistes.

In most mountain and some coastal resorts, rough wet weather may be frequent or prolonged. In dull weather visual pleasure has to be sought mainly in close surroundings – from riverside paths to paved walks among admirable or at least sympathetic architecture. During heavy showers, occasional shelters or shop awnings may afford welcome refuge – except in places such as Gstaad, where these awnings purport to be sunshades and automatically recoil as downpours increase, even if people are sheltering beneath them. Even in clement conditions, there may still be as many people enjoying the built-up surroundings as are walking in the hills or valleys. Within the resorts, pedestrians deserve priority or exclusive use of movement space; looking outwards they appreciate vistas into wilder tracts beyond; from without they look back at a comforting village skyline.

New development of tourist accommodation or sports facilities often extends the boundaries of built-up areas; for example, groups of holiday chalets arise on what were once green meadows. Besides ensuring that development suits the regional long-term land-use and landscape plans, there must be measures to ensure that the exclusivity that new enclosures offer does not impede routes past them. Near the points of access to these private enclaves, there should be public paths leading to the forest or hillside above or to any beaches or coastal cliffs below.

There is also an opportunity, where greenfield development is permitted, to tax that development. Taxation may discourage excessive opportunist development and could be used to support less easily financed enhancement of the landscape or local non-tourist economic activity.

Economic development

From an aeroplane over the western Himalayas, one looks down on deep green valleys, rocky precipices and snowfields quite as inspiring as those seen from the Alpine peaks by Edward Whymper and other Victorian explorers. At that time, Swiss peasants were described by travellers as 'miserably impoverished' (Schama, 1995).[6] One day, more of the people of India or ex-Soviet Central Asia may be able to join the modern Swiss in better enjoying their scenery. But meanwhile, in many regions, such as parts of Nepal, the landscape has become more densely populated and largely deforested. It is very unlikely that most of the population of the Himalayas can benefit by a transformation to a tourist economy of the sort that occurred in the much smaller country of Switzerland.

We have seen that mountain people typically live on economic margins (p254). Even when the population of the Himalayan foothills was less than it is today, many men had to seek semi-permanent employment on the plains in order to bolster their family incomes.

Natural resources of mountain areas are the mountains themselves, their weather (wind, snow, rain and run-off), sometimes minerals, often forest and grazing land and patches of more fertile soil. Storage reservoirs and hydroelectric schemes provide water and power for the benefit mainly of downstream urban communities, as do coal and ore mining or gathering of precious stones, activities that are site specific or employ few local people. Nor does forestry often offer intensive employment. So agriculture, much of it on steep slopes, continues to be the main source of livelihood – a hard enough option on the plains, more so in the marginal soils and steep topography of the hills. Somehow, by social and economic transformation, man's heritage of effort in steep country agriculture must continue to find reward.

There are two forms of such agriculture – rain-fed and irrigated. Both have been developed over centuries and have required great ingenuity and effort to produce and maintain the level ground and water channels that dominate some man-made landscapes. Indeed, some steep rain-fed terraces in the Philippines constitute a World Heritage Site, and irrigation channels have long supported communities that would not otherwise have been viable. Well-maintained terraces or canals and cropping activity reflect the health of production in many rural valley settlements. But there are ubiquitous problems of low market prices for farm produce; and agricultural work is insufficient to support the available labour force. The end of Chapter 12 foresaw a need to diversify rural employment and to include some of the activity, particularly that related to agriculture, which is presently undertaken in large towns. Hill districts need to do this just as much, balancing difficult topography and soils against any economic advantages that mountain resources can generate.

The highest, most scenic towns in the Himalayan foothills are the hill stations. In India, Darjeeling arose as a centre for highland tea plantations, Shimla as a hot season seat of government. Smaller agricultural settlements are more isolated, sometimes not accessible by mechanical transport. With improving communications, it can be hoped that more towns will arise with some of the employment opportunities of the hill stations and with similar scenic resources. In Utopian communities, more people may be able to live in Shangri-La and undertake most of their business there. Information exchange facilities, schools, mosques and garden centres could be architectural gems of new economic focuses within the finest landscape. If the total hill country population can be stabilized so that the majority live in small towns, then the pressure of inhabitants on the agricultural areas might be relieved to the extent that a hard life in farming could at least be viable.

River control works

The scenic effects of dams, reservoirs and diversion structures have been identified in Chapter 6. Major water storage sites have to be downstream of the highest mountain catchments. Reservoirs may provide clear and attractive foreground to mountain scenery, but dams intrude on narrower more intimate scenery – low-key intervention for an embankment dam in sylvan or gentle topography, more striking assertion for an arch dam in a steeper gorge. Below any water diversion site, the white water characteristics of natural streams may be spoilt.

Most weirs and mountain irrigation systems have been established through centuries of experience. Being constructed essentially from local resources, their provision requires a balance between robust construction and frequent maintenance or rebuilding after the ravages of annual floods and storms.

It is not always appropriate that essentially small-scale and simple water transfer systems should be replaced by more permanent structures of the sort incorporated in major water diversion projects. If a government provides such works, it becomes incumbent on that government to maintain and operate the system – this is much more difficult and usually more expensive than the farmers having to do it for themselves. However, farmers need authoritative advice and assistance in adapting irrigation and agricultural systems to meet the changes demanded by modern markets and technologies, for example in adopting sprinkler or drip irrigation.

Transport and pedestrian routes

Main roads and railways are aligned in accordance with principles outlined in Chapter 8. Their design can suit their surroundings by:

- modifications (planting or clearance, possibly earthwork) of adjacent landscape, to promote assimilation rather than intrusion, and to provide views from vehicles and trains (Photo 13.4); and
- design of structures, especially high bridges or viaducts (Photo 13.5), to form their own elements of landscape, in admirable contrast rather than shocking disparity with valley profiles or wooded hillsides.

Lesser roads and tracks across scenic country are the means of access to smaller settlements, remote installations and beauty spots. They offer opportunities for enjoyment, en route or at the destinations, to those travellers or walkers who can accept restrictions on the numbers of vehicles or people. On roads, there have to be adequate parking and setting-down points at each stage at which route capacity is reduced or where people have to alight from their vehicles to take alternative transport, perhaps on foot or using bicycles. The location and capacity of parking space is a key element in the regional route and movement pattern. Over time, adjustments to road capacity and connections may also have to be made to suit changing social demands and economic circumstances.

Footpaths allow people to enjoy the prime attractions of scenic country – physical exercise amid changing views, lolling by the wayside or

Photo 13.4 View of an otherwise secluded cove from a railway cut across a well-vegetated slope, Tor Bay, Devon.

Photo 13.5 A railway viaduct across a quiet valley in scenic Corsica (photo by Robin Carpenter).

exploring particular points of interest at the end of the trail. The more the people who wish to enjoy the routes, the more it is necessary that the paths should withstand wear and related erosion. In those inhabited mountain territories where there is no motor transport, carefully constructed and regularly maintained footways are essential. There may have to be careful stone paving, low retaining walls and drainage culverts, although the results are commonly perceived as ancient parts of the landscape since essential maintenance effort is not always visible. Maintenance can be combined with cutting vegetation to provide views and vistas from forested slopes.

Zones of more accessible less sacrosanct semi-natural beauty

Many large cities have spacious hinterland, within about 50km, which envelops both attractive countryside and pleasant residential areas. In the Chiltern Hills, northwest of London, valleys and plateaux covered by beech wood, fields or chalk downland surround small towns that expanded largely for London commuters but which are increasingly providing their own employment. In the English Midlands, Baines (2000)[7] describes commuter territory that 'still maintains its old rural integrity, with brick and timber villages and market towns, winding lanes, high hedges and a healthy mix of arable and livestock farming'. Near Greater Manchester, the country towns lie in steeper valleys, some among high moorland which is much wilder but still easily accessible on foot.

The *countryside* is a magnet for walkers and visitors to pubs and village greens, from the locality or from the cities. It encompasses the 'critical natural capital' for which protection Punter (1998)[8] says 'must be a vital part of our drive towards more attractive urban areas'.

The *residential areas* embraced in this happy situation are guarded by green belts and restrictions on development – for the benefit of those who enjoy the landscape or are fortunate enough to live there already. At the same time, they are under great pressure to release more land for newcomers to the region. The latter include those who can afford to move out from more crowded areas, particularly as, with improvements in telecommunications, these semi-rural centres become viable centres of employment.

The challenge for the future is therefore to find extra space for residential development whilst conserving the best aspects of the countryside. Punter (1998)[9] points out that critical natural capital 'does not usually occur in neat one to ten-mile belts around the built-up edges of our towns – it frequently takes a narrow linear form'. If there has to be new development beyond established city boundaries – and in most parts of the world this seems certain during this century – then there are some greenfield areas that are more suitable than others and where development would be less destructive than at other sites. But choice of new sites – for buildings, roads and services – should avoid:

- regionally scarce or nationally significant features, much of which may already be designated for conservation;
- the locally best-endowed areas of woodland;
- viable agricultural land units that are essential elements of recognized conservation areas;
- linear features of natural capital such as ridges or escarpments, long-established hedges and rural watercourses; and
- flood plains unsuitable for settlement; and wetland that may absorb excess flood flow.

At the same time, enjoyment of inhabited landscape depends ultimately on maintenance of the regional economy and the livelihoods of those who live on or tend the land.

Scenic attraction depends partly on wild vegetation but mainly on irregular topography. On land that has few or no natural undulations or viewpoints, recognizably attractive landscape has to be contrived by man. Such land is the subject of the next chapter.

Notes and references

1 Ruskin, J. (1856) *Modern Painters*, George Allen, London.

2 Denniston, D. (1995) 'Sustaining mountain people and environments', in *State of the World 1995*, Earthscan/Worldwatch, London.

3 Briggs, D., Smithson, P., Addison, K. and Atkinson, P. (1997) *Fundamentals of the Physical Environment*, Routledge, London and New York, p448.

4 Hackett, B. (1971) *Landscape Planning: An Introduction to Theory and Practice*, Oriel Press, Newcastle upon Tyne, p11.

5 Hackett (1971) as Note 4, p11.

6 Schama, S. (1995) *Landscape and Memory*, HarperCollins, London, p480.

7 Baines, C. (2000) 'Arden', in B. Bryson et al (2000) *The English Landscape*, Profile Books, London.

8 Punter, J. (1998) 'The household factor', *Landscape Design*, July/August.

9 Punter (1998) as Note 8.

14

Construction in Flat, Open Country

Chapter 13 concerned attractive country, much – but not all – of which is mountainous and much of which is devoted to human leisure. This chapter concerns flat or only gently undulating country, much – but not all – of which is scenically dull and much of which is put to economic use. Attention here is directed at inhabited but mainly rural land. It is generally productive, especially through agriculture, and has numerous settlements and networks of roads or waterways. Section 14.1 describes the uses of such land. Section 14.2 concerns the effects of construction on geologically stable ground, while Section 14.3 deals with river plains where the land surface is less stable and where engineering works may influence the effects of floods. Section 14.4 discusses how rural scenery may be visually enhanced by appropriate construction.

14.1 Rural land use

In totally rural areas, people use the land to produce food or industrial crops. In rural tracts of otherwise mainly urbanized regions, there are mixed economies in which agriculture still takes up most space but more livelihoods depend on commerce, small industry and services in country towns or on commuting into cities. Agriculture concerns growing crops or rearing livestock on ground watered by rain or with added irrigation. Trees provide shade, fruit or timber. Construction is involved in any substantial modifications of land surface or drainage needed for these and other productive activities, and provides the infrastructure of human settlement – accommodation, storage and communications as well as out-of-town facilities supporting urban communities.

Farming and forestry

Annual crops, such as cereals, are sowed and harvested at least once per year. To grow well they need fertile soil and an appropriate climate (see Chapter 1). Besides varying cropping patterns or adding fertilizers to restore nutrients, topsoil may be replenished by alluvium deposition (during flooding or from canal clearance), or it may be lost to water run-off or wind erosion when the surface is bare. Meanwhile, in some well-inhabited regions, fertile soil is being lost more quickly than it is being replenished.[1]

The main means by which dry climatic conditions can be countered is supplementary watering. Where rainfall is only marginally sufficient and irrigation with additional supplies of water cannot be afforded, then farmers need to maximize the benefits of such rain as does fall. This they can attempt by traditional techniques for capturing rain in the immediate vicinity of farmland – in ponds, on terraces or even underground, or by planting crops in hollows or shallow valleys to attract most immediate run-off before it evaporates.

Trees can grow on less fertile land and with less regular rainfall. They yield orchard crops, building or fuel wood, and provide field boundaries, windbreaks and shade. *Forestry* is large-scale managed cultivation of woodlands to yield timber for construction, furniture or pulp products. In flat well-inhabited

Table 14.1 Resource use in agricultural production

Resource	Risks	Opportunities
Human	Modernizing or intensifying agriculture in ways that reduce employment so far as to upset the viability of rural communities.	Increasing agricultural and agro-industrial production by means of technology, knowledge and communications – for optimum use of human effort and adequate reward for rural people.
Land	Excessive conversion of fertile land to less appropriate land uses.	
	Soil degradation – by wind or rain/run-off erosion, by salinization through excessive water application or by too intensive cropping and depletion of soil fertility.	Balanced production and conservation use of land.
	Excessive maximizing of cropped areas, eliminating features such as hedges, copses or fallow land.	Realizing planned priorities in allocation of land …
Water	Misallocation of irrigation water flow to those users that can most easily afford to waste it, rather than to efficient productive or conservation use.	… and water
	Similar mismanagement or wasteful use of rainfall, run-off, groundwater or effluent, or of controlled water supplies.	Optimum use of water according to its occurrence and quality.
Rural scenery (related to all of these)	Damage and intrusion.	Enhancement of scenery and natural habitat.

country it is practised mostly where the soils are poor and pressures for settlement are not intense, as in the Breckland of Norfolk. In wider zones unsuited to agriculture or dense settlement, such as much of Russia, forestry can be practiced spaciously and, hopefully, sustainably.

Livestock are pacified, captive animals. They are bred, fed and managed mainly for food – meat, milk or eggs. Some large mammals provide hides, some oxen are still used for haulage and horses carry people. Animals feed:

- by grazing on spacious rangeland territory, land at best marginally suited for arable agriculture, otherwise generally unoccupied;
- on farms capable of providing permanent, seasonal or occasional pasture within their own boundaries; or
- intensively, on bought-in feed, in concentrated areas or closed sheds.

Generally, animals are inefficient in converting vegetation into human food. So, on arable farms in well-populated regions where land is scarce, livestock should be kept only to a limited extent, mainly to meet local demand. Animals can graze fallow land, eat crop residues and supply manure as a crop nutrient.

Marginal land is that which is capable of producing traditional regional crops, such as grain, but only at low yields or with harvest failure in dry years. However, crop seeds exist, or can be developed genetically, that may become economic for more reliable crop production on land currently considered marginal.

For *agricultural land* in general, Table 14.1 identifies some risks and opportunities in attempting to meet demand for rural produce by means of land, water and human resources. Construction is most relevant when irrigation is required.

Rural settlement

Traditionally, villages comprise the homes of people who work the land. Such were vast numbers of settlements developed before the days of motor transport. But, with wider economic development during the last millennium, market towns arose in which trade and craftsmanship were undertaken, based originally on agricultural production. Over the last two centuries, in the energy-consuming mechanized communities, there has been a strong trend towards much smaller labour forces in organized farming. On extensive farms, one family or a small group live in isolated buildings, depending on vehicles for their communications with other people and commercial centres. At the same time, the main villages have become partly or largely inhabited by people employed in towns but with the same transport facility.

Most settlements arose near to farming land where there was a supply of water – from streams or springs – or at strategic locations adjacent to river crossings, seaports or mines. Today, planners tend to choose the less fertile, harder ground to locate buildings. But, when living space is in short supply, flat land is not likely to attract any priority for conservation even if it is intensely cultivated. It may be ripe for building on, if social needs exceed those of agricultural production.

14.2 Construction on stable plains, plateaux or gently undulating country

Stable, in this section title, implies topography that is unlikely to be altered by natural forces such as floods. Nor is the land so steep as to need the construction of terraces. Some extensive flatlands, such as the Punjab, have been flood plains in recent geological time but are mostly stable today. Such territory occurs in most soils and climates. Besides many of the intensively populated and farmed regions, it also extends over vast empty landscapes like the Central Asian steppes.

Locations for construction

Siting of new structures may concern:

- a requirement to be located in the vicinity of the settlement which it is intended to serve, for example, as a water tower;
- a route for a road between two settlements, or a water channel from an abstraction point to an irrigation system or municipal reservoir; or
- the need for a particular advantage – lying beside a waterway, on a high bluff, or in seclusion.

Environmental issues can then be considered in examining how construction will affect the landscape. Ensuring that any sort of future development best suits the wider locality requires comprehensive regional planning. The boom of ribbon development along main roads in the early 20th century was only stemmed by introducing strict delineation as to where development could take place and where it could not. Binney (1996)[2] refers to the success in Britain of planning for 'protection of the countryside from sprawl and spasmodic development' resulting in some 'unspoilt landscape in virtually every county'. By contrast, 'you can drive for two or three days across vast expanses of middle America and there will always be a house in view'.

Existing natural features

Otherwise smooth terrain may occasionally be pierced by hard visible outcrops or chasms. Sometimes there are water features; commonly there is vegetation, perhaps unexpectedly a special or unique ecosystem. Construction on plains or gently undulating ground should respect or take advantage first of any geological features that protrude. These could be, for example:

- ribs of rock strata revealed across eroded peneplains;
- glacial deposits (moraine fronts, eskers and hillocks) or eroded furrows (lakes) as in the northern parts of the European plain; or
- harder shields of older rock which have rougher, less fertile surfaces and very complex drainage (Labrador, northern Scotland and some Arctic lands).

Hackett (1971)[3] refers to the pattern of post-glacial lakes in midwestern US and notes that, in such flat territory, 'it would be important to emphasize or at least not subdue this variation to the landscape'. Layouts of structures where geological features occur should:

- make these features evident, rather than hiding or disfiguring them;
- provide opportunities for people to see views from high points; and
- take advantage of any sloping ground to create attractive background.

Vegetation, as low ground cover and foreground in open views, might be tundra growth, heath, grazed pasture or managed vegetable crops. Thicker taller shrubs or forests eliminate views in flat country except where there are lakes or clearings. Much scarcer are oases in the desert or wildland in densely inhabited places. Any development in the vicinity of ecologically sensitive places should be planned with great caution.

Long escarpments or narrow lakes are prominent features, which – at least as seen from the air – might be compared with features introduced by man, such as:

- agricultural boundaries, comprising physical barriers (for example, rabbit-excluding fences in Australia or political borders – between Egyptian Sinai on one side and intensive Israeli/Palestinian agriculture on the other, or Canadian grazing land and US arable fields on either side of the 49th parallel), or dark green circles irrigated by radial sprays extracting groundwater in the Libyan or Jordanian desert;
- engineering structures such as for transport routes, waterways or reclamation works, for example, the 1932 Afsluitdijk which holds in the Ijsselmeer and its polders, the straight lines of drainage channels across the English Fens, or arrays of solar panels in US deserts.

Space and layouts for settlement

Homes and settlements have been introduced in Chapter 12. Here we are concerned with planning construction for settlements – villages, country towns, often city suburbs – that are not constrained by physical limitations on space in the same way as:

- steep hillsides mentioned in Chapter 13;
- high density city accommodation, considered in Chapter 15; and
- land prone to flooding, the subject of the next section (14.3).

Without such constraints, there are usually no particular construction difficulties in preparing land surfaces or providing access. However, actual allocation of space may depend more on the commercial value of land (particularly in developed countries) or the conditions of tenure (particularly in less developed countries).

The economy of more affluent communities is based on non-agricultural employment, much of it performed some miles from home using ubiquitous, possibly unsustainable, road transport. High building land and property prices have stifled wider ownership, perhaps as international banks replaced local mutual building societies. In less developed economies, rural employment opportunities must be boosted by transfer thence of more secondary production and service jobs.

Assuming that effective political and economic management can find equitable solutions to these problems, the allocation of building plots and provision of services can be planned to realize actual needs for accommodation. In a crowded world, individual opportunism has to be tamed by authoritative measures such as land use planning and strategically engineered infrastructure.

Actual layout of villages or country towns relates to the disposition of structures, the pattern of streets and any interface with semi-natural features such as water channels or trees. Land devoted to human residence and amenity extends well beyond the buildings. Besides the surface areas of streets and market places, there are public or private gardens or yards. Nearby there must be facilities such as water treatment or sewage disposal plants, as well as recreational space ranging from informal playgrounds to manicured golf courses.

Roads and other paved surfaces

Almost all human settlements are connected to some form of road. Hard surfaced vehicle parking areas adjoin industrial plants, railway stations or out-of-town retail outlets. Main highways cut swathes across rural land between cities. Airports land space can exceed 1000ha.

Rural roads are integral features of semi-natural landscape. They are boundaries as well as routes. Typically they may extend to 10m in width including verges. These verges are often green corridors but are interrupted where there are bridges, crossroads or roadside buildings. Ground level *parking areas* are certainly space consuming, but their configuration can be planned to suit the shape of space available amid established or higher priority land uses. If car parks are needed in the countryside, their obtrusiveness can be softened (see pp155–156).

Main highways separate land on one side from that on the other. In most flat country, this separation affects economic activity more than landscape. On new routes, alignment adjustments may be feasible to minimize disruption on either count. Otherwise road shapes relate to landform engineering and their layout should be conceived and managed to minimize intrusion and yet to optimize views seen from the highway. Across flat terrain, most main roads are slightly raised.

Airports, because of their extent and to alleviate aircraft noise over residential areas, are often located on rural land. Several European airports are approached by air and land over or through forests, isolating the airports from other settlements and affording them attractive access. Within airport perimeters there are areas of unpaved land, assigned as safety buffer zones, which can be conserved for nature or allocated to certain low-level low-density human activities. It has been suggested[4] that London's Stansted Airport is well suited for extension into the Essex countryside because the existing airport is better ecologically managed by the British Airports Authority (BAA) than what was described as 'agricultural wasteland' nearby. Heathrow Airport is in a more built-up area west of London, better suited to industrial activity and air freight facilities in its immediate vicinity.

Water channels

Villages and towns require communal or municipal water supply systems. These take water from run-off, streams or groundwater in the vicinity, or from pipeline or open channel aqueducts supplied from more distant rivers and reservoirs. In flat country, irrigation and drainage networks follow linear patterns. Between the channels grow profuse crops in season (Photo 14.1). A more dismal prospect is seen where ill-managed irrigation has resulted in abandoned saline or waterlogged land (Photo 14.2).

In addition to their vital role in transporting water, open channels with shady trees have always been essential features of Paradise, as conceived by philosophers in hot dry regions such as Arabia. So, still today, tree-lined canal roads offer pleasant prospects and shade from the sun.

Canalside trees must be planted, but their growth and that of nearby strips of marginal grass or shrubs have to be sustained by such rainfall as occurs or by seepage from the water channels. Paths and roads – for pedestrians, cyclists and occasional vehicles or animals – can be elevated on canal banks. Points of interest occur at canal regulators and offtake structures, especially where there is falling water. Other features in the landscape may be short-term reservoirs, incidental wetland or waste water treatment ponds – all elements of a balanced plan for irrigation and drainage.

Thus water, where it is scarce, can be a key element of pleasant environment, and its contrived transfer from wet to drier places is vital to land resource values. Where there is too much water, construction plays a quite different role.

14.3 Construction on flood plains

The role of floods in river geomorphology and the type of construction that attempts to cope with them are introduced on pp41–42. Flood protection embankments are described on pp54–55. The limitations and scope of flood control are further discussed for very large river basins on p88. The landscape of alluvial plains is explained on p94.

Characteristics, opportunities and risks

Except where river channels have been totally stabilized by man, river flood plains are dynamic landforms. Ground contours and channels change every season as fast currents scour the river beds and slower overflow deposits sediment on adjacent plains. In these conditions, ancient cities and settlements have been buried beneath the deposited alluvium or destroyed by scoured channels. Yet, as the global population continues to increase, so flood plains are still chosen as principal crop production areas and some are densely settled.

Photo 14.1 Precast concrete irrigation distribution channels, southeast Turkey. Water is transferred to the fields through portable siphons (water is released from earth channels by temporary hand-dug breaches). Excess water should be drained off at the bottom of the fields.

Photo 14.2 Soil degradation on irrigated land, Punjab Province, Pakistan. Most of the cropped fields are, so far, not seriously affected. In the lower middle bare patch, some groundwater has reached surface level, while the white areas have become saline.

Photo 14.3 Qara Tappeh, Iran – a village raised by centuries of building and rebuilding above the surrounding seasonally flooded plains and Caspian Sea coastal lagoons.

On land that is seasonally dry, cultivators have often taken advantage of floods to sow crops just before or just after inundation. In wetter countries, river flow may be diverted to complement rainfall in watering the fields. Construction of permanent buildings on flood plains carries risks both to the structures and in exacerbating the effect of floods by constricting the flow. In situations where water levels rise gradually and seasonally rather than due to sudden inundation, it has been possible for villages to stand permanently on stable raised ground (Photo 14.3). But large-scale settlement is only viable behind protecting embankments.

Landscapes adjacent to rivers are those:

- of plains subject to unimpeded seasonal flooding;
- of areas – much agricultural land and some settlements – that are now protected by embankments sufficient to withstand most annual high flows but which are occasionally overcome by exceptional floods; and
- of more densely populated or strategically important areas more securely protected by higher walls or embankments.

Figure 14.1 indicates ways in which man-made protection banks define these landscapes.

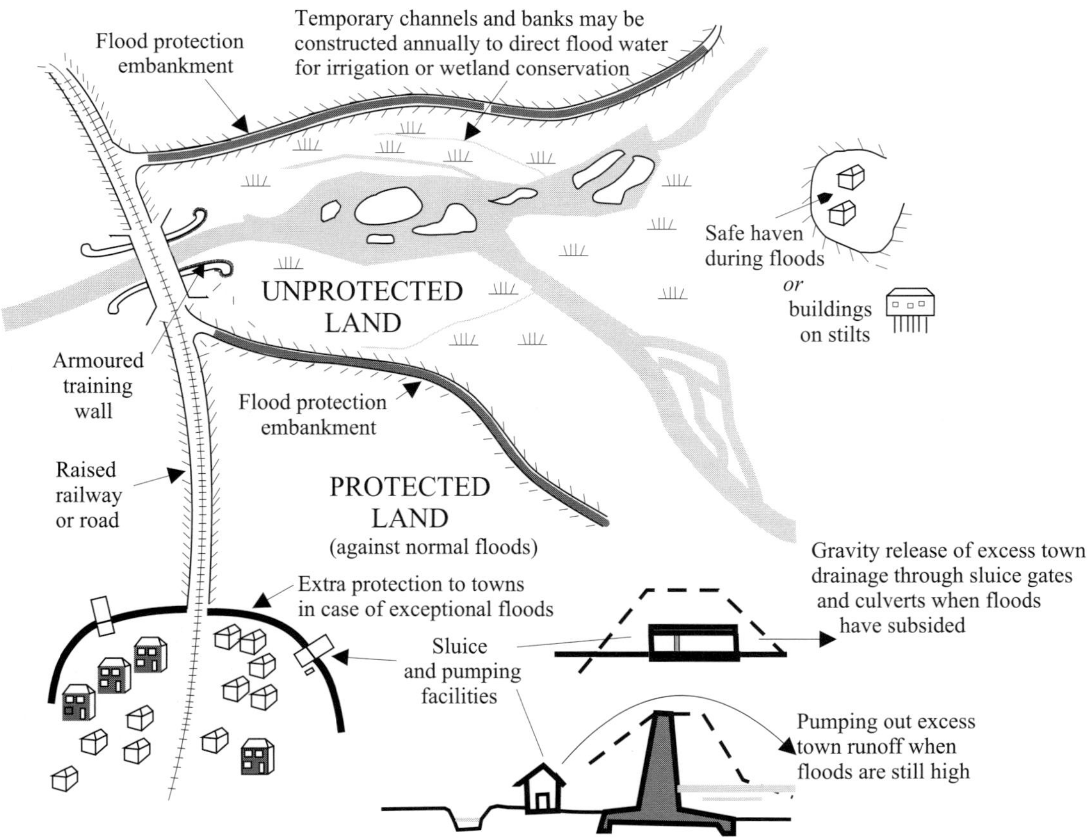

Figure 14.1 Flood plain land – protected or unprotected

Seasonally flooded plains

As sediment is deposited and river bed levels rise, so the chances of overtopping natural or artificial levees also increase. Hydraulic engineers in China, starting over 2000 years ago, strove to dig the channels deep but to keep the banks low. Today, rivers such as the Mississippi continue to build up their bed levels slowly between almost continuous man-made embankments. Perhaps there may come a time and circumstances when dredging main river channels becomes as effective as constructing embankments. Meanwhile, settlements should be located outside the confined flood plain and construction on that plain restricted to cross-river structures such as bridges and barrages. 'Armoured' training works – rock or concrete block-protected banks – are necessary to direct flow through cross-river structures such as barrages or bridges.

Thus the landscape features of unprotected flood plains are cultivated fields, perhaps woodland and, hopefully, some natural wetland. Road embankments, bridges or raised settlements are added man-made features. Photo 14.4 shows a railway embankment crossing the River Spey flood plain.

Apart from their agricultural potential, seasonally flooded wetlands are important in two respects. They are often significant wildlife habitat, and they have a major effect in soaking up river flow and mitigating the effects of floods further downstream. Marshes can also be silent backwaters of mist and solitude.

Where villagers have no option but to live on flood plains, they need to:

- build their houses to stand up above flood water; or
- construct, where practicable in earthwork, raised safe havens to which they and their animals can retreat during floods; and
- be given timely warning of floods.

In Bangladesh, the width and flows of the rivers are too great for embankment protection to

Photo 14.4 Railway embankment across Insh Marshes, Scotland. The two parts of this National Nature Reserve are joined by culverts through the railway embankment thus conserving the quality of one of the most important wetlands in Europe.

be viable except around cities. Therefore wide areas and many villages of that densely inhabited country are flooded every year, much more in years of exceptional flow. In these conditions, main river channels are left untouched by engineering, except where it is essential and practicable to stabilize their courses.

Most of the villages in the Ganga-Brahmaputra flood plain and delta are raised slightly above the surrounding ground level, but this is often still below frequent flood levels. The villagers make valiant efforts to construct and maintain earth banks in an attempt to bring their homes into at least the category that is flooded only in bad years. Men carry baskets of mud on their heads and weave long lengths of fibre to reinforce the banks. They cannot afford either to hire earthmoving machinery or to purchase hard protection material. On stretches of land where river courses change frequently, villagers have to dismantle their houses before they collapse into the water, carry away the pieces and reassemble them on safe ground.

Land protected by embankments

The scenery of flood-protected flatland is similar to that of more stable (non-flooded) plains in respect of most construction or land management. Differences are in the more formal linearity of the man-made features and in problems of drainage. The linear features are transport routes, built up on low embankments across the flat ground, and the excavated water channels and drainage ditches. The resulting patterns are striking, seen from the air or any high structure, as is the sight of any less formal features, such as nature reserves or pockets of wild land.

The main problem of drainage occurs where the construction of long flood protection banks prevents escape of run-off accumulating on the land protected by those banks. This occurs whenever the level of river water, kept out by an embankment, exceeds the level in the protected area. The accumulated run-off either has to pond up – in reservoirs or marshes – until the outside level has receded and gates can release it, or it has to be pumped out. In the large part of the Netherlands that is permanently below sea and river levels, pumping has been a perpetual requirement for centuries. The English Fenland has been drained by pumps driven in turn by wind, steam, diesel engines and electricity as the land has sunk and the difference in water level has increased.

There is seldom serious concern about agricultural land that is only rarely flooded. Nor is it necessarily tragic if transport routes are cut for short periods. But these occasional effects may be more serious for towns. Engineering strategies can include:

- permanent protection where this is feasible;
- appropriately sited and adequately powered pumping stations;
- temporary protection walls that can be put in place whenever high water levels are forecast;
- acceptance of infrequent flooding if the cost of flood damage can be limited, for example, by raising floor levels and storing valuables and perishables upstairs, or by periodic reconstruction of houses; and
- demolition of unsuitably located buildings – which should never have been built in the first place – to increase the flood plain's hydraulic capacity.

The dangers of construction on flood plains are that structures may be built in areas where floods have become rare (for example, because of storage and diversion works upstream). High floods still occur, now less often, more unexpectedly and sometimes more disastrously. In addition, new built-up areas, paved surfaces and hydraulically-efficient flood relief bypass channels all contribute to rapid runoff and worsened flood conditions downstream.

Natural flood plain vegetation is well suited to soaking up floods and can be associated with valuable wildlife habitat. Fertile land can be adapted for agriculture. Robust engineering works to control or divert river flow or for flood protection can be provided if the benefits justify the cost. But permanent settlements are better located off flood plains where this is practicable.

14.4 Rural scenery

Perceived quality of flat scenery

Flat open views can be dreary if there is a bare or indistinct horizon with an absence of discernible structures. Equally dismal may be featureless foreground, particularly in monoculture, when crops or vegetation are not in flower or when there is no sign of animal life.

People at ground level have to be familiar with regional maps to sense the pattern in a flat landscape. Actual visual appearance is concerned with land/sky contrasts on wide horizons or specific features in more focused vistas. Only occasionally does one pleasantly encounter an oasis of wilderness.

An attraction in flat landscapes has been recognized by some artists and even some champions of nature. Dutch landscape painters depict rustic semi-natural countryside, man-made structures and human activity, and they take advantage of the few natural high features such as sand dunes.

David Bellamy (2000),[5] describing the Fens, denies that their flatness is boring and counts them among his favourite landscapes with 'hardly a tree to get in the way of each breathtaking view'. He explains his favour in terms of clouds and skies as well as 'spires and steeples in the furthest distance'. It may also be relevant that Bellamy is well aware of the wetland nature reserves and pockets of ancient fen that are conserved within the wider man-made landscape and which, although generally out of sight, perhaps offset what other people consider is visual monotony.

So, perhaps a perceived glory of open landscapes is the contrast between celestial features and isolated man-made towers pointing to the sky from a subdued horizon. However, in many places in less temperate climates, skies are dazzlingly bright and the sun sets behind a pall of dust about 5° above the blurred horizon.[6] Nor are the human settlements so permanent, resources so ample, or culture so fine as to encourage tower or spire construction.[7]

Nevertheless, there must be many dull plains where opportunities exist to manage interesting foreground crops, plant middle distance trees or build whatever structures are needed in attractive regional style and colour.

Farms and woodland

Some intensive cultivation of fruits, fungi or flora takes place in glass houses or sheds. Perhaps this is agro-industry more akin to the scenery of Chapter 10. Conventional farming scenery comprises:

- open fields for cultivation of grain, vegetables, grass and industrial crops, offering seasonal foreground (ploughed, growing, in flower, harvested) to any taller background scenery;
- hedges or walls, which constitute some of that background, and tracks or drainage ditches further forward;
- shrub crops – vineyards and tea or coffee plantations, seen as regimented rows on undulating hillsides; and
- orchards and copses.

Whether trees grow along field and roadside boundaries or at the edge of deep forest, they are the most common form of background to agricultural, grazing or heathland scenery. Tree-lined avenues or forest rides allow vistas of whatever lies beyond. Where new roads have to cut through woods, it may be possible to choose routes that minimize disturbance of viable units of special habitat or destruction of historic woodland edge features.

Structures in distant views

Location or alignment of new works should take account of local topography, existing buildings and any remnants of undeveloped land or water margins. Proposed structures can then be designed to be visibly acceptable in their surroundings, for example, as:

- elegant towers such as those described in Chapter 11 for remembrance (monuments), inspiration (spires) or function (high water tanks);
- regionally distinctive roof shapes on houses;
- slim power or telecommunications masts;
- bright horizontal strips such as extensive greenhouses in the Netherlands;
- interesting profiles of military or industrial structures; or ancient monuments, such as Stonehenge on Salisbury Plain; and

- cunning earthwork and sympathetic planting to soften any perceived intrusion of structures in distant prospects.

These considerations are mainly for the long term. However, on productive land there are many temporary structures that can confuse the situation, sometimes for years. Temporary ugliness is perhaps a symptom of careless opportunism, for example in the siting of sheds or car parks. Particular difficulties, even in medium term structures, occur where:

- people make use of random pieces of timber, iron sheets or bricks;
- the purpose of a structure is intended to be brief; or
- everything may suddenly be destroyed by a cyclone.

However, even in these cases, a degree of artistic ingenuity must be possible. Enforcement of aesthetic standards is no doubt easier around handsome structures.

High hedges may well hide small buildings, local power lines or road traffic in otherwise open views. However, there is seldom any point in trying to conceal tall structures or high voltage power lines. Elegant design of towers and pylons may mitigate any blatant intrusion.

Many people enjoy the sight of wind turbines, some greatly dislike them. Few people admire high voltage electricity transmission lines, many dislike them. Electricity generating stations, which often have to be located in open countryside and near to water, are commonly regarded as an unwelcome but necessary intrusion. All these are inevitable – where there is enough wind or enough people to justify power generation and transmission. In determining where any type of new or replacement structures should be built, their impact on the landscape can be optimized by selecting sites or routes well suited to other land features, such as by aligning power lines to run parallel to (rather than along) river banks. Structures can – at best – be attractive, often interesting or – at worst – their layout can hide their most offensive aspects.

Village scenery

The scenery of rural settlements comprises three elements:

1 Dominant buildings – such as churches or manor houses in English villages, grain silos on North American prairies, pagodas in Myanmar, minarets or domed bathhouses in the Middle East (Photo 14.5).
2 Ordinary houses, in traditional styles, tend to be functional and uniform in appearance – flat earthen roofs on thick walls in semi-arid countries (Photo 14.6), sloping thatched or tiled roofs on brick, stone or timber frameworks in wetter climates. Shape, symmetry or asymmetry, texture, colour and contrast should perhaps be less striking and more harmonious than they are in ever-developing cities.
3 Roadways and other shared space are the foreground of village scenery and may be best served by soft but rough rather than geometrically sharp hard features, for example in roadside kerbs or fences around property.

Some semi-natural scenic features – such as humps or holes, ditches, banks or ponds – are inevitably eliminated by development, but there may also be specific opportunities for their conservation or enhancement. Paved or gravel surfaces can be walkways and foreground as much as vehicle space. Rural byway priorities are different from those of main roads or city streets.

As explained in Chapter 12, the permanence of houses depends upon climate, available building materials and skills, funds and security of tenure.

Photo 14.5 A whitewashed domed public bathhouse is in bright contrast with the comparatively sober houses in most villages in Mazandaran Province, Iran.

Photo 14.6 Shush, a town on the plains of Khuzestan, southern Iran, within a mixture of comparatively short-lived buildings and one iconic conical monument – Daniel's Tomb, in the background. The town has grown considerably since this picture was taken in 1975.

North European village structures might not be appropriate in less affluent communities or in warm climates where utilitarian semi-permanent buildings are the established norm. However, it is in transition from simple communities to more sophisticated ones, or in rapid extension of previously integrated settlements, that village landscapes need to be secured against insensitive alteration. Constructional changes result from improved communications, economic adjustment and influx of people who work in towns.

In what were comparatively isolated rural areas, wider commerce has introduced manufactured construction materials such as cement, corrugated iron sheets, fibreboard and even solar panels to complicate the original rusticity, and small-scale private enterprise can bring a host of undisciplined overhead power connections and shallowly buried water pipes. This is where there is the main need for planned engineering – in constructing extended or replacement municipal infrastructure for both functional and visual coherence.

Thus land planners and builders can devise means whereby drab landscape or settlements may be transformed to inspire more enjoyable living. But meanwhile, in many countries, there are political difficulties in maintaining even basic livelihoods for rural people.

Viable rural landscapes

The quality and variety of rural landscapes reflects the way in which people manage the land – plough fields, graze livestock, tend orchards, construct drains – and how they fashion settlements. The cultivated land and the human settlements are no doubt part of the 'interwoven complex of habitats, together forming a scene of beauty' which Crowe and Mitchell (1988)[8] suggested define a 'good' landscape.

The attraction of English countryside, for example, depends fundamentally on the temperate, moist climate and often on undulating relief. But the rich tapestry of fields, pasture, hedgerows, woods, lanes and villages was made by long-established, often feudal communities in which a stable rural population engaged in agriculture in support and close liaison with local market towns. With 20th century social egalitarianism and mechanized farming, a large proportion of people living in villages moved to towns or travelled there to work. To balance preservation of country landscape features with a viable rural economy has become a prime objective in regional planning.

The landscape of the North American prairies was monotonous when it was grassland grazed by bison. So it remains under arable crops or as managed ranch land, relieved only by planted windbreaks, farmsteads and water towers. There is space for more people to live in these regions, but most must necessarily reside in towns. Not many extra people can be employed in agriculture even in more intensive or more carefully husbanded farming.

In much of the overcrowded farmlands of the less developed world, there is an imbalance between the dynamic and demanding economies of the towns and the struggling overmanned agriculture of the country.

Engineering and construction should contribute towards any solution by sustainment of fertile land by erosion control, provision of adequate drainage and careful water management. Then they must continue to provide, adapt and extend the infrastructure for villages (whose inhabitants could add more value to agricultural produce) and rural towns (where people could take over more of the remunerative employment currently monopolized by the larger cities).

Notes and references

1 Most comments on land degradation concern global trends. In some crowded countries, losses may be worse than official data show because governments are loathe to report depleting areas of fertile land – see, for example, Young, A. (1998) *Land Resources: Now and for the Future*, Cambridge University Press, Cambridge. Based on later FAO data, 24 per cent of the world's 15 million square kilometres of cultivated land suffered degradation in the years 1981–2003 (compared with an earlier United Nations Environment Programme estimate of 10–20 per cent in the same period).

2 Binney, M. (1996) in *Times* supplement on New Towns.

3 Hackett, B. (1971) *Landscape Planning: An Introduction to Theory and Practice*, Oriel Press, Newcastle upon Tyne, p9.

4 A comment by Germaine Greer in a BBC programme, c.2006.

5 Bellamy, D. (2000) 'The Fens', in Bryson, B. et al, *The English Landscape*, Profile Books, London, p267.

6 Here the author is referring to the hot flat plains of the Punjab where he lived for three years. At best tree-lined canals, water regulation structures or archaeological sites are interesting *landscape*. For holidays, one travels to the outstanding *scenery* of the Himalayan foothills.

7 There are outstanding exceptions. In the Indus-Ganga plains these are lasting monuments of Mogul cultural resource such as Jahangir's Tomb near Lahore and, of course, the Taj Mahal near Agra.

8 Crowe, S. and Mitchell, M. (1988) *The Pattern of Landscape*, Packard, Chichester.

15

Construction in Urban Landscapes

15.1 Keeping cities civilized

Civilization is any advanced stage of social development – the betterment in human condition that can result from large-scale collaboration. The word civilization is derived from the Latin *civitas* (a city) and *civis* (a citizen). In Rome, the city offices were the *munia*, and *municeps* was a citizen with privilege. The privileges of citizens have to be adapted, shared or relinquished, especially where there is a need to correct the urban/rural economic imbalance – explained at the end of the last chapter – or to accommodate the continuing influx into cities of people from rural areas or poorer countries.

Correction of economic imbalances involves transfer of certain employment from urban to rural centres, adjustment of urban and suburban employment to suit modern techniques and methods of communication, and perhaps lower urban wages due to competition both from rural competitors for non-agricultural jobs and from more people settling in the cities. Lower wages may afford less spacious accommodation and less extravagant but possibly healthier lifestyles.

Accommodation of city people has to provide for their increasing numbers in the land space available. This space is needed for:

- buildings in which people live, work or meet for communal activity;
- open ground, such as suburban gardens or public parks; and
- the infrastructure of roads and municipal services.

The form of a city 'has been and always will be a pitiless indicator' of man's civilization (Bacon, 1978).[1] Therefore, to accommodate change, cities have to be dynamic; their layouts, structures and services have to be continually adapted, rationalized or extended.

This chapter describes the role of construction first in particular aspects of urban landscape, such as:

- natural features, within or around which cities are constructed (15.2);
- views that are possible in urban situations (15.3);
- city heritage features (15.4); and
- urban ecology – the fauna and flora of towns and suburbs (15.5).

Section 15.6 then examines ways in which construction planners are involved in city development and the best use of urban land resources.

15.2 Natural features in city scenery

Natural scenery that may be a part of city landscape includes:

- rock outcrops – robust examples of raw geology;
- lakes and rivers – water surfaces that can be created, drained, raised or reshaped by engineering works; and

- trees and other vegetation – including transitory elements of scenery and depending on how these are managed.

Rock outcrops occur occasionally as cliffs or ravines within cities or, on a large scale, as mountain background. The cities of Edinburgh, Fribourg and Luxembourg (Colour Plates 7 and 8 and Photo 15.1) are traversed by gorges. These give rare opportunities for construction to take advantage of the low and high points, slopes, cliffs and views. Cliffs and rock pinnacles are rare in urban situations. However, on the smaller scale of human building, cliffs can readily be created, whether in cuttings for flat land space or quarries to win building materials.

Backdrop scenery may be high sometimes snow-covered Central Asian mountain ranges behind Tehran or Almaty, or the American western cordillera beyond Vancouver, Calgary or Denver. There are lesser but steep escarpments close to Islamabad or Bogota and more spectacularly shaped mountains behind Cape Town, Rio de Janeiro, Naples or Edinburgh (Photo 15.2). Even if there are no crags, nearby hills offer sites for structures deserving prominence (Photo 15.3).

Hills may also afford fine views of the cities from without. London's skyline can be seen from the open space on Hampstead, Greenwich and Epsom Heaths, while Florence can also be viewed from its surrounding hills. Just as these distant prospects chiefly concern the silhouette of the higher, more prominent buildings, so the rockier features of neighbouring hillsides and especially their crests need to be kept undefiled by structures that might mar their beauty as city backdrop.

Lakes and rivers contribute to the scenery around them by means of the low level open space which they occupy and the views afforded across them or reflected in the water, as well as riparian features such as lakeside greensward, rougher wetland habitat or promenades along engineered riversides. Photos 15.4–15.11 illustrate various waterside situations in towns and cities.

Bankside streets or promenades are the pride of towns that turn their faces to their rivers. Bewdley, for example, faces onto its river (Severn) as do most French towns, and Amsterdam faces onto its canals.

Photo 15.1 A railway viaduct, lawns and ancient fortifications decorate the gorge entering the city of Luxembourg.

Photo 15.2 Salisbury Crags, part of the volcanic mass of Arthur's Seat, soar steeply above eastern Edinburgh.

Photo 15.3 Montmartre, Paris. The bright profile of the Sacré-Coeur church crowns its hill as seen across the River Seine and the Tuileries Gardens.

Cross-river views of city buildings are renowned visual assets of Paris, Budapest or Prague. So they are in London where redevelopment has gradually provided walkways and river-facing buildings with fine new aspects. More widely, waterfronts have become spearheads for regeneration of former wharf and warehouse areas.

Meanwhile, few cities can display outcropping rock, and by no means all have waterfronts. But people can plant *trees and lower vegetation*. These mature within a few years after which their growth can be managed. Besides encouraging avian wildlife, the urban function of trees is:

- to provide summer shade within which to enjoy walking or sitting in the scenery;
- to define the edges of vistas; and
- to show up as green strips when seen from a distance.

Lawns surrounded by bushes or flower beds provide more open foreground and amenity space.

Photo 15.4 River Seine, Paris. Fine bridges, riverside promenades and water traffic are the principal attractions.

Photo 15.5 The Thames and its bridges provide foreground for views towards the City of London.

Photo 15.6 Contrasting architecture as London's towers rise behind the Thames Embankment.

Photo 15.7 Office buildings on London's South Bank. These high blocks have been kept well back from the river, leaving space for various lower developments that started with the Festival of Britain in 1951. An open car park remains which shows that there is still scope for further development.

Photo 15.8 The sheer high shapes of London's Canary Wharf buildings can be appreciated across a former dock. Barges carrying construction materials were still using the waterway in this 2002 view.

Photo 15.9 Some of London's warehouses originally backed directly onto the river. Where these buildings have been demolished or converted there are now walkways overlooked by apartments, for both of which the river is a prime attraction.

Photo 15.10 The Regent's Canal in north London originally provided direct access to warehouses, factories and gas works. Now there are apartments …

Photo 15.11 … and houses, some with gardens running down to the water.

15.3 City views

Built scenery and how we see it

Buildings are the core of urban landscape. With characteristics introduced in Chapter 12, they may be seen as individually splendid, as elegant rows of a particular style or period, or as deliberately eclectic or wantonly random architectural mixtures. Streets, squares and waterways provide foreground across which buildings can be admired. Linear horizontal features – such as pavements or flower beds along boulevards – or vertical ones – such as colonnades, lamp standards, signposts or shade trees – then compose frameworks for vistas and views.

Urban form comprises the frontage (elevation) of buildings and the layout (plan) of the spaces between them. Civil engineering is one of the inputs to urban form. It influences the location, determines the type of foundation and sometimes the shape of major structures, and fixes the layout of the associated infrastructure.

Street and building layouts

Streets contained by tight centuries-old patterns are often little more than narrow walkways with wider space only in front of the more prestigious buildings. Beyond the sights and sounds of one's immediate surroundings, the sights are buildings at corners or ends of streets (Photos 15.12 and 15.13), there may be a glimpse up a side alley or waterway (Photo 15.14), or, if the land slopes steeply, one may look up to a variety of styles at different levels (Photo 15.15).

Wider layout patterns for city streets can encompass:

- straight streets and right-angled intersections – convenient for traffic management but visually dull unless some road or building lines are offset at junctions;
- curved streets (as Regent Street, London) or crescents (particularly impressive if foreground space allows a wide frontal view, as at Bath);

Photo 15.12 Canterbury, England. Vistas up narrow streets show buildings at corners ...

Photo 15.13 ... or terminal structures such as the cathedral gatehouse.

Photo 15.14 A glimpse up a narrow canal in Venice reveals a more accessible and friendly prospect beyond.

Photo 15.15 A stepped walkway in Fribourg, Switzerland.

- spokes from a central boss, such as the 12 streets running radially from the Place d'Etoile, Paris – profoundly satisfying seen from the Arc de Triomphe above, less evident at ground level except for the appearance of wedge shaped buildings; and
- wide boulevards, such as from the Louvre Palace to the Arc de Triomphe in Paris and now beyond on the same line by the Avenue de Neuilly to the striking modern complex of La Defénse.

The shape and width of streets determine how well building facades can be seen. The configuration of buildings affects what can be revealed along or beyond the street vistas. The height of structures and the slope of the ground are also relevant in planning higher buildings behind those in front, for example in cross-river views[2] (Photo 15.6).

Elevated walkways, parallel to or across roads, offer welcome views provided their own structure does not obstruct street vistas or building facades. Chicago and Hong Kong have high walkway networks in planned environments of marble, steel and glass, but a similar 1960s 'Pedway' scheme in London emerged only in fragments as the concept was swept away by rampant commercial redevelopment (Stanley, 2004).[3]

Open space

Rivers have already been cited as natural features that influence urban scenery. But city waterfronts are invariably artificial engineered structures. The construction of walls, landing stages, bridges, riverside roads, promenades and verdant strips provides spacious viewpoints, foreground and scenery itself. River traffic complement throngs of people and vehicles in the streets; and variations in stream flow reflect quirks of nature like the changing light on building facades.

Photo 15.16 Sir Christopher Wren planned, in vain, that long wide streets would stretch out from St Paul's Cathedral in late 17th century London. This late 20th century vista achieved the effect, but might better have been wider.

Besides riverside promenades and traffic-free strips on boulevards, open ground on city land includes:

- shafts of space, provided by open-ended courtyards or avenues, private or public – such as Queen's House seen from the Thames between the wings of Greenwich Palace, or the view from the Millennium Bridge towards St Paul's Cathedral (Photo 15.16);
- city squares, such as some Spanish or Latin American plazas, Trafalgar Square and some more limited openings in London, and in many other old cities or new towns (Photos 15.17–15.22); and
- other space that is not built on, whether empty with rough growth or neatly cultivated and mown.

Meanwhile, many city parks and the spacious *maidan* lawns of Calcutta or Delhi, areas of potentially enormous real estate value, have fortunately been preserved as public amenity space. Most parks are less open because of the many trees they accommodate, creating islands of peace divorced from awareness of the city surroundings except by distant traffic noise.

Photo 15.17 City squares – paved in stone blocks in Brussels ...

Photo 15.18 … in bitumen in Moscow …

Photo 15.19 … or concrete slabs in Kathmandu.

Photo 15.20 Many London squares are now pedestrian areas, including the north side of Trafalgar Square ...

Photo 15.21 ... Paternoster Square in the City ...

Photo 15.22 ... and in grassy terraces in this quiet haven in Paddington.

Occasional vistas can be seen, such as that across the lake in St James's Park, London, towards the neo-Gothic spires of Whitehall.

Tidiness or wilderness, sterility or niches of nature

Particularly in urban surroundings it is necessary to designate which are to be tidy clean places and which should allow a more balanced mixture of natural and cultivated forms. Where people gather in large numbers, pass by or enter buildings, there is little scope for wildlife other than sparrows or pigeons. To discourage public nuisance, lurking criminals or bomb hazards, security-conscious municipalities tend to eschew construction of narrow passages, dark subways or even litter bins. In small parks or along boulevards there may be mown grass, tended flower beds and trees but bushy shrubs are discouraged. In larger parks or along canals, more separated from intense population, wilderness is more appropriate. On derelict land or anywhere comparatively remote, where human access is restricted or difficult, unkempt greenery or rough structural remnants are better suited as refuges for flora and fauna, if also to human miscreants. In railway cuttings or on private ground, habitat management can be both functional and conservational.

City skylines

Features of built skylines include:

- towers (such as in San Gimignano), spires (Oxford), domes and minarets (Istanbul);
- sloping roofs of sombre slate, brightly coloured tiles, iridescent glass or solar panels;
- groups of uniformly dimensioned high residential blocks with plain flat roofs, sometimes cluttered with extraneous extras, sometimes profuse in greenery; and
- less uniform but striking office blocks and skyscrapers (Photo 15.23).

In 1950, the view from Singapore harbour revealed little of the metropolis, yet 25 years later there was a packed skyline presenting the emergence of a prosperous city-state.

Photo 15.23 The skyscrapers of South Manhattan, 1964 (Photo by Elizabeth Carpenter).

Municipal height limits have been imposed in some cities. At the low end of the height limit range, Preece (1991)[4] reported that 'views of the skyline of historic Oxford from surrounding hills are protected by the banning of buildings within defined "view-cone" areas identified in statutory plans'. In less intensively historic cities, the maximum height of buildings continues to be under frequent review in each district. Once limits are raised, competitive prestige takes over up to the limits of engineering practicality and finance.

Total assessment of a city's scenery

The scenic character of a city as a whole can be comprehended by moving through or resting in its various landscapes and by absorbing each one in its changing diurnal and seasonal moods. Urban ambience is also associated with stimulating sights and sounds of people, traffic or city flora and fauna.

Bacon (1978)[5] describes city layouts and architecture in terms of 'movement systems'. Visual experiences of a pedestrian include turning a corner, entering an old city through its gateway, approaching buildings or just experiencing the intervening space. More rapid, perhaps less intricate, impressions may be gained from the top of a double-decker bus or an elevated metro.

15.4 City heritage

Preservation within inevitable change

Urban features of recognized historic or cultural significance encompass many of those listed in Chapter 2, including individual classic structures

– such as temples, public buildings, walls or monuments which are of high quality and have endured the ravages of time – and 'old city' quarters or other extended groups of historic buildings that have survived in their entirety.

Yet cities as a whole have to accommodate changes in population, economic activity and social norms. Buildings become redundant, neglected or ill-suited to new purposes. On the one hand, new activities or travel patterns call for changes in routes and structures such as new or wider roads. Heritage structures in the way may be protected, adapted, reconstructed or demolished. On the other hand, constraints on development have to be imposed to conserve the character of whole areas of old layout and structures. Furthermore, policies that protect historic structures may have to be extended to promote such economic activity as is essential to maintain the ambience and viability of a heritage district.

Preserving historic buildings

The best way to preserve a structure is to maintain and use it for its proper purpose. As far as city landscape is concerned, visual appreciation concerns the facades of houses or offices and the window displays or arcades of shops. Inside, certain rooms may justify conservation measures, while much may have to be rebuilt to serve modern needs and thus to afford upkeep of at least the shell. Some historic groups, such as the merchants' houses in the Red Sea port of Suakin, were abandoned in the mid-20th century and were gradually reduced to rubble (Photo 15.24).

Some very old structures remain complete after more than a thousand years, for example the Pantheon in Rome, topped by its pioneering concrete dome. Most other Greek and Roman edifices, particularly temples, have long been bereft of their roofs. They have become picturesque ruins; their columns in particular display fine capitals and fluting and have been robust enough to survive;

Photo 15.24 The port of Suakin on the Red Sea coast had been replaced by Port Sudan for most shipping by the 1920s. This 1967 picture shows that, by then, the roofs and floors of former merchants' buildings had been lost. Twenty years later, little remained.

Photo 15.25 Baalbek, Lebanon (1995). Unkempt surroundings to an urban heritage site.

and tiled floors have been conserved beneath coverings of earth. But a ruin remains picturesque only in a sympathetic setting. Edward I's Welsh castles are towers, walls and passages now enclosing lawns or grassy banks, but some Roman structures suffer from neglected surroundings (Photo 15.25).

Open air space is also an important attribute of well-endowed city properties. The forecourts, cloisters, residential closes or gardens surrounding palaces or cathedrals are islands of peace, but commercial land values are such that, without democratic building controls, all ground might profitably be built over and the heritage features destroyed.

Individual buildings in special styles may be randomly located, for example, Art Nouveau buildings created by Victor Horta in Brussels or by Antoni Gaudi in Barcelona. When threatened by local redevelopment, buildings of isolated character have to be judged individually as well as within their setting. If they are to be reconstructed elsewhere, the setting of the new location should be more aesthetically considered even than the, perhaps fortuitous, choice of the original site. Can they continue to fit in where they are or would they be better removed or replicated elsewhere?

Reconstructed historic forms

Where the boundaries of historic quarters are less well defined, there may be single historic buildings, or small groups of them, which have been isolated by, or stood in the way of, modern development. In some circumstances, it has been possible to remove such buildings, bodily or in parts, to more suitable settings. Where this is impracticable or too expensive, complete reconstruction or reproduction of old buildings can now be so expertly undertaken as to conceal their modernity. Some old cities such as Dresden or Warsaw were almost totally destroyed during World War II and have since been painstakingly reconstructed (Photos 15.26 and 15.27). Buildings dismantled centuries ago can be reconstructed, such as the Shakespearean Globe Theatre in London (Photo 15.28).

Preservation of 'old city' districts – extensive groups of historic buildings

Old cities are separated from modern extensions and commercial centres by the lines of ancient walls, by water (in Venice) or by modern ring

Photo 15.26 Semper Opera House, Dresden, rebuilt more than 50 years after it was destroyed by bombing.

Photo 15.27 The Old City of Warsaw was also totally reconstructed after World War II.

Photo 15.28 The Shakespearean Globe Theatre was reconstructed at its original site on London's South Bank some centuries after the original had been demolished. Behind it is the former Bankside electricity generating station, a vast brick building and chimney. The power station is now the Tate Modern Art Gallery and its riverside coaling berth is a promenade.

roads. Within these defining borders, the needs of conservation are served by limiting motor traffic, while preserving viable communities. Such measures encourage tourism and visits to old buildings, small shops and restaurants, thereby funding continuing maintenance and restoration of old structures and infrastructure in picturesque settings.

Since the 18th century, many buildings and streets, now recognized as heritage, remain in full commercial or traffic use. The furniture of squares and wide streets can be triumphal arches, statues, obelisks, fountains, well heads, continental 'Morris' columns, metro stations entrances, even bus stops. Where road traffic is permitted, it is a challenge to provide signposts that are clear to the vehicle drivers but discreet in the scenery at large.

Fortresses and city walls

Some cities are dominated by castles or palaces (Stirling, Lhasa). Others are contained within walls that have for long encircled old or more recent structures. Examples are York (Photo 15.29) or Tallinn or fully fortified towns such as Carcassonne and Aigues-Mortes in France or Conwy, Caernarfon and Tenby in Wales. The visual quality is in their own structures and the architecture of gateway towers (barbicans), as well as the fine views and exhilarating promenades afforded by walkways on the walls. City walls have been prone to demolition for road widening. Possibly they can be reconstructed to accommodate an appropriate opening.

Photo 15.29 The walls of York, once a stout defence, now a heritage walk.

Layouts for future urban heritage

There is little order or composition in the competing styles of 20th century skyscrapers. Historically, opportunities to create elegant but homogeneous groups of buildings have occurred, somewhat rarely, where:

- completely new cities were created by engineers and monarchs or political leaders, such as 18th century Saint Petersburg;[6]
- cities have been devastated by war or fire;[7] or
- municipal authorities were powerful enough to commission completely new concepts, such as Haussmann's grand design for Paris or, on a lower key, the quiet streets of East London that were cleared of slums in the 1950s.

Such opportunities still occur from time to time. Perhaps there is most to gain by creating new cities with today's best architecture on layouts that will suit citizens for centuries ahead.

15.5 Urban and suburban wildlife

City buildings and pavements are certainly unsuited to the wilder forms of fauna and their vegetative habitat. Early man hunted wild animals and then, having cleared their forest habitat, converted some to beasts of burden. Where formal human settlements were created, and eventually in medieval cities, people excluded all big animals except those they used for transport or put to graze on pasture outside. They kept cats and dogs that chased rats and other nuisance species.

Then and now, peopled settlements attract birds, rodents, foxes and feral animals. The attractions comprise:

- waste food that people throw out or which is dumped in landfill sites and which supplements sometimes precarious natural food cycles;
- habitat for birds in lofty roosts under eaves of roofs or on chimneys, electric power stanchions or monuments;

Photo 15.30 Street and roof greenery in Lausanne, Switzerland.

- habitat for ground animals in lower nooks, in space beneath sheds, in temporary structures, or even in machinery;
- pockets of neglected or spare ground or watercourse verges where wild vegetation can take hold, offering refuge routes for animals; and
- similar pockets created within more formally laid out and managed gardens.

Urban garden habitat – in public parks, at private houses or even on roofs (Photo 15.30) – can be richer in its own biotic species than many rural agricultural areas.

Parks – from those of the kings of Babylon to those of European or Oriental monarchs – were once the leisure grounds of the rich. English mill and mine owners of the Industrial Revolution built mansions in verdant surroundings on the outskirts of the towns where most of their workers lived in more crowded rows or tenements. Then philanthropy and broader political awareness brought municipal parks or less formal but official open space for the recreation of ordinary townspeople. There, trees (for bird and small animal perches and nests), any rough grass or waterside features (for smaller mammals, amphibians, insects and water fowl) and even extensive lawns (for blackbirds) were a firm wedge for subsequent profusion of urban wildlife. Often exotic fauna have been introduced. A variety of wildfowl live unconfined beside lakes such as that in St James's Park, London. Others have escaped and roam the countryside. Thus Canada Geese are as well-known in London as they are in the empty northern lands where they originated; and rose-ringed parakeets, traditional in the parks of cities such as Lahore, now fly between the treetops of Brussels or the Thames Valley.

New Towns, in England, were an extension of the concept of earlier 20th century garden cities. Most new towns comprise a number of basically self-contained residential villages, each a kilometre or so from a town centre complex, where main commercial and municipal activity now takes place, and from accessible industrial areas or 'business' parks. Among these diverse functional areas and the walkways and roads that connect them, space can be designated and set aside for recreation and nature conservation. Bracknell, designated a New

Town in 1949, today supports a population of about 50,000 and has 80 recognized wildlife sites. Telford supports 123,000 people in a landscape, once largely devastated by mining, which now includes the historic Ironbridge Gorge. Milton Keynes (161,500)[8] has 165ha of lakes (*The Times*, 1996).[9]

However, it is *suburbs* that provide the most extensive shared kingdom of people and wildlife. Typically, in England, suburban houses have individual private gardens and face on to quiet tree-lined streets. As habitat for birds and small land animals, these small islands of greenery and cover combine their attraction with human gifts of food, intentional or incidental. It is probable, in many well-populated countries, that there will not be enough space for the majority of people to be accommodated at the low density usual in new towns or suburbs; but, provided the quality and connectivity of the wildlife habitat is well planned, its actual aerial extent may be less critical.

Surface *transport routes*, starting with late 18th century canals, became catalysts for wildlife immigration into urban areas. Baines (2000)[10] points out that 'together with railways, streams and minor rivers of the region, the inland waterways of Birmingham make a marvellous green network of wildlife habitat'. Along navigation canals in European cities – or irrigation canals in more arid climates – it is the riparian water features and vegetation which provide a haven; along railways and orbital highways it is the verges and the slopes of embankments or cuttings where human intrusion is prohibited that, notwithstanding the noise and hazard of adjacent traffic, provide comparatively undisturbed habitat.

Empty or derelict city land may comprise:

- city areas where slum housing or factories have been cleared or railway yards and docklands made redundant; or
- suburban or conurbation land where industrial or extraction processes no longer take place.

Dereliction can result in exposed ground that can be burrowed into, ponds that can be drunk from or bathed in, coarse brambles that provide hidden roosts, ornamental weeds like rosebay willow herb, and other edible growth. However, the very arbitrariness of neglect is neutral whether or not it may result in viable regionally valuable ecological communities.

There can be more positive benefits if structural development and human and wildlife habitat management are planned together. Thus, ground space, contours, vegetation, drainage and water features can be shaped into patterns that permit the movement, nesting, breeding, hunting and feeding of such creatures and plants as warrant encouragement in each type of landscape.

Leafy residential suburbs are complemented by city parks and 'a marvellous mosaic of wild green official open space, neglected cemeteries, worked-out quarries, long-absorbed landfill sites and here and there a patch of relic woodland, a railway bank of cowslips or heather' (Baines, 2000).[11]

Historically, urban wildlife has adapted to take advantage of human settlements and structures. Today, in a world where it is evident that man threatens to overexploit remaining resources, it is incumbent on us to take positive steps to encourage diverse wildlife so as to make city conurbations into as rich a landscape as the options for planning, construction and management can offer.

15.6 Construction planning for urban development

Urban development can involve expansion of cities or creation of new metropolitan areas; or it may comprise redevelopment. Commonly, in industrial or post-industrial cities, there need to be changes in land use and physical reconstruction to suit new circumstances. Planning for urban development includes:

- recognition of existing or more appropriate layout patterns;
- fitting local development into that of any greater conurbation;
- identification and resolution of engineering issues affecting location and type of structures and infrastructure;
- allocation of land space for conservation, reconstruction or new development; and

- contriving attractive landscape in a manner which balances existing scenery and cultural traditions against options for economic and social reconstruction.

Urban layouts

City expansion requires new building, often on agricultural land. Layouts have to take into account land topography, drainage patterns and climatic factors as well as any particular needs of the forecast population. For new or substantially redeveloped cities, three basic types of city layout can be described.

Concentric layouts originate from feudal strongholds with scattered farming and trading communities in the surrounding countryside (Wheeler, 1964).[12] In today's cities, ring roads rather than defensive walls define the circles. In London, the central Square Mile (actually 3.15km^2) of the financial hub is defined mainly by the buildings, but the Circle Line underground railway, built by cut-and-cover construction beneath the streets, encircles an area of 15km^2 and defined the limit within which 19th century railways were not permitted to penetrate further into the city. A dual carriageway North Circular Road ringed northwest London since the 1930s at about 12km from the centre, from the 1980s, the whole of what is approximately Greater London, formerly within recognized Green Belt zones, is now effectively bounded by the M25 motorway, within 20–30km of the capital's centre.

Perhaps a variation on concentric layouts is what Hilling (1996)[13] describes as a Latin American model for developing countries. A conventional central business area is encircled by mature residential areas, then a zone for 'in situ accretion' (presumably on land first kept in reserve) and finally by squatter settlements (which may or may not be controlled, see p238).

Radial layouts are focused on roads and railways leading outwards from city centres toward the environs. Undeveloped countryside tends to remain, at least initially, between the radial corridors and some of these green fingers have been deliberately preserved. Ribbon development may have taken place along main roads and been followed by concentric spread around railway stations or village centres on the radial routes. Particular classes of housing or types of economic development tend to concentrate on particular routes from the centre rather than on zones at various distances from that centre. London is so large a city and has developed as such for so long that it has radial as well as concentric elements. Meanwhile, radial and other urban transport facilities can be improved to 'encourage people to travel in a manner that best suits the urban geography that we have inherited' (Sherlock, 1991).[14]

Rectangular grids may constitute the detailed layout of a suburban estate, or the whole plan of a new city, such as Islamabad, capital of Pakistan. The Indus Civilization's main cities (3000–2000 BC) were planned on a regular street grid with modular houses and unified sanitary facilities. Greeks build residential extensions on precise street block patterns. These were usually on flat ground, a felicitous situation according to Wheeler (1964),[15] which did not apply at Rome where 'the rugged hills and woodlands … stimulated the Roman genius for engineering (roads, services, aqueducts, water pipes)'.

Peking (Beijing) has been described by Bacon (1978)[16] as 'possibly the greatest single work of man on the face of the earth'. It is a grid city on several scales – the central palace complex of the Forbidden City contained by a moat, then the Imperial City defined by its outermost streets, all within the considerably larger collection of blocks known as the Inner City, and its southern extension – also within walls and with a railway along its inner boundary line – known as the Outer City. In 1981, the outer boundary was consolidated by completion of the 'Second Ring Road'. Since then more ring roads – up to the Sixth – have defined subsequent expansion.[17] But, through all these formal divisions, watercourses are still allowed a meandering course.

Grid pattern layouts have to be modified at the edges to suit irregular perimeter boundaries compelled by the topography (for example, contour-defence lines around ancient settlements) and within and beyond the boundaries, to suit drainage, water supply and other civil engineering.

Industrial areas have not been mentioned specifically in these layout descriptions. Their location, as it relates to business and residential areas, has changed historically. Until well into the 20th century, industry was noisy and polluting. So early planned improvements separated it from housing, at best with public transport to link the two. Now there is less heavy industry but some quieter and cleaner manufacturing or assembly. Planning strategies might return to the concentrated multi-use city centre. As likely may be development in multiple-nuclei patterns, in which various housing areas surround or lie between a number of business, manufacturing or wholesale centres.

Meanwhile, there is probably a majority of towns that would be difficult to classify in terms of any layout model. However, each rough pattern has elements – such as existing routes, watercourses, corridors or skylines – which are relevant in recognizing the shape of urban landscape and then in extending, modifying or enhancing it to meet social needs. Perhaps this is even truer in the urban spread into rural areas that constitutes conurbations.

Conurbations

Conurbations are extensive enough to encompass both economic planning for large regions and physical planning at a local scale. On the largest scale, the US has a number of multi-state economic development regions, known, for example, as Upper Great Lakes, New England and Appalachian, which, notwithstanding that they encompass considerable unbuilt areas, effectively constitute very extensive conurbations (Sendich, 2006).[18]

Megalopolis was an early, possibly 19th century, concept for a very large urban community. Forman and Godron (1986)[19] described the same as 'an enormous suburban landscape ... within which cities are scattered'. They anticipated that each megalopolis would have to be sustained by massive amounts of fossil fuel, would be dependent on other leverages and would be slow in responding to disruption.

Alsop (2003)[20] described a different concept of supercities and suggested three for England, each about 100 miles long and 20 miles wide:

1 '*Coast-to-coast*' near the M62 motorway from Liverpool to Hull;
2 '*Diagonale*' near the M1 and M6 motorways and West Coast Main Line railway between London and Birmingham; and
3 '*Wave*' near routes along the South Coast from Hastings to Poole.

Alsop envisaged, or we may surmise further, that these concepts should incorporate:

- medium distance passenger transport, for example buses between railway stations and combined motorway service areas and park-and-ride facilities;
- a number of self-sufficient very-high-rise (25-storey) community towers;
- reconstruction of crowded cities with underground services;
- emerging villages with new houses, services and transport connections; but also
- plenty of green agricultural and amenity land space, including surviving gems of rural heritage (but not national parks which should continue as extensive protected areas outside these conurbations).

Each of the focal settlements within the greater megacities needs its own detailed physical, social and landscape planning. But the economic and socio-economic consequences for the whole region, and the extent to which parts of the greater landscape is conserved or modified, is largely dependent upon transport infrastructure and controls on different types of traffic.

Engineering issues in city planning

Earthworks, slope stabilization works and avoiding or using existing foundations have been considered in Chapter 4 and frameworks for tall buildings in Chapter 11. Branches of civil engineering required in planning and providing essential city infrastructure include:

- surface drainage and watercourses;
- water supply, waste water and solid waste disposal;
- electric power lines and district heating or cooling systems; and
- roads and railways.

Surface drainage has to cope with rainfall. Some of the run-off is absorbed in vegetation or infiltrates into the ground. But in urban areas much of the land surface is built upon or paved by an impervious covering – such as concrete or asphalt. As a result, surface run-off is much more rapid and rainstorms can overload even large drains. This increases the risk of floods, and health hazards can arise if the capacity of combined foul drainage and stormwater sewers is exceeded.

Sustainable urban drainage systems (SUDS) use porous paving, pervious strips, swales (widened, often grassed, ditches), detention or retention ponds and wetland. Infiltration trenches or basins can encourage natural recharge of groundwater, thus promoting biological cleansing as well as reduction in downstream flooding and erosion. Provided there is enough land space and they are maintained responsibly, these systems constitute an important assigned use of the land surface and one that may well enhance the landscape.

However, SUDS strategies do not work where the ground is impermeable or if the groundwater level is high, as it is in low-lying areas and floodplains. Some cities are subject from time to time to extraordinary hydrological events – for example, in 2005, exceptional rainfall in Mumbai and hurricane surges around New Orleans – which can only be coped with by adequate stormwater culverts and sea/river defences respectively. Where defence against common hazards is not practicable, urban settlement may be unviable and the land should be reallocated to agriculture, conservation or flood water storage.

Water supply solutions depend on the availability of water at source, costs of its procurement and treatment, and the ability of the population to pay for its supply – whether at communal standpipes or from taps within their houses. Options involve charging for water by quantity, quality or both. Perhaps untreated or only partially treated water may be separated from potable supplies.

Water supply through shallow pipes under sidewalks may readily reveal points of breakage or leakage and be accessible for repair. However, pipes laid more deeply, at the time of road construction, may have to last for many decades until it is time to renew the complete water supply system. Accessible culverts carrying all services alongside roads may be a satisfactory solution. *Waste water* is both a potential pollutant and a source of water for reuse.

Sewage treatment for cities usually takes place on a large scale. Each plant requires space, for example, for sedimentation or aerobic biological treatment ponds, that may not be available within the municipal boundaries. Sewage treatment works may then become part of the urban perimeter. Disposal of sludge and other effluent for agriculture may suit agricultural crop production in nearby fields. In some climatic conditions and where land space permits, natural waste treatment systems may use lagoons, soils and aquatic plants in managed wetland.

Solid waste, in enlightened municipalities, can be recycled or burnt as fuel in power stations. More often recycling or reuse is limited and much of the waste is disposed of as landfill – itself a landform. Disposal of waste in landfill requires wide areas of land space where holes are being filled or hills created; and it entails considerable engineering in preparing or sealing off the subsurface, in preventing groundwater contamination, in compacting each layer of fill, in controlling methane and other gases, and in capping the completed fill.

Electricity is a wonderful form of power for static domestic or industrial applications or metro railways. Electric power lines in cities are usually medium or low voltage, so the cost of subsurface cable installation and insulation should be affordable for permanent distribution systems. In fact, to supply unplanned urban spread, rapidly erected cables are suspended from randomly sited poles.

Heating and cooling for buildings often makes wasteful use of energy – because thermal design is suboptimal or through not taking advantage of energy sources more immediately available. For example, in London, Head (2003)[21] proposes that use of solar energy derived on the buildings themselves should be increased 100-fold, while for cooling London Underground trains and stations, it has been proposed that groundwater should be pumped through pipes in the railway tunnels. Combined heat and power (CHP) systems are well suited to dense housing close to power stations in cold climates. District heating in ex-Soviet Bloc

Photo 15.31 Elevated highway over local road over commercial waterway and pedestrian space in west London.

cities is provided through highly visible steam or hot water pipes. Pipe routeing is directly related to road and building layouts, and the brightly coloured insulated pipes – often elevated above the ground – are strong if inelegant elements of urban landscape.

Roads, railways and waterways – with their crossings and earthworks – are basic elements of urban civil engineering. One of the requirements of their routing is to negotiate and make use of differences in ground levels with provision of elevated sections to make use of air space (Photo 15.31) or submerged underpasses (Photo 15.32), or by constructing viaducts across valleys for roads (Photo 15.33) or railways (Photo 15.34).

Great improvements have been made in providing city centre pedestrian space where motor vehicles have been excluded. In many cities, underground railways have always provided the main public transport system, in others, buses have been given priority on the streets. However, except in very cold climates, people prefer to travel in trains or road vehicles above ground level rather than in darkness below. Construction of *elevated railways* appears to be socially acceptable in new developments in Hong Kong, where they are planned and constructed at the same time as new high-rise housing. On the other hand, disused elevated railways have successfully been converted to parks or promenades in Paris and New York (Dyckhof, 2004).[22] *Elevated roads* have been known to blight the residential areas around them. In San Francisco, two road viaducts collapsed during one of that city's occasional earthquakes. Instead of rebuilding them, the ground was cleared, the residential areas recovered their former prestige and the car drivers found some travel alternative (Nash, 2002).[23] Perhaps paramount in urban road pattern planning is conservation of quiet safe residential streets which accommodate only limited motor vehicles.

Besides their associated walkways, canals are routes for barges including those carrying constructions materials and wastes.

Photo 15.32 City highway underpass through the Parc Cinquantenaire, Brussels.

Photo 15.33 Road-over-road viaduct crossing in Lausanne, Switzerland.

Photo 15.34 Railway viaduct at Folkestone, England (photo by Robin Carpenter).

Allocation of land space for urban development

Spatial planning for development needs to be long-term and geographically inclusive. Stages in space allocation may include:

- identification of regional opportunities – areas that might be suitable for urban growth;
- recognition of particular features of the natural and artificial topography and their implications for construction;
- delineation of various corridors which need to link any development zones; and
- allocation of space for those zones, in different categories and with priorities for managed development.

These processes are not necessarily consecutive. Much relevant data and previous plans may already be available. The latest aerial photography, remote sensing and ground survey data can be taken into account together with information about the nature, capacity and operation of existing and planned infrastructure. Determination of priorities and allocation of development zones can then be based on fresh assessment of all these physical issues and on medium and long-term social and economic targets.

Identification of opportunities for urban development can be for any broad geographic region, such as a river basin upstream of an existing city, so as to establish the area's long-term capability and the features that must be conserved. McHarg (1992)[24] described how suitability for urban development can be mapped, for example for the Potomac River basin for expansion beyond Washington, DC. This basin assessment was undertaken in stages starting with the exclusion of most floodplains, woodland controlling erosion and steep slopes.

Natural topographical features that affect suitability for development include sloping ground and surface drainage characteristics. Data can be assembled and analysed concerning geology, hydrogeology, natural vegetation and microclimate. McHarg (1992)[25] drew a series of map overlays depicting the many aspects relating to ecological and scenic conservation.

Artificial features are existing settlements and their related infrastructure. Adaptation of existing drainage, roads and utilities or construction of new ones has to be examined in the light of topography and the engineering issues already described.

Corridors, for land planning purposes, are rivers and watercourses, transport routes, paths and wildlife links. They are generally linear and continuous. So they are susceptible to interruption and may need better connections or protective buffer zones. The critical role of rivers and man-made drainage courses is to accommodate floods. Adjacent wetlands may be able to absorb some of the flood water.

Highways and railways are through routes, sometimes linked to adjacent communities but sometimes regarded as barriers or sources of noise. Pathways, from which all traffic is excluded, are welcome features in any but the most hostile environments.

Wildlife corridors have been introduced on p8. They lie along watercourses, hedgerows, highway verges or wider brush or woodland. Planning for wildlife conservation should identify specific biological purpose in encouraging particular species or habitat, even in cities.

Zones are areas of land set aside by planners for specific use. In 'master planning' for development or reconstruction they may designate space suitable for residential, community, commercial or industrial buildings. The buildings themselves may then be subject to restrictions, for example concerning density, type or height. Alternatively, zones may define mixed development for communities where many people will live within walking distance of their work and most of their amenities. Within these zones, planning authorities may stipulate forms and layouts or restrictions may be relaxed to encourage creativity. In the US, Planned Unit Developments (PUDs), 'homes and other buildings may be arranged freely. They need not face upon a public street. They may instead front upon off-street courts, plazas or dedicated walkways – with parking bays or compounds located beside or underneath' (Simmonds and Starke, 2006).[26]

Simmonds and Starke also note that zoned areas tend to be oversized, nominally 'to be sure of sufficiency' but actually 'in response to political pressure by landowners who buck for the highest potential sales value of their property'. It may therefore be better that any 'sufficiency' – if that includes provision for future development by future populations – should be provided for in separate zones in areas currently unbuilt on or ripe for redevelopment in the more distant future.

Zones or permanent reservations are also needed for open space, whether for human amenity or for nature conservation.

Urban landscape planning

Within any allocated land space and preliminary building plan, the optimum visual landscape has to be designed. This can be achieved by assessing the existing landscape, identifying fragile landforms and deciding what aesthetic goals or structural themes should be encouraged.

Assessment of the existing observed landscape usually involves identification of:

- natural scenic features – such as cliffs, other sloping topography and water features, providing both viewpoints and scenery;
- existing man-made scenic and heritage features and their characteristics, for example, in terms of regional typicality and vernacular styles or as outstanding exceptions; and
- open spaces and corridors, views, vistas and skylines that will result from the development.

Risks to landform character – and integrity of structures – may arise from natural causes (such as earthquakes, fires or hurricanes) but are more likely to result from man's interference with natural processes. Thus there are greater risks of landslides where vegetative cover has been cleared or hillside geology disturbed by excavation for buildings, or of flooding where paved surfaces and lined drainage channels in the catchment area accelerate run-off from rainfall, much of which would once have seeped into the ground.

Pregill and Volkman (1999)[27] suggest that 'contemporary urban landscapes that are planned for public benefit fall into two categories, those that attempt to reinforce the identity of a historic district, and those that seek an identity through the imposition of contemporary forms and materials'. Perhaps this is a simplification of many more complex situations. Exterior design can emphasize the concepts that show up best in whatever views can be contrived. It may be practicable to select architectural themes or model layouts that can be applied to each type of zone, building or service, bearing in mind particular local conditions.

Few architects, civil engineers, town planners or landscape planners ever design a new city layout. However, from time to time, these people may have opportunity to influence how popular or official goals for urban improvement are translated into projects. Such influence – for example, to ensure good landscape design – may be brought to bear:

- in recognizing the options when preparing a preliminary layout for a project;
- in negotiating approval for the project with the developer; or
- in preparing full designs for construction.

As the population of the Earth has soared, the proportion of that population living in towns and cities has increased even faster. If living in cities, suburbs or urban conurbations is to be the norm, then urban landscape should be created as much for the enjoyment of living or moving about in as for successful economic activity.

Notes and references

1 Bacon, E. N. (1978) *Design of Cities*, Thames & Hudson, London, p13.
2 According to his obituary in *The Times* on 7 May 2003, Sir Philip Powell designed three-storey houses on the Thames Embankment 'to allow the tall structures behind fine views over them'.
3 Stanley, R. (2004) 'Taking a walk in the clouds', *The Times*, 24 August.
4 Preece, R. A. (1991) *Designs on the Landscape*, Belhaven, London, p85.
5 Bacon (1978) as Note 1, p64.
6 The evolution of St Petersburg is described in Bacon (1978) as Note 1, p196.
7 Even after the Great Fire of 1666, the interests of London landowners were too powerful to permit Wren's vision and the haphazard pattern of the city's streets continued as an aspect of its own peculiar heritage.
8 The plans for construction of Milton Keynes were described by E. H. Brown and J. Salt (1969) 'New City in the Oxford Clay', *The Geographical Magazine*.
9 *The Times* (1996) supplement on New Towns, 11 October.
10 Baines, C. (2000) 'Arden', in Bryson et al, *The English Landscape*, Profile Books, London.
11 Baines (2000) as Note 10.
12 Wheeler, M. (1964) *Roman Art and Architecture*, Thames & Hudson, London, p25.
13 Hilling, D. (1996) *Transport and Developing Countries*, Routledge, London, p201.
14 Sherlock, H. (1991) *Cities are Good for us*, Paladin, London, p183.
15 Wheeler (1964) as Note 12, p29.
16 Bacon (1978) as Note 1, p244.
17 http://en.wikipedia.org/wiki/Beijing, accessed 20 January 2010.
18 Sendich, E. and the American Planning Association (2006) *Planning and Urban Design Standards*, Wiley, Hoboken NJ, p85.
19 Forman, P. T. T. and Godron, M. (1986) *Landscape Ecology*, Wiley, New York, p309.
20 Alsop, W. (2003) *Supercities*, a three-programme series on UK TV Channel 4.
21 Head, P. R. (2003) 'Better building through sustainable development', *Proceedings of the Institution of Civil Engineers*, vol ES4.
22 Dyckhof, T. (2004) 'New York's West Side is back on track', *The Times*, 17 August.
23 Nash, A. (2002) 'Tearing down the freeway', *Land and Liberty*, autumn/winter.
24 McHarg, I. L. (1992) *Design with Nature*, Wiley, New York, p115.
25 McHarg (1992) as Note 24, p110.
26 Simmonds, J. O. and Starke, B. W. (2006) *A Manual of Environmental Planning and Design*, 4th edition, McGraw-Hill, New York, p329.
27 Pregill, P. and Volkman, N. (1999) *Landscape in History: Design and Planning in the Eastern and Western Traditions*, Wiley, New York, p10.

16

Built Landscapes in the Future

This book concerns the landscapes in which we live. It has suggested how construction, deemed necessary for civilized development, can best be adapted to make optimal use of land resources. Allocation or conservation of certain types of land should be pursued vigorously. But it would be socially callous or blatantly hedonistic to expect that our own regional landscapes should be preserved everywhere in their traditional shape in a world of still increasing human numbers and stark inequality of opportunity in use of resources.

To meet our long-term obligations we should be concerned at this time:

- to recognize and, where practicable, quantify the challenges posed by dwindling resources and rising populations and deduce the implications for civil engineering;
- to identify political issues and options in meeting these challenges;
- to assess the ways in which construction techniques can make the best use of land resources; and
- to plan long-term allocations, for at least a century ahead, of land that should be economically developed, conserved in a more natural or semi-natural state or reserved for needs that may arise yet further into the future.

16.1 Daunting challenges for civilized development

Stark problems

In the early 21st century:

- the world's population continues to rise; it increased from 1.6 to 6 billion between 1930 and 2000 and may reach 10 billion by 2100;[1]
- there are problems in providing food, accommodation and livelihoods for the extra people and for at least a billion who are already undernourished; most of the population increase is occurring in poor countries where locally-produced food is not always adequate to feed the population;
- global warming is taking place – some experts fear adverse, possibly catastrophic, effects of continuing climate change;
- there are severe water shortages in some regions;
- there will soon be shortages in petroleum oil and gas even though some communities have only recently begun to use these extensively; yet response to imminent energy shortages is often hope of unproven alternatives rather than comparison between realistic demand forecasts and supplies that might be sustained from established options;
- there is flagrant inequality between peoples and considerable waste of human resources; and
- there is continuing war and strife.

The challenges to 21st century politicians, planners and engineers are to solve these problems as quickly as possible. Most of them are severe already, yet we must plan how to cope with still further increases in population and more restricted resources.

Burgeoning population

We have two particularly onerous obligations consequent upon the population explosion.

Firstly, we must try to slow down and eventually reverse the unsustainable global rate of population increase. Relationships between high birth rates and poverty have been well argued, for example by Sachs (2005).[2] But elimination of poverty is proceeding very slowly. Even containing world population to 10 billion by 2100 is unlikely to be attained without political measures of the sort undertaken in China in the latter part of the 20th century. Education of boys and girls ultimately encourages stable families as well as opportunities for more rewarding employment. Secondly, we must cope with the surplus people seeking food, shelter and livelihoods.

Feeding people

Perhaps 1 billion people are undernourished. Uncounted numbers starve every year and millions rely on outside food aid in both emergency and long-term situations. To provide enough food, we have to assess:

- diets – what people eat and what is adequate;
- food crops that are best suited to local soils and climate and that best meet dietary needs; and
- available fertile land.

Many people could eat food that is easier to produce and on less land than their present choice. New varieties of food crops can continue to be developed to cope with marginal or adverse climatic conditions.

Fertile land is that with soil of the right texture and nutrients in a climate of sufficient temperature, sunshine and water to promote plant photosynthesis and growth. But the more sunshine there is, the less the rainfall may be, while, in cooler climates, a common temperature requirement is that there be no frost during the growing season. Some desert or semi-arid land can be rendered fertile if additional water can be supplied for irrigation at the appropriate season, either conveyed from a distant water source or drawn from groundwater. But most of these opportunities are already being exploited in many of the drier inhabited countries. Soil productivity can then be increased only by nutrient-replenishing cropping patterns and more efficient use of water.

At the same time, many of the excess and poor people of the world live in the countries where there is little spare cultivable land left. Indeed, much cultivation takes place on hillsides by sustained arduous effort in conditions which might be considered uneconomic in less crowded lands, and lands of marginal agronomic capability may be overcropped or overgrazed. Young (2003)[3] suspects that the extent of uncropped land still available in developing countries has been grossly overestimated in national data. Officials are loathe to report, for example, that soil conditions have got worse.

In the less crowded countries, such as Russia and much of North or South America, it is commonly assumed that cropping could be intensified and new land planted for crops if the market demands it. However, some of the land may prove of only marginal potential or capable of successful cropping only of particular varieties by particular – not necessarily intensive – farming methods.

Meanwhile, the main contribution of civil engineering to crop production is construction of systems to supply and drain irrigation water and of dams to store water for timely release. On the flood plains of large rivers, semi-arid land can be fed with extra water gathered during flood flows at different seasons in wetter regions upstream. However, many of the best river storage opportunities have already been taken and very-long-distance water conveyance – requiring pumping rather than gravity flow – is generally unlikely to be economic at foreseeable 21st century energy prices. So, for the present at least, we can only make better use of such water as is available regionally.

Accommodating people

The greater part of the net increase in the world's population arises among poor rural communities. Coupled with the fact that efficient agriculture can only provide livelihoods for a modest proportion of the people, this results in a strong pattern of migration of rural people seeking work in already crowded cities.

Accordingly, the only ways of accommodating more people are:

- to restructure and expand rural economies so that more small or medium-sized towns arise

in rural areas, performing many of the activities currently undertaken in cities; but meanwhile
- to allow expansion of cities to continue; and if necessary
- to allow or manage migration of people to other countries.

Types of housing (described on pp234–238) have to suit climate (as shelter from rain and cold temperatures), an appropriate degree of permanence and available ground space, as well as what the inhabitants can afford and their attitudes to privacy and communality. Civil engineers and architects will continue to:

- collaborate – with planning authorities and promoters of residential development or redevelopment schemes – in planning buildings, layouts and open space to suit the extent of land available;
- provide the necessary infrastructure for the new or improved settlements including safe water supply and sanitation where this does not already exist; and
- devise types of structures, layouts and services to cope with particular site restrictions or difficult geographical conditions such as on flood plains, unstable ground or in extreme climates.

Livelihoods

Every community has to maintain itself and to produce something to sell in exchange for the goods or services that it receives. As far as planning construction of human settlements is concerned, there is likely to be increasing scope for multi-unit light industrial or office estates reasonably close to residential areas. These estates might well be combined with schools, shops and community centres. The aim is to limit the need for regular travel without stifling mobility for activities such as maintenance or construction. As for employment itself, open free-enterprise labour markets are not always effective or equitable in changing situations. So there may be a case for political direction, at least for short periods, as to who does what – fulfilling understaffed public needs and widening perceived opportunities for job seekers.

Climate change

Global climate change occurs naturally, usually over thousands of years. However, anthropogenic release of greenhouse gases has been associated with the more sudden changes that have taken place in recent decades. The cause-and-effect relationships are not yet fully established, but many scientists and some governments accept that some of the consequences may be adverse and that action, such as limiting carbon dioxide emissions, should be taken now to reduce the impact.

Recently global average temperatures have risen and many experts expect the trend to continue. While this may ultimately be of benefit in some cooler climates, problems may arise in countries that are already uncomfortably hot and especially in regions where effective rainfall may be reduced. As yet, it is very difficult to predict precisely what changes will take place and where the effects will be most serious.

Meanwhile, in all regions, there is an urgent need to better understand the issues, for example, in growing the most suitable crop varieties in different soils and rainfall conditions and with different cropping methods, applications of chemicals and provision of irrigation. Climate change, as it occurs, is only one of the variables to be considered in the continuing research.

It would be pessimistic to believe that man will never be able to influence local weather, for example by successfully seeding rain clouds; but the sun-derived energy in weather is so colossal that man's puny, even nuclear-generated, effort is only likely to be effective in marginal situations – such as at the border between drought and precipitation.

Coping with climate is a much more practicable matter. Man has always built shelters to keep warm in winter and as cool as possible in summer. People can work more at night in very hot weather. Thermal mass architecture, use of new insulating materials and selective inputs of available energy can better contain or transfer heat, enabling equable climates to be created in buildings. Hopefully these benefits may be extended to more people in the future as changed climatic conditions arise in particular regions.

One of the undoubted consequences of global warming is that of sea level rise due to thermal expansion of water and melting of the polar ice caps. The sea rose about 18cm in the last century (Wikipedia, 2010)[4] and further rises have been variously predicted before 2100. Along steep coastlines this may imply no more than adjustment of coastal protection strategies, which may already have to cope with rises of 1m above tidal levels during storm surges. On flatter coasts, including the Netherlands, consideration is already being given to allocating flood plain land that can be sacrificed temporarily during extreme flood events. In Bangladesh, such areas might have to be extensive, some lands even permanently abandoned. Meanwhile, some Pacific atolls have already been evacuated.

Coping with water shortage

Clean and adequate drinking water is a fundamental necessity for everybody. Outside assistance in providing supplies to the poorest communities remains vital, but people should pay for their water as soon as they can afford to. Thus, in temperate countries with frequent rainfall, most of us can afford to use water to satisfy all our needs without any onerous restriction. In drier countries, people have to be more frugal and make more use of 'grey' water for non-potable uses. In places where sea water has to be desalinated, people's ability to pay for it may ultimately be one of the causes limiting the population of desert cities.

Large water storage and control systems are often publicly owned. This is to ensure long-term responsibility for maintenance and to allocate or vary amounts released and sold for often competing, sometimes changing uses.

Shortage of petroleum and mineral resources

Various analysts – individuals, interest groups or companies – are currently making their own estimates of future energy needs and supplies. But this is seldom with sufficient certainty to publish conclusions valid for more than a few decades ahead, thus leaving room for a 'something will turn up' philosophy and short-term political solutions. Attention has been drawn strongly to greenhouse gas (GHG) emissions and the perils, rather than impending shortages, of fossil fuels. Meanwhile writers have explained imminent shortages (for example, Heinberg, 2003),[5] technical aspects of various energy sources (Hinrichs and Kleinbach, 2002)[6] and the numbers and relationships involved (MacKay, 2009).[7]

These commentators recognize the rapid depletion of conventional oil and gas reserves, contention about the growing use of longer-lasting coal stocks and the inevitably greater difficulty and cost in gaining fossil fuels embedded more solidly, at greater depth or under less pressure than those so easily obtained and burnt profligately in the past. We appreciate the scope of nuclear fission and renewable energy harnessed from water, wind and the sun but also the physical and political limits on how far and how fast these can be developed.

This author's own speculations assume that no new cheap source of abundant energy will be fully developed and commercially viable on a global scale by 2100. Such sources commonly envisaged are nuclear *fusion* electricity generation and *cheap* solar energy converted directly to electricity that *may* one day prove to be practicable, affordable and sustainable.

The key, but often disappointing, factor in proving the viability of any new energy source is always the net output after all the necessary effort in harnessing and refining raw resources. Meanwhile, indications are that, even with restrained increase in fuel use by the developing countries, we richer people are unlikely in this century to be able to continue to consume *any* form of energy at our current cheap rates. Only political action can stem current excessive demand for fuels and energy, mainly by selective tough taxation.

The main role of construction in 21st century energy production will remain provision of comparatively clean efficient fossil fuel and nuclear power stations while adding the infrastructure for harnessing wind, tidal, solar and new forms of geothermal energy to a century of achievements in hydroelectricity.

There are regional shortages of other minerals such as:

- rock and earth components of concrete and other structural forms; in regions where rock or gravels are scarce there may still be scope for further improvement in using such soils as exist; and

- metal ores which combine in many forms to form high tensile or flexible structural elements, but even global shortages of a particular ore may not rule out an alternative combination.

Note that polymer (plastic) materials are mainly derived from petroleum. However, the amounts concerned, much of which may eventually be recyclable, are small compared with the amounts of oil and gas burnt off as fuel.

Flagrant inequality

'In 2000,' as Jay (2000)[8] observes, 'the world's output was greater than it has ever been. Average production … and average living standards were higher. Expectations of life were longer, infant mortality lower. At the same time inequality was greater; and there were probably more poor people, however defined, than ever before.'

Sachs (2005),[9] in his proposals for ending world poverty, draws particular attention to the plight of about a billion people who are fighting for survival in 'extreme poverty'. These people he believes are caught in a 'poverty trap' comprising 'disease, physical isolation, climate stress, environmental degradation and extreme poverty itself'. The trap is such that these people – whose number is increasing, especially in Africa – cannot gain even a foothold on the ladder of development without substantial basic investment that the people or their governments cannot afford. Sachs believes that such a foothold could be gained through rich country official development assistance within the level of aid (0.7 per cent of donor country Gross National Product) that has long been promised but not delivered.

The difficulties in providing aid effectively are formidable, particularly where adequate political will and competent institutions do not exist in the country concerned. Some commentators (such as Collier, 2007)[10] believe that advancement, even with outside help, must be promoted from within.

As far as land resources are concerned, only internal political initiatives can resolve perhaps the most pernicious forms of inequity – those of insecure land tenure and particularly the plight of 'landless' people in overcrowded rural areas and on urban fringes. The role of construction in resolving this issue is perhaps participation in the planning and provision of settlements that more equitably accommodate total populations.

War and strife

Discontent arises from injustice or inequality within communities on issues such as land, employment or housing. Serious antagonism between nations, including disputes over water or petroleum rights, can result in war. Crime is engendered or can thrive in either case. Provision of housing plots for the dispossessed or works implementing international water treaties can relieve some tensions although social imbalance can always lead to strife. The worst consequences of crime can be no-go areas and a breakdown of conventional law. War involves widespread destruction and death. However, in the aftermath there will be urgent needs and opportunities for clean up and reconstruction. The needs will concern very gradual recovery. The opportunities will relate to better social cohesion, perhaps ultimately in a more egalitarian landscape.

Recent huge advances in information technology and communications will doubtless continue to make everybody on earth better aware of resource and political issues. But these advances have not yet on their own succeeded in resolving strife in much of the Middle East or Central Asia or genocide in parts of Africa. Nor is well-established democratic government necessarily exempt from action leading to global disaster – such as launching missiles with nuclear or bacteriological warheads. Future stability depends on everlasting success of human collaboration and technical ingenuity, not least in preventing unfair resource exploitation.

16.2 Political issues and options

The nature and role of government

Government is the control of communities – setting and enforcing rules for civilized collaboration among people. Politics is the art and science of government; and politicians practice government or attempt to influence it. Government exists at national level, at various regional or local levels and in certain respects internationally, as in the United Nations (UN) or European Union (EU).

Governments are supported by civil servants, who undertake the administration of the existing rules and new ones that governments introduce, and advisers, who may or may not be public employees, but hopefully are competent to plan for the longer term and to guide the politicians accordingly. Additional public services can be rendered by organizations controlled directly by governments. These services usually comprise at least the armed forces and the police, often major components of education, health care and scientific or agricultural research. In the third quarter of the 20th century government 'public' ownership was extended often into other sectors such as transport, public utilities, sometimes industry and even agriculture. Many of these activities were recently sold back to the private sector but people may want another reversal if they believe that public ownership can achieve fairer control of increasingly scarce resources.

Meanwhile, private companies dominate 21st century economic production. Political choice arises as to how these companies may be monitored and influenced, for example, in respect of their prices, trading conditions and use of resources. Governments should also resolve how they can best collaborate with a third economic sector – that of voluntary work and non-commercial non-governmental organizations (NGOs).

Governments have to fund the whole public administration and most of the public services. The greater part of government income comes from taxes. These may be levied on people's assets, income, expenditure and activities.

In foreign relations, governments must continually strive to avoid war in other countries, except with strong consensus as to how to rescue a failed community or restrain a destructively offensive one, and to promote fair trade (about which there is also considerable contention). On resource issues, there have to be agreements and adjustments concerning the sharing of ocean resources or river waters and the migration of people.

Tackling human community problems

Governments, even with substantial help from voluntary organizations, must play a leading role in alleviating some of the social and human resource problems mentioned at the beginning of this chapter.

Demographic problems arise from high rates of population growth. Sustainable birth rates depend on education and related empowerment of women as well as measures to reduce poverty. Migration of people from overcrowded states has to be enabled and controlled by the governments of the receiving countries, bearing in mind inevitable pressure over at least the rest of the century.

Government roles in *feeding the world's population* comprise:

- supporting research into food crop varieties providing higher yields in various climates and marginal conditions, so as to make optimum use of fertile land to meet essential dietary requirements; and
- arranging temporary or longer-term food aid or trade where necessary in ways that make best use of actual global food production.

Adequate *accommodation* for surplus people has to be furnished by planning where and in what form housing shall be constructed or services provided, and how local livelihoods may develop. *Employment opportunities* can be widened by directing people into labour where this is necessary and helpful.

Inequality has to be tackled both within countries – by selective taxation, provision of social services, education and overcoming caste-type prejudice – and internationally by permitting free or specially targeted trade and aid. Foreign relations extend to *preventing war*, especially with modern means of rapid universal communication and dissemination of information and if organizations like the United Nations are allowed to be effective.

Control of resources

Jay (2000)[11] defines good government as that which 'allows mankind to live more closely up to the potential of the existing supplies of natural resources, including land, and of knowledge'. Resource use may be limited by legal restriction or people may be persuaded to be frugal. Persuasion may be reinforced by taxes designed to reduce demand; selective subsidy may make more sustainable alternatives attractive. Some valuable natural resources are so prone to exploitation that only effective national ownership may ensure their sustainable use.

Legislation may forbid the use of certain resources in certain activities. It may enable reform of land ownership and confirm or strengthen traditional land or water rights threatened by new predators.

Laws may set frameworks for allocations to be made, for example:

- of zones of land for specific uses – such as agriculture, mining or urban development – or for conservation;
- of water distributed from reservoirs for competing uses in downstream places;
- perhaps, if it works, of permissible total emissions of pollutants or GHGs; possibly issue of tradable carbon emission permits which might be effective if the financial institutions managing the markets do not use them perversely for their own enrichment; or
- even of the number of children per family, as in China.

Persuasion goes hand-in-hand with educated appreciation of intelligent propaganda.

Taxation can be directed most strongly onto sales taxes on scarce non-renewable commodities. Probably, if all countries had levied much higher taxation on petroleum fuels for the last two or three decades, global stocks would be higher today, some less fair taxes might be lower and we would be less dependant on these fuels.

Land resource use can be strongly influenced by targeted land taxes – on possession of land as well as on its use – as has succeeded in some countries (Andelson, 1997).[12] However, strong pressures by landowners have tended to discourage governments from interfering in this difficult and not always well-understood field. More easily, landfill taxes have been effective in encouraging recycling of wastes.

Subsidies are often allocated by governments to support public transport systems, sometimes on urban heating or power, water supply or even food, or for selected sections of these to help particularly needy communities. But problems with subsidies are that they tend to help the rich as well as the poor, that they distort markets and trade, and that they are politically difficult to withdraw later.

Maximizing *human resources*, certainly at national level, has always been dependent on good universal education. However, education tends to equip more people to undertake the skilled work than the jobs that exist in open labour markets – another reason why direction of labour and widening of opportunity in fair packets of work may have to be introduced.

Public ownership of resources may be the only way of ensuring long-term, equitable and sustainable use of water or land in critical situations.

In water resource engineering, the management of whole river basins and probably the ownership of dams and hydraulic control systems has to be within the domain of the whole community. Operation and conservation of water and associated drainage and land management have to make use of the reservoirs and the land in a way that will continue to benefit future generations of competing users. In non-water-resource control, local issues on which governments should undoubtedly collaborate with, if not direct, construction planners include:

- giving political or fiscal encouragement and allocating land for certain renewable energy projects;
- restricting access to conservation zones, to reduce illegal or unofficial land development and land degradation; and
- encouraging or prohibiting land reclamation from areas of shallow sea (including with regard to the marine and littoral resources that may be affected).

16.3 Opportunities in land resource engineering

This section addresses the type of engineered landforms that will be appropriate to 2100 needs – for construction on steep ground, in mining and landfill, in providing water storage in reservoirs and to manage flood plains. It also considers how the formation of earth structures should be aided by advances in the classification and use of soils.

Construction on steep ground

Many human settlements and transport routes are necessarily located on steep ground. This has always been done but planned development gets no easier as we attempt to find building space or squeeze in new or wider roads. There are two sorts of opportunity in:

1 Fully designed creation of terraces for buildings or new road lanes or railway tracks with continuing improvements in different types of retaining walls, ground stabilization measures or planting and managing of vegetation.
2 Constructing basic infrastructure for self-help housing (see pp246–247) but extending this strategy to steeper territory so that shanty towns are no longer built in dangerous ill-prepared conditions.

Mining, tunnelling and landfill

Efficient means of hard rock quarrying have been in use for a long time. Gravel and sand extraction and restoration of pits is also well established, and excavation of tunnels has advanced with the use of tunnel boring machines in the second half of the 20th century. Scope for further development lies in further improvements in underground excavation, particularly of caverns for secure containment of fluids or hazardous solid wastes, and planning opencast mining for subsequent land restoration.

Wider opportunities for underground storage will depend on continuing advances, for example concerning:

- use of ever more efficient yet accurate excavation and blasting processes to *avoid* cracking the surrounding rock;
- means of *sealing* cavern walls to prevent escape or ingress of contaminating fluids, or means to *encourage* cracking for storage of inert fluids within the shattered rock; and
- remote sensing to *monitor* what is going on in underground situations, both the state of the stored materials and any effects in the surrounding ground or groundwater.

Most vital is that the location of underground caverns should be accurately indicated for future generations, for example of those storing radioactive wastes.

Soft minerals will continue to be extracted by opencast mining. For metal ores, coal and petroleum fuels embedded in solid form, the main issues will remain how the overburden removal is undertaken and how the landform and drainage pattern is restored for long-term settlement or new uses long afterwards.

Landfill is filling holes as an opportunity to dispose of waste. It is also a way of restoring land levels to what they were before quarrying began. Sometimes, however, no sooner is a quarry completed than its conservation value is recognized – as a geological or biological site of interest. Because of this and particularly in well-populated areas, there are not many holes left to be filled.

The value of worked-out quarries as welcome features of the landscape will continue to depend on how they can best be subsequently exploited. Disposal of solid waste is still in its infancy in a global context. So there is great scope for advances in recycling wastes or burning them to generate energy. Even when consigning material to landfill there is scope for further separation of waste in terms of the degree of precaution in its confinement that is actually necessary.

Sustainable water storage

Most large city water systems and irrigation schemes rely on reservoirs to even out the supply of river water over the year. When all practicable reservoir sites are in use consumers must accept limits on their demand or pay more for supplies from special or distant sources. The long-term success of river storage and control schemes depends on:

- satisfactory and safe operation of mechanisms releasing water on an optimum basis to serve all purposes;

- ecological safeguards such as fish passes or release of timely stream flow to otherwise bypassed river beds or wetland; and
- further development of mechanisms for sediment removed from reservoirs to retain storage capacity and to restore the benefits of fertile silt downstream.

Most of the obvious dam sites have already been used except in the less inhabited huge remote basins – such as the Amazon, Congo or Arctic Ocean rivers – which have no thirsty downstream lands. Elsewhere, more storage may be enabled in the future by:

- extending the life of reservoirs by reducing sedimentation (as above);
- creating all-around dams on flat terrain where consumers can afford the comparatively high unit cost of storing water in this way (as, for example, in London's reservoirs or those filled from the Great Man-Made River pipelines in Libya);
- devising some means of reducing reservoir evaporation (unlikely); or
- creating underground dams and reservoirs.

Subsurface aquifers provide an important form of stored water. In fertile areas, if the farmers have pumps and fuel or electricity, the groundwater is often overdrawn. There are two ways in which, in the right conditions, the sustainable yield of aquifers might be enhanced:

1 Underground dams might be constructed, for example, by impermeable grout curtains across the permeable alluvium in the bottom of gorges.
2 The rate of recharge might be accelerated by constructing ponds or ditches to absorb a greater proportion of passing flood water.

Neither of these measures was widely adopted in the 20th century, mainly for economic reasons; but there remains scope in the right conditions, for example, where flash floods occur and run across coarse material in arid country and where any way of trapping this ephemeral flow could increase downstream supplies.

Flood plain management

Construction in flood plains has been explained in Chapter 14. Flood damage can be reduced by protecting towns or by restricting new building on the plains to structures that can withstand flooding when it occurs. Flood plain management can also make best use of the high flow and inundation that does occur – to maintain wildlife habitat, to practise flood recession agriculture or by directing water into specially constructed fertile polders.

The ideal solution is perhaps that adopted in the Netherlands where historic construction of dykes (embankments) and polders and more recent massive coastal hydraulic works provide protection from most of the Rhine and Scheldt flood plains and to resist the force of North Sea tides and storms; but, even there, new works will become necessary from time to time and there has to be continuous maintenance of the works and energy-consuming pumping to drain low level land.

Such degrees of control are not viable in the more extensive delta area of Bangladesh with the much higher seasonal flows of the Ganga and Brahmaputra rivers and the more severe storm surges in the Bay of Bengal. The 21st century will surely see partial solutions in which the conditions in such regions can at least be alleviated. This might involve:

- a number of favourably located river training walls and flood embankments, lessening the chances of river channel changes in particular directions or to prevent certain areas being so often flooded;
- constructed polders for more controlled or retained use of flood water in those areas where river control can be at least partially effective; or
- construction of more robust semi-permanent raised villages or safe havens.

The main obstacle to massive construction of training walls, flood embankments and raised settlements is the soft nature of the earth material locally available.

Further advances in soils engineering

Soft earthforms are naturally deposited and eroded in a somewhat random pattern. Methods of classifying soils have long been established whereby we can recognize which are suitable to construct on or which suit extraction for incorporation in earthwork elsewhere. Means will continue to be sought for making use of a wider range of soils so that less local material has to be rejected.

Particularly for small-scale earthwork – such as fill around foundations or coarser material behind retaining walls or in land drains – there are already opportunities for more selective use of manual or mechanical methods of soil placement and compaction. Many traditional approaches are still economical using modern equipment and experienced operatives, but the choice of tools or machinery could often be better related to the scale of work and the quality of material involved.

At a larger scale, new ways need to be sought for containing or strengthening a wider range of common soils. Impermeable geotextile (polymer) sheets can play an important role in containing otherwise semi-fluid materials such as silt. Permeable sheets or more robust geogrids can be used as a medium for drainage or to strengthen slopes, and their incorporation may eventually enable easier dismantling or adaptation of earth structures. If more clays, silts and gravels could be stabilized or cemented in situ so as to form harder material, this should obviate much of the need to import rock or conventional concrete materials for river training works or semi-permanent safe havens on flat flood plains. If the performance of trial structures is monitored carefully, then there can be more universal application of soils engineering techniques in a wide variety of conditions.

16.4 Landscapes for 2100

Construction affects land resources roughly in proportion to the extent that man inhabits them. Review of urban and rural landscapes is followed here by that of an emerging combination of both rural and urban characteristics in well-populated regions. Separate consideration is then given to scenic land and to the wide extents of mainly uninhabited territory. The sooner that firm but flexible long-term allocation of land is made for various uses or for conservation, the more likely it is that the patterns that ensue will remain suitable for 22nd century populations.

Patterns for urban living

Cities are where people live in dense concentrations and gain livelihoods in factories, offices or services not related to the land. Globally, rural regions tend to be those where populations increase naturally, but, because agriculture does not have the capacity for more employment, the surplus people have to move to the cities where they hope to find work. Even where national populations are stable, immigration from abroad may trigger urban expansion from time to time.

Because their populations are rising and social and employment patterns are changing, cities must be dynamic. City planning adapts existing communities for new conditions and creates patterns for new (formal or less formal) development that:

- suit physical and human geography;
- complement or extend existing infrastructure and public transport systems; and
- offer livelihood opportunities.

Just as many layouts completed before 1900 still suffice today, so some of the prototypes for the 22nd century already exist. Equally, there are mistakes that should not be repeated. These include layouts that are too spacious to be frugally effective as well as those that are too tightly confined to be socially acceptable.

Most cities arose in locations where they could trade in or process the products of considerable hinterlands or, if ports, from overseas. But, in the 20th century, cities such as Phoenix, Arizona, grew up in semi-desert conditions dependant partly on irrigated agriculture or mining but largely as elements of very large national economies. In recent decades, there has been spectacular city development in the once isolated desert emirates along the Arabian shore of the Persian Gulf. This was initiated, with the funds of oil-rich sheikhs, by the construction of large ports where everything for construction and living could be imported.

State-of-the-art skyscrapers, 'palm' shaped marine reclamation projects, hotel, office and housing blocks and seaside villas have followed with all the associated infrastructure of airports, roads, power supply and desalination plants. Latterly development has been for expatriate communities and largely funded by foreign investors. By 2008 in Dubai, 1.2 million people lived in a 'leisure and commercial hub' where there had previously been mostly empty desert, and expansion towards 4 million by 2020 was still envisaged. However, the investment and loan bubble burst in November 2009.

The limits of viability of Arabian desert cities may relate to how many people or companies will be able to afford future water and energy prices. One can only wonder how they can continue to thrive in competition with places such as Mumbai, located in well-populated regions backed by large mixed national economies.

Meanwhile, the majority of 2100 cities not built in the desert will be accommodating people:

- in densities of housing dictated by the area of land available or, if space is ample, by the viable size of socially-cohesive energy-efficient communities;
- in layouts that are suited both to the climate and the topography; and
- in community economies providing adequate employment.

However, such is the certain increase in urban population that it may be necessary to revise even the basic separation of cities and suburbs from adjacent countryside.

Rural economies

Rural land has been defined both as countryside – everywhere except built-up areas – and as pastoral and agricultural land used productively for feeding livestock or growing crops. In our context, rural areas are those where the land is primarily productive – mainly using the soil for managed vegetation, occasionally extracting minerals or fulfilling other industrial or military applications that take advantage of physical features or are best suited to out-of-town locations. Globally, most extensive fertile land is devoted to agricultural production. Such are the vast wheat lands of North America or the Ukraine and the irrigated plains of India or China. But it is in some crowded mainly warm countries where there is the greatest need to increase food production, where fertile land is most in demand and where there are often problems in equity concerning, for example, land tenure or groundwater abstraction.

Future prospects for the agricultural land that feeds the increasingly urban world need to be perceived in terms of:

- conserving the productive capability of land;
- maximizing production of those crops that can best meet regional dietary requirements; and
- sustaining or rescuing rural economies.

Conserving land capability concerns prevention of salinization of soils in arid conditions, maintaining the long-term fertility of soils generally and flood management. Globally and in recent times, soil salinization has probably destroyed irrigated land resources at a rate faster than new irrigable land has become available. It is one of the tragic consequences of ill-managed watering and inadequate drainage. Saline land can be reclaimed, but it is far less costly to manage the land sustainably.

Long-term fertility of soils depends on permitting time for their recovery after cropping, or on inputs to artificially restore elements depleted by growth or by erosion. The cost of chemicals will probably remain too high for all but intensive cultivation. So traditional use of organic fertilizers and resting of land must continue. Meanwhile, allowing land to lie fallow (perhaps as pasture) and diversification and rotation of crops produce varied colours in the landscape as well as biodiversity not possible in monoculture.

Flood management to conserve land capability should be directed both at positive gain for agricultural production and at wetland preservation with consequent benefits in attenuating flood flows and their effects downstream.

Maximizing production of optimal crop mixes depends upon:

- introducing more effective farming in those regions where this has not already happened;

- improved crop varieties to increase yields, tolerance to difficult soil or climatic conditions and resistance to pests and diseases; and
- other inputs, such as chemicals where they can be afforded, to best combine with particular seeds and techniques.

All the while it must be kept in mind that most improvements have drawbacks which may reduce the net benefit.

Sustaining rural economies in overpopulated countries is primarily a political problem. It is not likely to be solved without persuading the majority of non-agricultural people to pay more for their food and possibly to adapt their lifestyles. Rural people must be able to afford at least the basic food supplies judged adequate for poor urban people, including the cost of any necessary transportation from distant places.

It may be necessary to somehow transfer more of the processing and marketing of agricultural produce to the rural areas, in other words, to produce finished goods in the communities that produce the crops. Urban and industrial interests may claim that this is retrograde reversion, but for too long now arguments based on economies of scale and cheap fuel for transport have encouraged concentration of large central factories and distribution depots in ways that are wasteful in resources and to the disadvantage of rural economies. So future rural landscapes may encompass more agro-industrial (factory) and agro-commercial (office) buildings bringing employment to rural centres. Almost certainly there will be more rural towns.

Extensive urban/rural combinations

A distinctive boundary between urban and rural areas suited medieval towns and is still evident around many established cities in the more spacious regions of continental Europe. However, 19th century industrial development based on coal mining followed valleys like the Ruhr. Corridor or ribbon development followed.

The rapid rise in urban populations in the second half of the 20th century saw the consequent growth of already very large cities (such as Mexico City, Saõ Paulo, Shanghai, Calcutta, Istanbul and Cairo), the birth of lesser but completely new cities (Brasilia, Islamabad, Chandigarh and Milton Keynes) and the coalescence of separate towns into great conurbations (Philadelphia/New York/Boston, Nordrhein-Westfalen, the English West Midlands).

Many of us live in these urban and suburban situations and most of us are sure to do so by 2100. The landscape will have to accommodate more people and not all can be contained within existing towns or their presently planned extensions. Some rural land will have to be released for building but it should be possible to retain viable units of countryside in a mixed land-use pattern. Some of the categories of landscape now classified as rural or urban will become combinations of the sort introduced as conurbations in Chapter 15. They could incorporate strips or pockets of open space amenity, nature conservation and heritage features as well as productive farming land – all within a matrix of towns and suburbs.

Instead of random urban expansion to meet immediate local population pressures or medium-term projections, regional combinations need to be planned specifically to accommodate more people (typically twice as many) in 100 years time. Ideal combinations will differ from recent part-planned part-random developments through updated concepts foreseen for 22nd century land use, housing and transport. Achieving long-term objectives needs not only planning parameters suitable for 2100 but also a strategy for transition from distinctive urban and mainly productive farming land into a more densely populated region retaining inherent green elements in towns as well as distinctive unbuilt-on 'breathing' zones beyond.

Land use in these circumstances has to encompass all activities inherent in urban civilization as well as the agriculture, public infrastructure and amenities normally associated with rural areas. A viable layout must accommodate features, mainly linear, that define the pattern of the landscape. These include watercourses, escarpments and other natural boundaries as well as railways and major roads, but the patterns and hierarchy of lesser roads may be altered to suit planned development. Then land space between the linear features or boundaries will contain or be allocated for:

- factories, waste recycling facilities, electric power stations or substations, service depots, major commercial or sporting complexes and any other facilities larger than are normally suited to residential areas, preferably of sufficient extent to accommodate these activities for 100 years; even if the types of facilities change, buildings or surfaces can be adapted or reconstructed;
- residential areas including those already notionally assigned for medium-term redevelopment or new construction and allowing for ample facilities (shops, schools, amenity space) appropriate to each settlement; and
- famland and woodland, in viable units, but with some assigned priority as to which particular land might eventually be transferred to built development later in the century, depending on actual demographic changes and the success of various medium-term housing and development policies.

Recognized nature conservation or amenity land will fall partly in urban or suburban and partly in agricultural land. Exceptional (sacrosanct) features could be close to but outside the boundaries of any urban/rural combination zone.

Scenic landscapes

Scenic regions can be assessed in zones, such as: land essential to the local (non-tourism) economy; conservation areas; and a wider class of less spectacular or biologically critical landscape that can, if necessary, be considered for human settlement and such structures as dams, windfarms or tourist facilities. Strategies can then be identified whereby certain zones may overlap, be shared by different interests, be firmly assigned for specific development or be preserved permanently.

On conservation land, construction is limited to that necessary for any permitted access – erection of simple footbridges or manually built and maintained pathways capable of withstanding pedestrian wear. Prevention of increased hillside erosion or water contamination should be paramount. Outside, but adjacent to the boundaries of the conservation areas, there can be buffer zones where the landscape merges into mixed woodland, pasture, arable fields, farmhouses, simpler buildings and rural roads. Specific sites can be located perhaps in woods or at riversides, for visitor centres, cafés, picnic areas and vehicle parks. These sites can be attractive destinations for less intrepid tourists, an introduction and enticement to those who wish to venture into wilder scenery and yet a firm barrier between economic production on one side and conserved landscape on the other.

In the future we should assume that the numbers of people attracted to scenic places will increase; but a reduction in the profusion of motor vehicles cluttering the approaches or parking in the scenery itself will result from a probable decrease in convenience and increase in cost of private car use, at least within the environs of resorts.

Both trends may be complemented by a reversion to the popularity of healthy walking, as practiced on the moors surrounding northern English towns in the first half of the 20th century and by workers who were encouraged to explore the mountains of eastern Europe in the communist era in the second half of that century. More Indians may visit hill stations, or the inhabitants of Rio de Janeiro, Cape Town or the cities of southern China may enjoy more days off in their stunning hinterland. Meanwhile, long-term strategies for tourism must ensure maintenance of local livelihoods and sharing of land resources between economic production and leisure amenity, as well as provision and control of the means of access.

Ideally in the poorer, comparatively densely inhabited but more extensive landscape of some Asian or Latin American mountain or lake terrain, there will, by 2100, be many more small resorts. These should bolster what may still be precarious local economies by attracting more, if generally less affluent, visitors than at present. Enjoyment of fine scenery is universal among people who have had opportunity to walk about in it.

Empty or remote territory

Regions that remain comparatively unexploited by man can be assessed in four different geographic types that are:

1 cold, often with infertile ground or very short summers; those not permanently covered in

snow are in tundra or sparse boreal forest, or in middle latitudes at high altitude (such as Tibet and Bolivia);
2 somewhat less severe more fertile northern zones;
3 dry deserts; and
4 warm rainforest and extensive wetland, which – in contrast to the other types – exhibit luxurious growth, support great biodiversity and have survived so far only where they have been inaccessible.

The *cold lands* have been the habitat of such people (Inuit Canadians, Siberian trappers or Mongolian nomadic herdsmen) and wild animals (deer, beavers, migrating wildfowl) as can survive by gleaning the sparse resources of great tracts of open or lightly forested land. Main human settlements have been centres for mineral extraction (Klondike gold, Alaskan oil) or its processing (Siberian bauxite) or as ports (Archangel'sk). Pollution of land by waste chemicals and fuels and devastation of land by strip mining could be avoided by relatively inexpensive precautions, enhancing whatever future settlements may be feasible or necessary.

The most substantial changes to northern landscapes have resulted from dams and reservoirs created for hydroelectric schemes, for example in Quebec and Labrador in Canada or at Bratsk in the headwaters of the Enisei River in Siberia. Total river flow into the Arctic seas is considerable, but it is seasonal and often ceases completely during the winter freeze. Nor is there much regional consumptive use of water.

However, the hard geology of the catchment regions upstream of most northern hydroelectric dams means that there is little risk of sediment filling the reservoirs – a major hazard in less stable catchments. The Canadian reservoirs, exceeding 500km^2 in area, do impede animal migration routes but, in such areas of extremely old Lewisian gneiss worn into irregular rock and water landscape, any adjustment in migration patterns is likely to be no more than an incident in the animals' long-term routine within their vast otherwise unspoilt habitat. Less vast and much more accessible at their fringes are the numerous glaciated rock outcrops and lakes that characterize the same Lewisian formation where it was originally named in northwest Scotland. In most remote wild territory there is scenic landscape of appeal to adventurous recreation or extreme tourism – adventurous because of the difficulty of crossing it on foot, extreme where people travel far on ships to watch glaciers calve their icebergs into the sea.

Because remote settlements are in such spacious surroundings, there is all the more reason that their location and layout should respect any existing features of scenic or ecological significance; but there is opportunity, in their development, to incorporate the best of industrial functional forms or regional architectural styles to establish a visual identity.

All isolated settlements need energy, ideally in a non-fossil fuel form and therefore mainly as electricity. Hydroelectric plants are for bulk supply to major load centres and do not operate when the rivers are frozen. But nuclear power plants may prove appropriate to remote northern towns.

Tundra regions have neither the climate nor the soils to suit agriculture. Nor do most of the trees in the northern forest margins yield commercial timber.

Somewhat less severe northern zones are more promising. The cultivated prairies of southern Saskatchewan or steppes of central Russia, the productive forests of Quebec or Siberia, as well as the industrial regions of the Urals, lie in the same bands of latitude as western Europe albeit with extreme rather than temperate climates. Harsh winters and relatively short growing seasons probably account for the still relatively light population of these regions, but population pressure or land shortage elsewhere may eventually compel denser occupation. The lands may still be empty enough to afford spacious development opportunities – including for outdoor summer leisure – if needs can be fulfilled for efficient winter heating, adequate transportation and appropriate nature conservation. The opportunities should not be taken until the needs can be economically met, the resources of the land are fully understood and the pressure of overpopulation elsewhere requires it.

Deserts are of use to mankind:

- at oases – but these are so scarce, so ecologically important and so easily spoilt by more than

occasional camel-trains that any which remain pristine should be strictly conserved roads; in particular, should avoid them;
- for occasional mining opportunities; also some ancient groundwater resources such as are being exploited in southern Libya; the sooner data is revealed about the true capacity of these sources, the sooner limits can be placed on the rates of abstraction; and
- as sites for solar power arrays – possibly a vast source of renewable energy depending on the effectiveness and cost of the equipment.

Tropical rainforests are hugely prolific, but many tracts of them are sensitive and destructible. The forests of the Amazon and Congo basins were remote to entrepreneurs from the avaricious mercantile world only while they remained inaccessible. On the one hand, these warm, well-watered jungles are the most biologically diverse regions on Earth, there are thousands of only partly identified ecosystems and species whose biotic significance is not yet well understood. On the other hand, most of these unspoilt lands are regarded as ripe for human expansion and development at a time when crowded inequality is making life increasingly difficult in the inhabited lands bordering the primeval forest. Perhaps the best that one can plan for tropical rainforests in 2100 is:

- a key core of strictly conserved forests, inevitably reduced from even their present extent but in areas individually large enough to be viable for the indigenous species;
- assigned areas for further commercial forestry, rangeland or agriculture and consolidation of those that already exist – but expansion being severely limited to allow for gradual extension over the next 100 years and the limits being controlled as much by restricting the construction of roads, bridges and settlements as by enforcement of conservation plans; and
- rehabilitation of semi-wild land that has been degraded by overgrazing, intensive agriculture, mining or pollution.

Wetlands occur in great variety but usually on a smaller scale than forests. They are even more susceptible to anthropogenic destruction. Assignment and protection should therefore be more detailed.

Allocating land now to achieve future goals

Landscapes may remain sustainable in the future only if we allocate purpose to broad areas of space now. These areas should encompass enough land to meet the demands for each use anticipated for 2100. Each unit of land should have the characteristics, shape and size to be viable for its prime purpose; and reserve areas should be allocated for subsequent use (beyond 2100), including on land that is presently considered marginal.

16.5 Safeguards and aspirations for the next millennium

One thousand years is not an extraordinary length of time in speculating how civilized humanity can extend its survival. It is more than two thousand years since a culture of philosophy, arts and citizenship flourished in Greece, almost as long since Roman civil engineering made practical use of emerging science. Since then, there have been periods of barbarian destruction, famines and social turmoil interspersed with centuries of regional brilliance – in Arab medicine, surveying and architecture, massive Chinese hydraulic works and the European Renaissance. Episodes of remarkable achievement – such as today's revolution in information technology – have alternated with 'dark ages' such as the one currently stifling parts of Africa.

In construction, our inheritance includes the substantial surviving structures of ancient Egypt, Greece, Rome and Mexico, but the irrigation schemes of ancient Mesopotamia succumbed to the perennial risk of salinity, as did Arabian dams to that of reservoir sedimentation. In the year 3000, the Eiffel Tower or the Millau Viaduct may survive as heritage monuments; some transport routes which no longer follow Roman routes may continue along the corridors created for 19th or 20th century railways.

Populations and their civilized survival

Historically, there have been periods of regional population increase, episodes of plague and times of economic stimulation or migration. In the 20th century there was a population explosion, mainly in the less developed countries. By 2100, the global population will be 10 billion plus or minus 1–2 billion. Thereafter, there is great uncertainty, but it could reasonably be speculated that at the end of the third millennium the total might have changed by a factor of ten either way, resulting in a global population somewhere between 1 billion and 100 billion.

The lower figure of 1 billion people is equivalent to the world's population as recently as 1830. It was only 2 billion in the 1940s, within living memory for some of us and at a time when, certainly in the developed countries, much of today's social conditions and infrastructure were in place. A reduction to such numbers in the future could conceivably result from assiduously applied family planning and universal changes in cultural attitudes, or it might result from a holocaust of war or disease. Either way, there would be a need and opportunity to establish a more stable relationship between mankind and his surroundings.

A population of 100 billion would present a huge demand for food, water and land space. Survival at such a level is perhaps conceivable with huge migration, radical changes in diet and some enormous source of cheap energy capable, for example, of desalinating great quantities of sea water or transporting water over long distances or over mountains. Without such energy inputs, periods of excessive population growth are more likely to be followed by famine, unrest and disastrous local depopulation than by painless socio-economic adjustment.

Roles for construction in meeting human aspirations for the year 3000

Key functions for construction in the centuries ahead may include:

- on the largest scale – massive hydraulic works, ingenious landform engineering and effective infrastructure necessary to optimize land resource allocation for however many million or billion people there are in each part of the world;
- at the smaller local scale – provision of simple but measured and controlled means for good husbandry of land, water and available energy; and creation of pleasant environments in all human settlements with secure conservation of whatever wildlife forms are judged to be most deserving.

In conjunction with both of these would be social and political safeguards for civilized sharing of human and material resources.

We may see engineering capability harnessed in national and international groups also furnishing humanitarian aid and military force, discipline and logistics. When these combined groups are not engaged in the relief of sudden natural or man-made disasters, the construction teams in particular might be assigned to particular new works or reconstruction, which the countries concerned cannot complete with their own resources.

Last words

Returning to population speculation, perhaps a stable compromise of 5–20 billion people may emerge. The number will depend on social, political and constructive ingenuity in devising a reasonably peaceful if still imperfect world. There may still be some respect for regional cultural difference within strongly enforced global strategies for sustainable use of resources and application of human ingenuity and effort.

But that is another story.

Notes and references

1 The world's population reaches 7 billion in 2012 and, according to various forecasts, 9 billion by 2040–2050, the latter being UN's medium case forecast. UN warns that projections beyond 2050 are little more than guesses (UN Population Division (2004) *World Population to 2300*, UN, New York, p3). This author assumes that 2050 is too soon for global population to peak and there could be modest continuing increase to a round number of 10 billion at the end of this century.

2 Sachs, J. (2005) *The End of Poverty: Economic Possibilities of Our Time*, Allen & Lane, London, pp323–324.
3 Young, A. (2006) 'Spare land – a challenge to official elements', www.uea.ac.uk/gov/landresources/news-spareland, accessed c2008.
4 Wikipedia (2010) 'Current sea level rise', http://en.wikipedia.org/wiki/Current_sea_level_rise, accessed 20 August 2010.
5 Heinberg, R. (2003) *The Party's Over: Oil, War and the Fate of Industrial Society*, New Society Publications, Gabriola Is, Canada.
6 Hinrichs, R. A. and Kleinbach, M. (2002) *Energy: Its Use and the Environment*, Harcourt College Publishers, Orlando, Florida.
7 MacKay, D. J. C. (2009), *Sustainable Energy – Without the Hot Air*, UIT, Cambridge.
8 Jay, P. (2000) *Roads to Riches: The Wealth of Man*, Weidenfeld & Nicolson, London, p223.
9 Sachs (2005) as Note 2, pp18–19.
10 Collier, P. (2007) *Why the Poorest Countries are Failing and What Can be Done About It*, Oxford University Press, Oxford.
11 Jay (2000) as Note 8, p310.
12 Andelson, R.V. ed (1997) *Land-Value Taxation around the World*, Robert Schalkenbach Foundation, New York.

Subject Index

Index of Places, Projects and People